Sustainable Development Through Food and Nutritional Security

The Editors

Dr. Deepika Baranwal (Ph.D in Foods and Nutrition) working as Assistant Professor in Department of Home Science, Arya Mahila P.G. College affiliated to Banaras Hindu University, Varanasi, U.P., India. She has been awarded with Gold Medal in B.Sc H.Sc from Dr. R.M.L. Avadh University. Ayodhya, She has done M.Sc in Food Science and Nutrition from CSAUAT, Kanpur, Ph.D from MPUAT, Udaipur, Raj. She has qualified many exams like UGC NET and RPSC SET , ICAR JRF and SRF examination. She has received many awards for her contribution in Teaching Field and Research like Excellence in teaching, Global Teacher Award, Best Paper, Best Poster and Best Oral Presentation Awards in National and International Conferences in India and Abroad. She is a Councellor in IGNOU programms like MSCDFSM, DNHE, CFN, CNCC etc. She possesses five years teaching experience in the area of Food Science and Nutrition. She has many National and International publications a reputed journals, books, conference proceedings and magazines. She is a life member of many reputed societies like Nutrition Society of India, Hyderabad, A.P., Indian Dietetic Association, TAPASYA, AIASA, Sri Aurobindo Society, Pondicherry, International journals Trends in Biosciences and Advances in life Sciences etc. She is associate editor of an International journal of Advances in Obesity, Weight Management and Control. She has been attended many National and International conferences in India and also presented papers in the seminars. She is also member of Women Cell and Health and Hygiene committees of the college. She organized various events like National Nutrition Week, World Health Day, International Literacy Day, World Mental Health Day and Lectures on different topics in the College. Media also covers her name in various activities of Department of Home Science. She has presently taken charge of AuroYouth Co-ordinator, Sri Aurobindo Society, Varanasi Center and conducted various youth camps in UP.

Dr. Ram Chandra M. Sc. (Ag), Ph.D Professor in the Department of Mycology and Plant Pathology, Institute of Agricultural Sciences, Banaras Hindu University, Varanasi. He possesses vast research experience in the area of mushroom production technology, plant disease management and microbiology. He has 20 years teaching and research experience in the field of fungi, microbiology and plant pathology. He has been honoured with many prestigious awards which include; CST Young Scientist Scheme Award (2006), Bioved Young Scientist Associate Award (2012), Bioved Fellowship Award 2014 and ICAR, Best Teacher Award 2014. He is life member of several professional societies such as Mushroom Society of India, Association of Plant Pathologist of India, Doctor's Agricultural and Horticultural Development Society. He is also member as editorial board in Indian Journal of Scientific Research and Tatvanveshan Research Journal and Journal of Ecology. He has authored book on Mushroom and their Cultivation Techniques, edited book on Microbial Diversity Modern Trends, Microbial Biodiversity in Sustainable Agriculture, three practical manuals on Mushroom Cultivation, Microbiology and Mushroom Production Technology respectively. He has published more than one hundred research papers, review papers, book chapters and popular articles in National, International journals, books and magazines. Twenty five students have been awarded M. Sc. Ag. / Ph. D. Degree under his supervision. He has initiated and established Mushroom Spawn Laboratory in 2006 as Nodal Officer in Seed Production of Agricultural Crops and Fisheries under Mega Seed Project. He has completed and handling several project sponsored by ICAR, UPCAR, DBT, DST and Dhanuka as Principal Investigator and Co Principal Investigator. He has actively participated in All India Radio and Doordarsan by fifty talked.

Sustainable Development Through Food and Nutritional Security

Deepika Baranwal

Ram Chandra

2020

Daya Publishing House®

A Division of

Astral International Pvt. Ltd.

New Delhi – 110 002

ISBN 9789390371532 (Int. Edition)

This book brings together the edited manuscripts of the papers based on the syllabus of Ph.D and M.Sc course of Food and Nutrition of Indian Council of Agricultural Research, New Delhi. The manuscript of this book is prepared by eminent academicians, from various scientists, teachers, researchers of universities and institutions.

Authors will be responsible for any queries regarding materials of particular book chapters.

Published by : **Daya Publishing House®**
A Division of
Astral International Pvt. Ltd.
– ISO 9001:2015 Certified Company –
4736/23, Ansari Road, Darya Ganj
New Delhi-110 002
Ph. 011-43549197, 23278134
E-mail: info@astralint.com
Website: www.astralint.com

Preface

India is the world's second biggest populated country. Thus, the country requires more food production so as to mitigate the starving conditions. Since independence, our country is continuously facing the emerging challenges *i.e.* to achieve food security as well as to eradicate poverty. Nutrition Security is the strength of any nation. Sustainable development is a new type of human strategy that meets current needs, without compromising the ability of future generations to meet their own needs. Nutrition is an important environmental factor that influences health and well-being. Consumption of diets adequate both in quantity and quality is a prerequisite for the maintenance of good nutritional status. Agricultural production that determines food availability is, therefore, an important determinant of food consumption, though not a critical one if food imports can be assured. Self sufficiency in food production is of particular importance for developing countries, not only because they tend to have high rates of population growth, but also because such countries have malnutrition as a public health problem. The quantitative aspects of food production are undoubtedly of primary concern, but it cannot be forgotten that the qualitative aspects are extremely important, if optimal nutrition is to be provided. In the qualitative aspects food processing, agricultural practices, warehousing, storage, transportation of food all things are to be concerned in this book. This book is based on the syllabus of Ph.D coursework at agricultural universities (ICAR) in the subject Foods and Nutrition. The manuscript of this book is mostly invited one from the eminent academicians, from various scientist, teachers, researchers of renowned universities and institutions. I am thankful to all the contributors for their authoritative and thoughts providing review papers for this volume. This book Sustainable development through food and nutritional security is intended for the undergraduate, post graduate, research scholar, senior research fellow, research associate, academicians and scientist in the area of Home Science, Food and nutrition, food processing, Food science and technology, Agriculture and Biotechnology. I am also thankful to publisher, Astral International (P) Limited New Delhi.

Deepika Baranwal

Ram Chandra

Contents

Part B
Food Processing and Technology

Part C
Miscellaneous

List of Contributors

Sl.No.	Chapter Name	Author Names and Adress
1.	**Food Situation and Food Insecurity in India and Abroad**	**Neelam Basera and Arpit Huria** *Agricultural Extension and Communication. G.B. Pant University of Agriculture and Technology, Pantnagar, Uttarakhand*
2..	**Role of Nutrition for Agriculture Planning and National Development**	**Shailesh[1] and Puroshattam[2]** *[1]ICAR-Central Research Institute for Jute and Allied Fibres, Barrackpore, Kolkota – 700120, W.B* *[2]ICAR-Indian Institute of Pulses Research, Kanpur, U.P*
3.	**Impact of Physical Resources, Farming Systems, Cropping System and Agricultural Inputs on Food and Nutritional Situation**	**Ragini Ranawat** *Krishi Vigyan Kendra, Udaipur, Rajasthan*
4.	**Food Production, Food Distribution and Nutritional Status**	**Neha Nagda, Upendra Singh and Anuj Yadav** *Deptt of Agril Engg, SKNCOA, Jobner – 303329, Rajasthan*

Sl.No.	Chapter Name	Author Names and Adress
5.	**Food Distribution System and Food and Nutrition Security at National and Household Level**	**Tanu Jain** *Amity Institute of Food Technology, Amity University, Noida, U.P.*
6.	**Poverty and Vicious Cycle of Low Food Productivity**	**Amita Yadav** *Deptt. of Economics and Extension, School of Agriculture, Lovely Professional University, Phagwara, Punjab*
7.	**Food Price Control and Consumer Subsidy for Sustainable Development**	**Sripad Bhat and Puroshattam** *ICAR-Indian Institute of Pulses Research, Kanpur – 208024, U.P.*
8.	**National Nutrition Policy for Health and Food Safety**	**Uttara Singh[1] and Sadhna Singh[2]** [1]*Government Home Science College, Sector-10, Chandigarh* [2]*Narendra Dev University of Agriculture and Technology, Faizabad, U.P.*
9.	**Contribution of National and International Organization for Agricultural Develop-ment**	**Vishakha Sharma** *Department of Foods and Nutrition, Maharana Pratap University of Agriculture & Technology, Udaipur*
10.	**Nutritional Impact of Agricultural Programmes**	**Tanu Jain** *Amity Institute of Food Technology, Amity University, Noida, U.P.*
11.	**Impact of Agricultural Marketing System and Post Harvest Processing of Foods on Nutrition Security in India**	**Ragini Ranawat** *Krishi Vigyan Kendra, Udaipur, Rajasthan*
12.	**Post Harvest Losses, Food Spoilage and its Causes**	**Anurag Singh, Ankur Ojha and Ashutosh Upadhyay** *Department of Food Science and Technology, National Institute of Food Technology Management Entrepreneurship and Management (NIFTEM), Sonipat, Haryana*

Part A

Food and Nutritional Security

Chapter 1

Food Situation and Food Insecurity in India and Abroad

Neelam Basera and Arpit Huria

Introduction

At the 1974 World Food Conference, government leaders proclaimed that *"every man, woman and child has the inalienable right to be free from hunger and malnutrition"* (Gebremedhin, 2000). The 1974 Conference set an ambitious goal: to eradicate hunger from human society within a decade. At the time of the 1996 World Food Conference, two decades and two years later, many children still went to bed hungry, many families still feared for their next day's bread, and many individuals' potential continued to be stunted by hunger and malnutrition. Although world food production, overall and per capita, has risen, the goal of the 1974 Conference has not been fulfilled. In fact, today, the world faces a renewed food crisis, one at least as formidable and life-threatening as occurrences of the past (Gebremedhin, 2000).

Current trends in food and agriculture are increasing stress particularly on developing countries in a way that has implications for the future of food security, socio-economic development and sustainable peace. According to Food and Agriculture Organization, 2006 the highest number of 642 million hungry people live in Asia and the Pacific. India specifically accounts for 16.7 per cent of the world's food consumers. With the exception of China, India's size in terms of food consumers is many times larger than the average size of the rest of the countries. India has achieved food security at the national level but food security at the individual or the family level has not been adequate. With little reduction in the number of undernourished and poor people, the country is far behind in achieving the Millennium Development Goals and the targets set at the World Food Summits.

Meaning of "Food Security"

Food security concept has temporal and spatial dimensions which underpin that food must originate from efficient and environmentally benign production technologies that conserve, enhance and sustain the natural resource base of crops, farm animals, forestry, and inland and marine fisheries. Accordingly in 2006, Food and Agriculture Organization has defined food security as *"it exists when all people, at all times, have physical and economic access to sufficient, safe and nutritious food that meets dietary needs and food preferences for an active and healthy life"*. It includes the following dimensions:

- **Availability:** The availability of sufficient quantities of appropriate quality;
- **Access:** Access by individuals to adequate resources for acquiring appropriate foods for a nutritious diet on a regular basis;
- **Utilization:** Utilization of food through adequate diet, clean water, sanitation and health care to reach a nutritional well-being where all physiological needs are met;
- **Stability:** A population, household or individual must have access to food at all times and should not risk losing access as a consequence of sudden shocks or cyclical events.

Further, food security has three inter-dependent dimensions: (a) National or Macro Food Security, essentially amounting to the local (domestic) availability of food (b) Household Food Security and (c) Individual Food Security, the latter two largely determined by the physical and economic access to the available food.

Food Situation in World

FAO's Crop Prospects and Food Situation report of 2016 estimates that while global stocks remain at adequate levels, a total of 34 countries are in need of external food assistance due to a combination of civil conflict and El Nino linked weather conditions. Twenty seven of those countries are in Africa, where adverse weather resulted in significantly lower 2015 cereal harvests. Crop production prospects for southern Africa have been particularly affected, and this is likely to have severe food security implications in the region, according to the report. Crop prospects in Latin America are generally favorable, despite El Nino effects in Central America and the Caribbean. Higher than average cereal production in South America and Mexico are expected to fill the gaps left by decreased production in the rest of Central America and the Caribbean. Similarly, production prospects in Asia are favorable as well due to increased cereal harvests in China and Turkey.

Further, FAO estimates that global wheat production for 2016 will reach 723 million tons, down slightly from 2015's record outputs. The early indications suggest that El Nino driven weather effects will cause significantly reduced maize outputs in Asia, Europe, North America, South America and Southern Africa. Rice production estimates for the Southern Hemisphere are mixed, with lower outputs expected in Indonesia, southern Africa, and Australia as a result of weather conditions and water shortages. Rice prospects are positive in Ecuador and Peru, but are less so in Bolivia, Guyana, Argentina, Uruguay, and Brazil.

Food Situation in India

A Brief Account From Past to Present

At the time of independence in 1947, India was in the grip of a serious food crisis, which was accentuated by the partition of the country. The demand for food far exceeded supply; food prices were high and more than half of the population living below the poverty line with inadequate purchasing power. With high rates of population growth, the dependence on imported food increased further. However, the situation improved considerably after the mid-1960s, when new agricultural development strategy and food policies were adopted. The production of staple cereals increased substantially, mainly contributed by productivity improvements. The dependence on food imports decreased and the country became a marginal net exporter of cereals. There was also an improvement in physical and economic access of households to cereals and other nutritive food products. The proportion of households reporting hunger went down and the incidence of economic poverty has reduced.

During the last about one decade, however, the food grain production has almost stagnated and huge production gaps exist. The natural resources, particularly land, water and biodiversity, have degenerated, population pressure continues to intensify (although the population growth rate has come down from over 2 per cent during the 1980s to 1.5 per cent now but this still means addition of nearly 16 million people every year). In order to bridge the supply-demand gap, the Government has initiated several programmes, including the National Food Security Mission (NFSM) and the Rashtriya Krishi Vikas Yojana (RKVY) (GoI, 2007).

FAO Global Perspective Studies Unit, 2006, has projected that food consumption levels in India from current average level of about 2,400 kilo calories per capita per day will increase to about 3,000 kilo calories per day in 2050. As is typical of countries with rising incomes, the share of calories derived from cereals is declining in India, and is projected to fall below 50 per cent by 2050. Conversely the share of calories derived from higher-value foods like fruits and vegetables, vegetable oils and livestock products is projected to increase, with the most rapid growth projected in the case of chicken. In aggregate terms, projected demand for cereals (for direct human consumption) is projected to rise from 159 million metric tonnes in 1999-2001 to 243 million tonnes in 2050, an increase of 53 per cent, or 0.9 per year. This growth will help decline number of undernourished from 221 million people (about 22 per cent of the population) now to 70 million people (about 5 per cent of the population) by 2050.

Prominent Issues to be Addressed

There is a perception that China has, by and large, solved its 'food problem', whereas India has not (Timmer 2014). This rings true in a very specific sense. The crux of India's food problem today pertains not so much on increasing food availability or production but with the distribution of food. This is not to suggest that the challenges associated with ensuring food availability in sustainable ways is not a policy concern, but rather, in terms of the immediacy of challenges, ensuring food access would appear to score over concerns over food availability. For example, despite flagging growth rates in the agricultural sector relative to targets, India has seen impressive growth in food grain production in recent years.

According the Narayanan (2015), The National Food Security Mission (2007) has played a key role in augmenting production in cereals and pulses. Much of this has come from yield increases in the eastern regions in the country where the Green Revolution did not take place. At the same time, there has also been a strong and continuing trend for diversification into non-cereal and high-value commodities such as dairy, fruits and vegetables, which implies the possibility of higher quality diets. Investments in the agricultural sector have been especially strong after 2004–05, both public and private, with private gross capital formation accounting for an increasing share of all investment. Despite the large increase in production, access to food continues to be a serious issue especially in the context of extraordinarily high-inflation rates in food commodities in recent years and limited access in parts of the country to high-quality diets.

Recommendations for Ensuring Food Security

Based on a survey of India's food policy, approach to food security, its food situation, important policy instruments and strategic initiatives; it was obvious that there are quite a few lessons and issues, which need attention for achieving the goal of food security for all. The best assurance of food security in agriculture-dominant countries as in India can be provided through the accelerated growth of food and/ or other agricultural products and the introduction of cost-reducing technological changes in agriculture through a judicious combination of investment in agricultural research and technology transfer system, the creation of rural infrastructure and the provision of an incentive framework for the farmers (Acharya, 2002a).

Further, as the agricultural sector is dominated by marginal and small farmers, their participation in the growth process is quite critical. Most of the rural poor and food-insecure households either own small land-plots or work as farm labourers. The strategy to alleviate hunger and achieve 'food security for all' must, therefore, involve advancing the production and incomes of small farms. This in turn implies broad-based efforts to develop agriculture, animal husbandry, fisheries and forestry. Farmers must continue to receive adequate incentives to produce food and other farm products. National policies must ensure that small and poor farmers have access to land, water and animal grazing resources. Property rights of small farmers and fishermen must be secured. Institutional reforms are needed to ensure the rights of the poor to common property resources of water and grazing lands on an equitable basis.

India's food security policy is built on various livelihood entitlements, *viz.*, production based-, exchange based-, labour based- and transfer based-entitlements. A combination of policies and programmes was intended to help different food-insecure communities and sections of people. These included the supply of farm inputs at lower prices, price support to farmers, keeping the food prices at reasonable level through procurement, building buffer stocks and the public distribution system, food-for-work and other employment oriented programmes, and direct food assistance schemes. The rationale of the food and input subsidies has also been viewed in this context (Acharya, 2001).

Several issues relating to the financial and environmental sustainability and consequently to the economic costs of the package are, however, being raised. The debate revolves around the question whether the food security of the poor can be assured at a lower cost by phasing-out food and input subsidies? It is now being increasingly realized that although farmers must continue to receive adequate incentives for producing food and other agricultural commodities, the emphasis should shift from general input subsidies to the provision of specific targeted subsidies, the continuous introduction of pre- and post harvest cost reducing technologies, support for more efficient food processing methods, and the development of institutional mechanisms to reduce the prices to farmers and the elimination of risks regarding yield and income (Acharya, 2002c).

As the livelihood of a large section of population depends on the production of food, great care is needed in the liberalization of the food product trade. Studies have shown that unrestricted trade liberalization in food grains may expose both the small producers and poor consumers to the high volatility inherent in international food prices (Chand and Jha, 2001). Several safeguards available in WTO's Agreement on Agriculture should be prudently used.

Some studies (Patnaik, 2003) have shown that economic reforms have adversely affected the access of the poor to food and have forced many families into the trap of food insecurity. For instance, liberalized imports of edible oils with low tariffs have resulted in low prices for the oilseeds that are mostly grown by resource-poor dryland farmers. The low oilseed prices have induced the closure of a large number of tiny rural processing units, eliminating in the process the employment opportunities of many families. Despite the gains to consumers of edible oil by way of lower prices, the net social welfare of liberalized imports has been estimated as being close to zero (Chand and Jha, 2001). This implies a redistribution of welfare/income from the mostly poor oilseed growers (whose livelihood depends to a large extent on oilseed production) to the consumers (whether poor or not so poor, but for whom only a small proportion of their total expenditures goes to cooking). Therefore, the need exists for caution in deciding the speed and sequencing of trade liberalization and economic reform policies.

Liberalization of domestic markets must precede the opening up of trade and the reduction of import duties on agricultural commodities. An efficient marketing system significantly contributes to the alleviation of hunger and the improvement of food security. The policy environment and development initiatives in developing countries require several changes for reducing avoidable costs in marketing and for encouraging private sector investment and participation in marketing (Acharya, 1998).

Governments should reformulate market-related policies and make complementary investments in rural marketing infrastructure for attracting private investment to value addition and food processing (Acharya, 2003). So long as poverty persists and transient food insecurity occurs at frequent intervals, direct food assistance programmes will continue to be important in the fight against hunger, food insecurity and malnutrition (Acharya, 2002b). Chronic food insecurity is becoming increasingly concentrated in certain regions and communities, which makes the targeting of food assistance programmes a viable proposition. Furthermore, the success of programmes aimed at reducing food insecurity depends on an environment with active and vibrant civil society organization, voluntary groups and cooperatives.

Thus, attempts such as food assistance programmes should include a combination of (i) distribution of subsidized food grains in pre-decided quantities to targeted families; (ii) supply of adequate food grains free of cost to the poor, old and destitute; (iii) provision of one meal a day to poor school children during school hours; (iv) supplementary nutrition dispersion (including micro-nutrients) to infants and expectant/nursing mothers; (v) food-for-work programme for willing able-bodied adults; and (vi) an employment guarantee programme in all areas.

Experience has shown that whenever India needed to enter the world market to buy food grains (wheat), the price was higher than the world market average (Chand and Jha, 2001). This is because in the world wheat market, the characterization of India as a small country is not favorable to India. Therefore, a large populated country like India should continue to maintain a reasonably high degree of self-sufficiency in foodgrains.

Conclusion

The world has ample food and there is a growth trend in food production. Yet many poor countries and millions of impoverished people do not share in this abundance. There is widespread agreement that a leading cause of hunger and malnutrition is poverty. Although poverty alleviation and food security have long been recognized as among the most central challenges facing the human condition, today–in an era of unprecedented plenitude–there are probably more human beings suffering from deprivation than ever before in history.

We live in a world characterized by hunger, poverty, and increasing disparities. It is a world of disturbing contrasts–with hunger in some lands and waste of food in others–and with the disparity between many of the rich and poor nations widening constantly. We also live in a world of interdependence. This global interdependence has inescapable political and economic dimensions because boosting agricultural production in developing countries can stimulate commercial trade for developed countries. Likewise, security and peace for developed countries also depend on peace and stability in the developing countries. This is one planet. What happens to the least of us happens to the rest of us. It should therefore be acknowledged that poverty is no longer a problem contained within the borders of the developing world alone. The problem has substantial effects on domestic policies, international markets, and world peace.

It is also evident that the globalization of the world's economy and the deepening interdependence among nations can present challenges and opportunities for sustained economic growth and development, as well as risks and uncertainties for the future of the world economy. However, unless there are dramatic new and appropriate strategies for bringing about collaborations among all countries, rich and poor, on a much larger scale and through better approaches than have existed in the past, the global problems of food insecurity and resource degradation, particularly in poor countries, will be considerable and enduring.

However, long range solutions to poverty are multifaceted, with appropriate strategies required for each country based on its unique situation and location-specific characteristics. Moreover, alleviation of world food insecurity must come through such specific measures as (*a*) better distribution of control over food-producing resources, (*b*) instituting people-oriented public policies, and (*c*) sustainable and broad-based development at the grass-roots level which is socially just, economically efficient, and ecologically sound (SID, 1988).

The experience of India provides a major lesson for developing countries that are characterized by large segments of the rural population depending on food production for livelihood and by the high incidence of poverty, food insecurity and malnutrition: the strategy to improve food security must encompass programmes to increase food production that combine improved technology transfer, price support to food producers and supply of inputs at reasonable prices to the farmers, improvements in food marketing system, employment generation, direct food assistance programmes, and improvement in the access to education and primary health care.

REFERENCES

Acharya, S.S. (1998). Domestic Market Reforms: Should They Precede Agricultural Liberalization. Paper presented in the Annual Conference of Agricultural Economics Research Association of India, 21-22 September, Hisar: Haryana Agricultural University.

Acharya, S.S. (2001). Domestic Agricultural Marketing Policies, Incentives and Integration, In S. S. Acharya and D. P. Chaudhry (Eds), *Indian Agricultural Policiesat the Crossroads*. New Delhi: Rawat Publications, 129-212.

Acharya, S.S. (2002a). Experiences in Rice Policies, Strategies and Programmes in South Asia Region. Paper presented at the Workshop on Harmonization of Policies and Coordination of Programmes on Rice in ECOWAS Sub-Region, 25-28 February. Accra: FAO Regional Office for Africa.

Acharya, S.S. (2002b). Towards Sustainable Food Security at State Level – A Case of Rajasthan with Special Reference to Role of Food Assistance Programmes. Paper presented in Consultation on Food and Nutritional Security, organized by World Food Programme, 25-26 June, Jaipur.

Acharya, S.S. (2002c). Towards Hunger Free Asia: Some Facts and Emerging Issues. Plenary paper presented in IV Conference of Asian Society of Agricultural Economists, 20-22 August, AlorSetar (Malaysia).

Acharya, S.S. (2003). Agricultural Marketing in Asia and the Pacific: Issues and Priorities, In *Proceedings of the Mini Roundtable on Agricultural Marketing and Food Security in Asia*, FAO, Bangkok, RAP Publication 2003/02.

Chand, R. and Jha, D. (2001).Trade Liberalization, Agricultural Prices and Social Welfare, in S. S. Acharya and D. P. Chaudhry (Eds), *Indian Agricultural Policy at the Crossroads.* New Delhi: Rawat Publications, 7-126.

FAO, (2006). Global Perspective Studies Unit, FAO, Rome

FAO, (2016).Crop Prospects and Food Situation. June, Issue No.2: 1-38.

Gebremedhin, T.G. (2000). Problems and Prospects of the World Food Situation. *Journal of Agribusiness,* 18(2): 221-236.

GoI. (2007). National Policy for Farmers. Department of Agriculture and Cooperation, Ministry of Agriculture. pp24.

Narayanan, S. (2015).Food Security in India: The Imperative and Its Challenges. *Asia & the Pacific Policy Studies,* 2(1): 197–209. doi: 10.1002/app5.62.

Patnaik, U. (2003). Food Stocks and Hunger: The Causes of Agrarian Distress. *Social Scientist,* 31 (7-8): 15-40.

Society for International Development (SID). (1988). Objectives and terms of reference of the South Commission. *Development: Journal of the Society for International Development,* 4: 95-98.

Timmer, C. (2014). Food Security in Asia and the Pacific: The Rapidly Changing Role of Rice. *Asia & the Pacific Policy Studies,* 1(1): 73–90. doi: 10.1002/app5.6.

Chapter 2

Role of Nutrition for Agriculture Planning and National Development

Shailesh Kumar and Purushottam

Introduction

Agriculture is an important sector in India where nearly 60 per cent of small farm holder's categorized below poverty line and earn their livelihood. Sound and healthy physical status of them will help in driving maximum agricultural output from their farm land. For that balanced nutrition is very much essential for their body's dietary needs. Deprivation of it is generally manifested in various symptoms of malnutrition such as reduced immunity, increased susceptibility to disease, impaired physical and mental development. As per national Sample Survey report consumption of calories as well as protein is falling in rural India (in the year 2004-05, 8 per cent decline from the base year 1983). According to National Family Health Survey-3 extent of under nutrition amongst state it is found that underweight amongst children was highest in Madhya Pradesh (60 per cent) followed by Jharkhand (57 per cent) and Bihar (56 per cent) and least in Mizoram, Sikkim, Manipur and Kerala. However, even in these states, level of under nutrition is unacceptably high. As per report underweight adults can have lower productivity compared to adults of average weight (World Bank, 2006). Similar study from African country stated that for every 1 per cent increase in height, adults can experience a 4 per cent increase in agricultural wages (Haddad and Bouis, 1991). Thus fate of Indian agriculture can be imagined with these manpower.

Several decades ago the "green revolution" helped alleviate the problem of food scarcity by introducing high-yielding varieties of wheat and rice in India. Now it is time to address dietary quality. Millions of people do not get enough nutritious foods in their diets, especially foods that are rich in vitamins and minerals. According to the 2011 census, farm families constitute the majority of India's population. A high proportion of marginal farm families as well as landless labour families suffer from under-nutrition, because of inadequate income and many of these foods, such as pulses are too expensive for the poor.

Agriculture, nutrition and poverty are tightly interrelated. Poverty and under nutrition reinforces each other (Figure 1). Agricultural growth also tends to enhance poverty reduction more than other forms of economic growth.

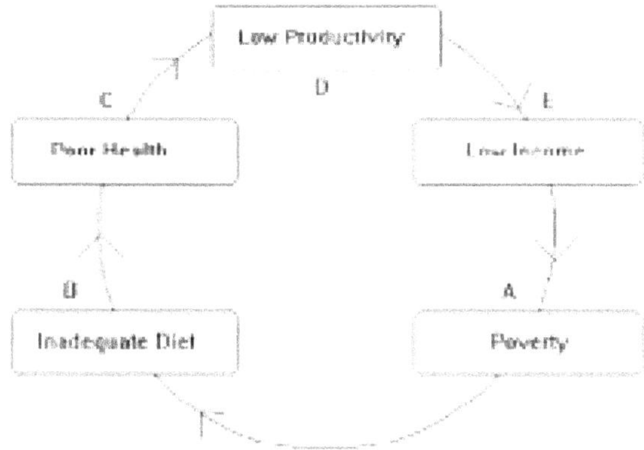

Figure 1: Figure Showing Reinforcing Nature of Poverty, Nutrition.

Poverty reduction from growth in agriculture is on average 2 to 4 times greater than from equivalent growth in other sectors of the economy, which can be attributed to a greater level of poor labour participation in this growth (Shetty, 2015).

In this back drop, National Nutritional Policy (1993) had rightly advocated a comprehensive inter-sectoral strategy between 14 sectors (which are directly or indirectly affect dietary intake and nutritional status of the population for improving nutritional status of all sections of the society. These 14 sectors are namely, agriculture, Food, Civil Supplies & Public Distribution, Education, Forestry, Maternal and Child Health, Food Processing Industries, Health, Information & Broadcasting, Labour, Rural Development, Urban Development, Welfare, Women and Child Development. They are implementing various direct short term (nutrition intervention for especially vulnerable groups, fortification of essential food items *etc.*) and indirect long term (like food security for improved availability of food grains, improvement of dietary patterns through production and demonstration *etc.*) nutrition interventions. Existing programmes and schemes of Government of India are target group specific such as: pregnant and lactating mother, children 0 to 3 years, children 3 to 6 years, school going children 6-14, adolescent girls 11-18 and adults.

Till its existence, 2016 Planning commission played a major role in agricultural planning in India taking into account all factors that were related to the rural sector with sole objective to enhance the total output of agriculture and boost the economic growth of the country through development in human resources. Now Niti Ayog has taken over this role.

In order to understand and address the failure of economic and agricultural growth to make significant inroads into reducing malnutrition in India, **Tackling the agriculture–nutrition disconnect in India** (TANDI) was facilitated by the International Food Policy Research Institute, with funding from the Bill & Melinda Gates Foundation during 2010–2012. As per the study agriculture–nutrition pathways (Figure 2) in our country operate in following ways (Kadiyala *et al.*, 2014):

1. Agriculture as a source of food: Farmers produce for own consumption.
2. Agriculture as a source of income for food and nonfood expenditures: As a major direct and indirect source of rural income, agriculture influences diets and other nutrition-relevant expenditures.
3. Agricultural policy and food prices: Agricultural conditions can change the relative prices and afford-ability of specific foods and foods in general.
4. Women in agriculture and intra household decision making and resource allocation may be influenced by agricultural activities and assets, which in turn influence intra household allocations of food, health, and care.
5. Maternal employment in agriculture and child care and feeding: A mother's ability to manage child care may be influenced by her engagement in agriculture.
6. Women in agriculture and maternal nutrition and health status: Maternal nutritional status may be compromised by the often arduous and hazardous conditions of agricultural labor, which may in turn influence child nutrition outcomes.

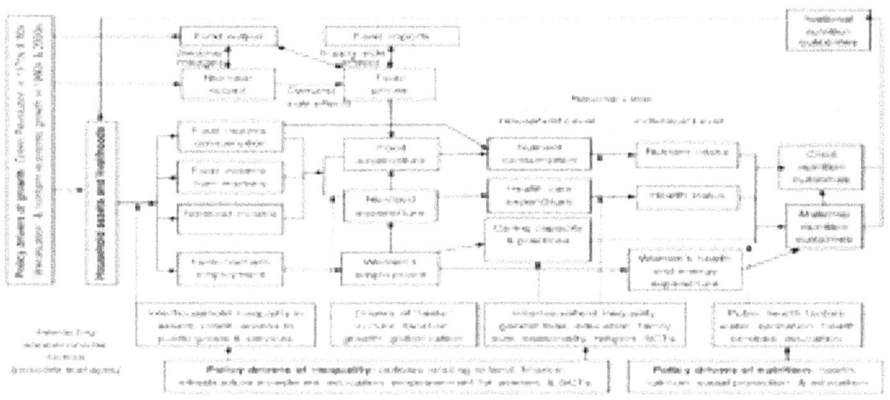

Figure 2: Mapping the Agriculture–Nutrition Pathways in India.

Pathways 1–3, consider impacts of agriculture as an economic sector, whereas pathways 4–6 assumes gender influence nutritional matter. Higher incomes predict increased higher calorie consumption in cross-sectional studies, but calorie consumption has seemingly been declining over time. Field data supports interactive influence of farm sizes and irrigation on the diversification of crop production and food consumption, as well as income generation from livestock rearing. It not only contributes in income generation, but also directly influences diets.

Different Phases of Agriculture and Extension Policies and Programs in India in Relation with Nutrition

In general, agriculture and extension policies and programs in India can be broadly categorized into following five different periods:

1. Pre-independence period
2. Independence to green revolution
3. Green revolution to liberalization reforms
4. Liberalization reforms to contemporary
5. Contemporary period.

An overview of different phases of agriculture and extension policies and programs in India in relation with nutrition reveled that during **Pre-independence period** extension and agriculture development programs (such as Gurgaon project, Shantiniketan project, Martandam project *etc.*) were sporadic in nature, initiated by either by influential persons or the social organization mainly guided guided by the colonial interest rather than need and interests of natives.

During Independence to green revolution phase institutional reforms like abolition of zamindari system; enactment of tenancy laws and ceilings on Landholdings, Community Development, and National Extension Program provided a basic platform of rural development. Panchayat Raj institutes were started for decentralization of decision making. During 1962 Intensive District programmes was started followed by high yielding varieties programmes. The Indian Council of Agricultural Research (ICAR) began its participation in agricultural extension through National Demonstrations in 1964. The Food Corporation of India (FCI) and Commission on Agricultural Costs and Prices (CACP) were established in 1964-1965 to oversee the country wide procurement, distribution and stocking of food grains and recommend minimum support prices of agricultural crops to the Government, respectively. All these efforts led to increase in productivity of rice and wheat significantly, what we called it as Green Revolution. But nutritional aspects were not touched.

In Green revolution to liberalization reforms phase self sufficiency in food production caused by green revolution remained restricted to the well endowed, irrigated areas of the country and rice and wheat crops only. Of late 1980s, ill effects of green revolution were also experienced such as nutrients mining; increased risk of agriculture-associated diseases and environmental degradation. Government launched Krishi Vigyan Kendras (KVKs) or Farm Science Center's, Lab to Land programs, and Operational Research Programs by the 1970s. To meet the gap between research and extension, Training and Visit (T&V) extension management system (1974) was started for rapid dissemination of broad-based crop management. Till the end of the 4th plan, India's main emphasis was on the aggregate growth of the economy and reliance was placed on the percolation effects of growth.

In the face of continuing poverty and malnutrition, an alternative strategy of development comprising a frontal attack on poverty, unemployment and malnutrition became a national priority. From the beginning of the 5th plan, there was a shift in strategy with initiation of a number of interventions like Integrated Rural Development Programme (IRDP) and employment generation schemes like Jawahar Rozgar Yojana, Nehru Rozgar Yojana and Development of Women and Children in Rural Areas (DWCRA) to increase the purchasing power of the poor, to improve the provision of basic services to the poor and to devise a security system through which the most vulnerable sections of the poor (*viz.* women and children). The era of 1980s saw regional spread of agricultural growth and crop diversification. There was shift from cereal to other food crops. Accordingly, government started technological mission on oilseed in 1986 and strengthening of Operation Flood (which was initiated in 1970) to bring about white revolution.

In liberalization reforms to contemporary stage (1991 to 2014) observed yield plateau or technology fatigue in agriculture sector. Government started shift in policy emphasizing to liberalizing trade in agriculture. The National Agriculture Policy (2000) highlighted the need for restructuring agricultural extension ignoring the nutrition aspect (Dev, 2012). In 11th Five Year Plan (2007-12) nutrition has assumed a central role and gained highest attention and priority in general and improving nutrition of working population in agriculture in particular. The monitorable targets of 11th Five Year Plan included key nutrition goal:

- ✰ Reduce malnutrition among children in the age group of 0-3 to half its present level and
- ✰ Reduce anaemia among women and girls by 50 per cent by the end of 11th Five year Plan

In 12th Five year Plan (2012-17) Rs. 1, 0175 Crore has been allocated for major areas (Multi sector nutritional programme in 200 high burden district, additional support for intensive action to programmes, strengthening of systems and strengthening of institutions).The 12th Five Year Plan Approach Paper indicated that output of fruits, vegetables, and protein rich food items needs to grow at a faster pace than production of cereals, to meet the rising demand in these items. Rashtriya Krishi Vikas Yojana (RKVY), National Horticulture Mission (NHM), integrated scheme of oilseed pulses oil palm and maize and National Food Security Mission (ISOPOM) were started for improvement in Indian citizens' nutrition security, especially in districts with a substantial overlap between poverty and under nutrition. RKVY encourages effective integration of livestock, poultry and fish farming with the crop sector. NHM resulted in increasing production and area under horticultural commodities. Along with RKVY, National Rural Livelihoods Mission and state-level nutrition missions are worth serious exploration other programs, like the Mahatma Gandhi National Rural Employment Guarantee Act (NREGA), also have an impact on nutrition of women and children. Disadvantaged workers participating in the NREGA shown positive correlation between program participation and anthropometric indicators of health outcomes. National Commission for Enterprise in the Unorganized Sector identified several positive externalities created by NREGS, including-reduction in distress out-migration, improved food security with wages being channeled into incurring expenses on food, health, education and repaying of loans, employment with dignity, greater economic empowerment of women workers and sustainable asset creation (NCEUS,2009). In extension also, new initiatives like National Agricultural Technology Project (NATP), Agricultural Technology Management Agency (ATMA) and National Agricultural Innovation (NAI) program were started. Unlike earlier programs, these were characterized by bottom up approach, demand driven, location specific, women participation, private sector participation and diversified agriculture. Diversification and women empowerment are linked to nutrition implicitly.

The Biovillage Paradigm is considered as 'Eco-technology in Action', was initiated in 1993 where the sustainability is ensured through appropriate interventions/technology dissemination that is environment friendly and provides opportunities for ensuring livelihood security. Its Pro-women orientation is ensured through empowering women with skills and access to credit facilities/finances and reducing the drudgery of work at home and outside. Thus the Biovillage is an approach, which strengthens the capacity of the rural community to blend sustainable natural resource management with livelihood security through economically feasible, socially acceptable, ecologically viable and gender-sensitive interventions. This approach encourages a value addition process within the system, to generate sustainable eco-jobs and income in the village, based on principles that are pro-nature, pro-poor and pro-women. Biovillage is eco technology in action Biovillage model focuses on:

- Enabling the community to understand the potentials of sustainable natural resource management.
- Introducing various livelihood opportunities in on-farm and non-farm sectors, blending traditional knowledge with frontier technology.
- Strengthening the human resource development through skill and knowledge empowerment.
- Building grassroot institutions such as Self-Help Groups, Farmers Associations, Federations, *etc.* which will take up the development

Contemporary phase or post planning commission phase, is marked by new initiatives taken up by Government of India. Government has initiated the steps for doubling the income of the farmers, diversification of agriculture by starting fund for livestock, balanced use of fertilizers through soil health cards, e NAM-e market platform, Pradhan Mantri Shichaiyi Yojana, Paramparagat KrishiVikasYojana, Fasal BimaYojana, Digital India, health insurance for poor, different schemes for the women and child. These all are expected to contribute to the nutrition security through better health, more income, income transfer and availability and access nutritious foods to the consumers.

Now there is a recognition that at the national, state, and district levels agricultural extension has the potential for integrating nutrition extension through its network of established outreach channels that connect with farmers (Babu *et al.*, 2016). In this respect multipluristic extension approaches such as :collective approach, market extension (Market led extension), farmer led extension, women empowerment and education, use of ICT, focus on diversified agriculture, farming systems approach has great potential to bridge the gap of nutritional imbalances in different sections of Indian society.

Collective approach such as Self help group model, cooperative model, contract farming provides a cushion against national and international price volatility. Thus affordability component of nutrition security can be also fulfilled by collectives of farmers. Issues of hunger and malnutrition through strengthening capacity of the public agency, supporting private sector marketing extension through extensive use of media, internet and IT in information and technology dissemination to the farmers is the need of hour. Farmer led extension can be started by organizing farmers into functional groups, such as Self Help Groups (SHGs), Farmer Interest Groups (FIGs), Commodity Associations (CAs) and/or other types of Farmer Organizations (FOs). These FOs can provide an effective channel for both dissemination of technology to smallholder and feedback to research and extension (Sharma, 2002). Maharashtra State Grape Growers association (MRDBS) is excellent example of such approach. This potential of social capital building can be utilized for reducing under nutrition and poverty. More than 80 per cent of rural women engaged in the labor force work in the agriculture sector strengthening their position both within the agricultural sector and within the household can significantly improve households' nutrition and health. There is evidence that women's employment does have beneficial effect on household nutrition, female education also has a strong inverse relationship with IMR, educated women have greater roles in household decision making, particularly those relating to nutrition and feeding practices (National Nutrition Policy, 1993). ICT can play important role in awareness creation various issues of nutrition and need for diversified diet, change in dietary habit of the consumers and production pattern of the farmers. Focus on diversified agriculture such as livestock, poultry, fishery, fruits and vegetables *etc.* will ensure easy availability of micronutrient rich foods. The Farming System (FS) approach considers the farm, the farm household and off farm activities in a holistic way. In order to integrate nutrition and agriculture or livelihood security Dr. M.S. Swaminathan and colleagues have designed the Farming System for Nutrition (FSN) model dealing with problem of malnutrition (Das *et al.*2014, Nagarajan *et al.* 2014).

The Indian budget (2013-2014) outlines a plan with Rs 200 crores to develop Nutri-farms in malnutrition affected districts, where iron-rich pearl millet & finger millet, zinc-rich rice and wheat, and protein-rich maize will be grown. These Nutri-farms have been piloted in the 100 districts of 9 states in India that are most affected by malnutrition. It is a major initiative for marrying nutrition and agriculture by establishing Nutri-farms. The detail objectives of the programme are as follows:

1. Demonstration of improved production technology to promote cultivation of Nutri-rich crop varieties.
2. Encouragement of commercial cultivation of specified Nutri-rich crop varieties through self help groups (SHG).
3. Development of supply chain of Nutri-rich produces to vulnerable sections of population.

Conclusions

It is time for reorientation of Indian agriculture policy from productivity to diversity and from food security to nutrition security. Farm policies that encourage research on improving productivity and quality of the nutrient rich commodities would stimulate the production system resulting in high value commodities economically accessible to disadvantageous groups of the society. Public information campaign by government through mass media will likely to create awareness of farmers and consumers about the nutrition problem and how to secure. Implementing land reform (both tenural reforms and ceiling laws measures) and related issues would reduce the vulnerability of the landless and the landed poor. Ultimately, all these will help in strengthening of agricultural workforce for nation building.

REFERENCES

Babu S.C., Singh M., Hymavathi T.V., Umarani K., Kavitha, G. G. and Karthik S. (2016) Improved nutrition through agricultural extension and advisory services: case studies of curriculum review and operational lessons from India, World Bank Group Report Number 94887

Das, P.K., Bhavani, R.V., and Swaminathan, M.S. (2014). A farming system model to leverage agriculture for nutritional outcomes. Agric Res 3: 193-203.

Dev, S.M. and Sharma, A.N. (2010). Food security in India: Performance, challenges and policies. Oxfam India working paper pp.46.

Dev, S.M. (2012). Agriculture-nutrition linkages and policies in India.Indira Gandhi Institute of Development Research, Mumbai, India pp. 1-25.

Haddad, L.J., and Bouis, H.E. (1991). The impact of nutritional status on agricultural productivity: Wage evidence from the Philippines. Oxford Bulletin Economics Statistics 53: 45-68.

Kadiyala, S., Joshi, P.K., Dev, S.M., Kumar, T.N., and Vyas, V. (2011). Strengthening the role of agriculture for a nutrition secure India. Indian Policy Note, IFPRI, pp. 4.

Nagarajan, S., Bhavani, R.V., and Swaminathan, M.S. (2014). Operationalizing the concept of farming system for nutrition through the promotion of nutrition-sensitive agriculture. Current Science 107: 959-964.

National Nutrition Policy (1993). Government of India, Department of Women and Child Development, Ministry of Human Resource Development, New Delhi, pp. 1-25.

NCEUS (2009). The challenge of employment in India: An informal economy perspective, Volume I-Main Report. New Delhi, India.

Sharma, R. (2002). New policy framework: Reforms in agricultural extension. Rev Agric 37.

Shetty, P. (2015). From food security to food and nutrition security: Role of agriculture and farming systems for nutrition. *Current Science*, 109: 456-461.

Visvanathan, N. (2013). Proceedings: Tracking hunger and malnutrition for food and nutritional security in India. A policy consultation, Jawaharlal Nehru University, New Delhi, pp. 1-178.

World Bank (2006). Repositioning nutrition as central to development: A strategy for large-scale action: Overview. Directions in Development, Washington DC, USA.

Chapter 3

Impact of Physical Resources, Farming Systems, Cropping System and Agricultural Inputs on Food and Nutritional Situation

Ragini Ranawat

Introduction

India's population of 1.3 billion (Worldometer, 2015) is around 18 per cent of the world's. It is further estimated to reach 1.6 billion by 2030 (Population Division, Department of Economic and Social Affairs, UN, 2015). Clearly, Indian agriculture systems have a huge responsibility to ensure secure access to food by every one of its citizens, now and for the future. Besides, with 58.2 per cent of the Indian population dependent on agriculture sector for its livelihood (Committee on Agriculture, 2013), the contribution of agriculture to the country's Gross Domestic Product (GDP), which is 14 per cent currently, will determine the economic benefits to the large section of the population.

The relationship between agriculture and human nutrition, or from food production to food consumption is intuitively direct, but in practice is quite complex. Increased food production should lead to greater food availability, access, and ultimately improved food intake and diets. Yet the persistence of malnutrition as a global public health concern despite the successes in increasing agricultural production belies any notion that malnutrition and under nutrition can be solved entirely from the supply side by increasing agricultural production. The question of how agriculture can more effectively contribute to improved nutrition outcomes therefore requires an answer that encompasses factors other than food supply, and that takes into account other sectors in addition to agriculture that contribute to nutrition.

India is a clear example of the failure of policies framed to achieve improved food and nutrition security. While it is estimated that between 1966 and 2007, the *per capita* availability of calories increased by nearly 25 per cent in South Asia, with India having an amount of food sufficient to feed its population, the prevalence of child malnutrition here is among the highest in the World. While the green revolution improved food productivity significantly, the role of healthcare, childcare, and diverse and quality foods for household food and nutrition security was less emphasized.

First Millennium Development Goal states the target of "Halving hunger by 2015". Sadly, the recent statistics for India present a very gloomy picture. India currently has the largest number of undernourished people in the world and this is in spite of the fact that it has made substantial progress in health determinants over the past decades and ranks second worldwide in farm output.

But there are ways to change the scenario. Modifications in farming and cropping systems along with targeted research on agricultural inputs can improve upon the production of nutritionally balanced food along with livelihood security for the population.

Relation of Physical Resources with Food and Nutrition Situation in India

India is rich in bio-diversity. India's geographical condition is unique for agriculture because it provides many favourable conditions. There are plain areas, fertile soil, long growing season and wide variation in climatic condition *etc.*

Apart from variation in landform, the country has varieties of climatic conditions, and soil types. These physical variations along with other factors like availability of irrigation, use of machinery, modern agricultural inputs like High Yielding Varieties (HYV) of seeds, insecticides and pesticides have played their respective roles in the evolution of different farming practices in India. Majority of farmers in India practice subsistence farming. This means farming for own consumption. In other words, the entire production is largely consumed by the farmers and their family and they do not have any surplus to sell in the market. In this type, landholdings are small and fragmented. Cultivation techniques are primitive and simple.

With availability of optimum resources, the farming practice generally shifts to commercial type. In this system, farmers use inputs like irrigation, chemical fertilizers, insecticides, pesticides and High Yielding Varieties of seeds *etc.* For example, rice farming in Haryana is mainly for commercial purpose as people of this area are predominantly wheat eaters.

Agriculture systems, being heavily resource-intensive, interact with natural resources and environment at a large scale. Around 50 per cent of India's total land area is under agriculture, using around 90 per cent of the total water withdrawals in the country (FAO, 2015). Agriculture sector is the third-largest consumer of power in India; it accounted for 19 per cent of the total power consumption in 2011.

Apart from the shrinking resource scenario, natural resources are also witnessing resource degradation due to various anthropogenic factors that affect the quality of resources available for practicing agriculture. India is losing 5,334 million tonnes of soil every year due to soil erosion because of indiscriminate and excess use of fertilizers, insecticides and pesticides over the years (Roy *et al.*, 2009).

Irrigation would continue to play an unquestionable role in achieving food self sufficiency, creating grain surpluses, stabilizing food prices, sustaining agricultural growth, absorbing labour force in rural areas, and alleviating rural poverty; all of which are vital for food security. Recent research by many scholars and institutions has shown that the future water supplies are going to fall short of the demand from different sectors. This will pose a threat to food security at the aggregate level. The problems would be acute in semiarid Gujarat, Tamil Nadu, Rajasthan and Maharashtra, which also experience ever increasing demand for water in all sectors. Mainly, food security of the poor will be at risk, as they would face severe resource constraints not only in accessing water, but also in investing in land and water management.

WASSAN (Maharashtra and Watershed Support Services and Activity Network, Hyderabad) and Pravah (Jharkhand) have demonstrated the ways in which communities can share their resources by forming teams for collective buying of inputs from market, develop water sharing, electric motor sharing groups and therefore using the available resources most optimally for maximizing profits (Bhamra *et al.*, 2016).

The resource sharing community models aim to efficiently and equitably use the resources at disposal by the community. These resources range from natural resources- water to physical and human resources available at community's disposal. The use of limited set of water resource in WASSAN's case that is shared by the community allows for increasing efficiencies and hence allowing more food production per drop.

Natural resource sharing models allows for efficient resource use that increases productivity and enhances income of the farmers. Other models studied that share human and physical resources allow the community and group, together, reduce costs of farming by efficiency in the division of labour and increase bargaining power and buying of inputs in bulk. All this allows for reducing the costs and increase in the selling price of commodities, allowing for higher income generation thus increasing the purchasing power of the farming community and ensuring better access to balanced nutrition.

Farming System's Link to Diet Quality

According to FAO, a farming system is defined as a population of individual farm systems that have broadly similar resource bases, enterprise patterns, household livelihoods and constraints, and for which similar development strategies and interventions would be appropriate. The classification of the farming systems of developing regions has been based on the following criteria:

- Available natural resource base, including water, land, grazing areas and forest; climate, of which altitude is one important determinant; landscape, including slope; farm size, tenure and organization; and
- Dominant pattern of farm activities and household livelihoods, including field crops, livestock, trees, aquaculture, hunting and gathering, processing and off-farm activities; and taking into account the main technologies used, which determine the intensity of production and integration of crops, livestock and other activities.

Globally, India is the third largest producer of cereals, with only China and the USA ahead of it. India occupies the first position in milk production and is the third largest producer of fish and second largest producer of inland fisheries in the world. The fisheries sector also provides livelihood to some 11 million people involved fully/partially in fisheries and on subsidiary activities connected with the sector. India ranks first in respect of cattle and buffalos and second in goats, third in sheep and seventh in poultry population in the world and nearly 90 million people work in the livestock sector.

There are two different pathways that can precisely link the kind of food produced and farming systems on nutrition situation of India:

According to Agriculture and Rural Development Department of World Bank (2007), Household production for the household's own consumption is the most fundamental and direct pathway by which increased production translates into greater food availability and food security. The different types of foods produced determine the impact of the production increase on diet quality. The production of more staples leads to mainly quantitative increases in energy intakes. Increased production of fruit, vegetables, dairy foods, eggs, fish, and meat can likewise raise macronutrient intakes, but with greater impacts on micronutrient intakes that can close dietary gaps in essential nutrients like iron, zinc and vitamin A. These more micronutrient-rich food sources can also make staple foods more palatable, and lead to higher still energy intake. Animal source foods in particular are energy and protein-dense. Given favorable intrahousehold processes of food distribution, these developments can greatly improve the food intake and nutrition of the more vulnerable members of the household.

Increasing market orientation brings a second pathway into play, one in which the relative importance of subsistence-related production and income-related production switch. Income now becomes the principal determinant of nutrition outcomes and household production for own consumption assumes a supplementary role. Technology becomes more important relative to the household's resource endowment, and the selection of crops to be grown is based principally on their tradability and the price they are expected to command in local markets. The same technologies that enable export also enable import, and the variety of food sources available to consumers is likely to expand, making possible more complex and higher quality diets. Intensification of staple food production, and opportunities for livestock, fish and fruit and vegetable production, can also generate employment for landless or land-poor individuals. Increases in income can of course be used for purposes other than and in addition to household food security, but the share that is used to purchase foods can have a variety of nutritional effects. Extra income may be used to buy more food, or to buy higher quality food that increases the consumption of micro-nutrients without raising caloric intakes at all.

Cropping Systems in India and the Nutrition Situation

FAO defined cropping system as a community of plants which is managed by a farm unit to achieve various human goals. The latter include food, fibre and other raw materials, wealth and satisfaction. Farmers are part of the system. They are able to set, or modify, their own goals, so two farms with identical climates and soils may be managed with different aims to achieve a different mix of outputs. Food security is the most basic output from a cropping system. For a subsistence household, all other outputs are secondary. Thus, the first property of a cropping system is its effectiveness; that is its ability to meet the needs of the goal-setter (the farmer). Production is a most obvious output and measure of the activity of a cropping system.

Figure 1: Sustainable Crop Production Intensification Overview (FAO, 2010)

From many perspectives, agriculture in the country today is in a state of crisis. A national survey some years back revealed that given a choice, 40 per cent of farmers in India would not like to be in farming. Wheat and rice together constituted 78 per cent of total food grains production in 2009-2010 (Economic Survey, 2013). Coarse cereals accounted for 15 per cent of the food grain production in 2009-2010. The growth rate in area of total coarse cereals comprising jowar, bajra, ragi, maize, small millets and barley, was negative in all the three periods 1980-1981 to 1989-1990, 1990-1991 to 1999-2000 and 2000-2001 to 2011-2012.

With regard to pulses, while during the 1980s there was negative growth in total area under pulses and growth in production and yield was 1.52 and 1.61 per cent, respectively, during 2000-2001 to 2011-2012 whereas area and production grew by 1.6 and 3.69 per cent, respectively, growth in yield at 2.06 per cent was almost stagnant. There has been progressive decline in per capita availability of pulses; it fell from 69 grams in 1961 to 32 grams in 2005. The requirement was estimated to be 21.3 million tonnes by 2012. The Economic Survey 2012-2013, reports the estimated production of pulses in 2011-2012 as 17.09 million tones, indicating a wide gap in demand and supply.

With such inadequacies in the production of pulses, oilseeds and coarse grains, the availability of nutritionally balanced food to every individual from any economic background seems very difficult. Our major thrust on growing rice and wheat indeed brings the option of optimum energy to our plates, but fails miserably when it comes to curbing the "hidden hunger". In India, the prevalence of stunting among less than five years old children is 48 per cent (moderate and severe); wasting is 20 per cent (moderate and severe) and an underweight prevalence of 43 per cent (moderate and severe) which is highest in the world. The status of anemia is more horrifying.

The nutrition situation in India can be improved by encouraging cultivation of pulses, oilseeds and coarse grains with equal importance to fruits and vegetables which can ensure not only availability but access (with upgraded supply chains) of healthy and nutritious options to its population. We have done it with wheat and rice and we can very well do it with other crops. The very change in farming systems of our country can indeed improvise the nutrition security situation.

Impact of Agricultural Inputs

Supply of key farm-inputs at reasonable prices has been another important instrument of food security policy in India. The twin and conflicting objectives of assuring remunerative prices to farmers and making available food to the consumers at affordable prices were reconciled inter alia by keeping the prices of inputs at reasonable levels. This led to the emergence of input subsidies. Input subsidies in Indian agriculture are of two broad categories, *viz.* direct or explicit and indirect or implicit. Direct or explicit subsidies are in the nature of payment to the farmers to meet part of the cost of inputs like seeds, plant protection chemicals, or machines. These are usually made available to specific target groups like marginal or small farmers and account for a small proportion of the total input subsidies. The indirect or implicit subsidies arise on account of the manner of determination of sale prices of inputs. There is no explicit payment of subsidy to the farmers. The inputs are supplied at a price or user charge lower than the cost of production, which amounts to implicit subsidization. Implicit or indirect subsidies on fertilizers, electricity for irrigation and canal water are the major input subsidies in the Indian agriculture.

The availability of inputs like high yielding and disease resistant varieties, fertilizers enhancing production and productivity; along with implements that can reduce the cost of cultivation has a big impact on food and nutrition situation as it increases the availability of food and aids in controlling price hike thus ensuring access.

Conclusion

Agriculture plays a key role in increasing food availability and incomes, supporting livelihoods and contributing to the overall economy, and is thus central to improving food and nutrition security. Ways in which agriculture can sustainably contribute to improving dietary diversity and nutrition outcomes include support for: agricultural extension services that offer communities information and improved inputs such as seed and cultivars for better crop diversity and biodiversity; integrated agro-forestry systems that reduce deforestation and promote harvesting of nutrient-rich forest products; aquaculture and small livestock ventures that include indigenous as well as farmed species; education and social marketing strategies that strengthen local food systems and promote cultivation and consumption of local micronutrient rich foods; and biofortification via research and development programmes that breed plants and livestock selectively to enhance nutritional quality.

Some of the most important emerging themes for nutrition-friendly agriculture, essential as part of a broader nutrition-sensitive development framework can be pro-poor food production systems which can provide direct support to rural smallholder production and urban and periurban food systems to expand, enhance and sustain people's ability to procure and use the amount and variety of food required to be active and healthy. In conjunction with this, improving agricultural production practices to address environmental concerns such as biodiversity, sustainable use of resources, and livestock sector reform must also be given priority to.

REFERENCES

Bhamra, A., Hazra, M. and Niazi, Z. (2016).Buidling Resilience in Agriculture For Food Security. Heinrich Boll Foundation. Development Alternatives.Pp 35.

Economic Survey. (2013). Ministry of Finance. Government of India.

FAO. (2010). An Ecosystem Approach To Sustainable Crop Production Intensification.

FAO.(2015). AQUASTAT. Retrieved December 21, 2015, from FAO: http: //www.fao.org/nr/water/aquastat/countries_regions/ind/index.stm

Roy, D., Chattopadhyay, P. and Tirado, D. (2009). Subsidizing Food Crisis. Greenpeace India.

World Bank (2007). World Development Report 2008: Agriculture for Development. Washington D.C. © World Bank. http: //openknowledge.worldbank.org/ handle/10986/5990 License: cc BY3.0IGO.

Worldometer. (2015). Retrieved December 21, 2015, from Worldometers: http: // www.worldometers.info/worldpopulation/india-population/

Chapter 4

Food Production, Food Distribution and Nutritional Status

Neha Nagda, Upendra Singh, Anuj Yadav

Introduction

India is the world's second biggest populated country. Thus, the country requires more food production so as to mitigate the starving conditions. Since independence, our country is continuously facing the emerging challenges *i.e.* to achieve food security as well as to eradicate poverty. To overcome this situation, Agriculture promotion came in practice. It started food production in the country to feed millions of people. This leads to acute food shortage and Indians have to borrow their wheat from United States under the PL 480 agreement. The Green revolution technique emerges the high yielding of seeds and efficient use of fertilizers and irrigation system. It continued the policy of expanding agriculture lands. The ICAR (Indian Council for Agricultural Research) under the ministry of agriculture played a vital role in Green Revolution era. ICAR developed the high yield variety of staple crops. The 'Green Revolution' resulted peak output of approx.131 million tons during 1978-79. This leads India to be one of the world's biggest agricultural producers. Increasing population leads to requirement of self-sufficient agriculture produce. This led to formulation of agriculture improvement, public distribution system and price support system for farmers. Public Distribution System aids to have efficient availability of produce.

Food Production in India and World Wide

Food production is the root for food security. According to FAO's State of Food Insecurity in the World (2001). Food security is defined as, "A situation that exists when all people, at all times, have physical, social and economic access to sufficient, safe and nutritious food that meets their dietary needs and food preferences for an active and healthy life".

The growth of food grain production during the 1970s and 1980s was mostly due to institutional efforts by uplifting the levels of technology used in agriculture through research and extension, investments to develop rural infrastructure and human capabilities, procurement at minimum support prices and the strengthening action of supportive institutions like the Food Corporation of India (FCI). The Economic Survey of 2012-2013 gives comparative results of the area, production and yield of crops during 1980-1981 to 1989-1990, 1990-1991 to 1999-2000 and 2000-2001 to 2011-2012. With regard to rice and wheat, while the compound annual rate of growth (CAGR) in area was marginal at 0.41 and 0.46 per cent, respectively during the 1980s, growth in both production and yield was above 3 per cent. The CAGR of area improved to 0.68 per cent for rice and 1.72 per cent for wheat between 1990-1991 to 1999-2000, but it fell for both production and yield in the case of rice and yield in the case of wheat. The subsequent decade 2000-2001 to 2011-2012, an improvement of area under wheat was observed but the CAGR of production fell for both rice and wheat. This suggests that in these two crops there is need for updated research to boost production and productivity. Further, it is necessary to note that these two crops together constituted 78 per cent of total food grains production in 2009-2010.

After a record harvest of food grains and other crops in 2013-14 with timely distributed monsoon rains, our country faced rainfall below Long Period Average (LPD) *i.e.* 12 per cent and14 per cent respectively. This results into declining global price of agricultural produce in response to record production. India's farm sector growth measured by GVA declined to 0.2 per cent in 2014-15 and is more or less remains unchanged in 2015-16. The data signifies that despite of sharp decline in rainfall, there is no such change observed in the agricultural production. The novel technologies in farming practices have reduce the impact of less rainfall over agricultural production. Timely intervention of government has mitigated the crop failure and thus, overcome the drought losses up to an extent. The drop in commodity price leads to increase the supplies of wheat and rice through Public Distribution System (PDS) from government stocks.

Current Scenario of Food Production

In 2015, according to UN update report, World Gross Product (WGP) expanded at a slightly faster rate at 2.8 per cent and 3.1 in 2016. The World Bank in its Global Economic Prospects Report (June 2015), 13 estimates global economic growth at 2.8 per cent in 2015-16 and expected to rise up to 3.2 per cent in 2016–17.

India stands as the third largest producer of cereals, than that of China and USA. Apart from cereals, India also grabs first position in milk production, third largest producer of fish and second largest producer of inland fisheries in the world. The dairy industry provides employment to nearly18 million people. Of these, 70 per cent are women and 67 per cent have no access to land, credit or technology. Per capita availability of milk rose from 124 g/day in 1950-1951 to 176 g/day in 1990-1991 to 290 g/day in 2011-2012, a figure obtained when compared globally. Despite of highest milk production in the world, the productivity per animal is extremely low compared to international standards. Considering around 4.4 per cent of the global fish production, the fisheries sector gave employment directly and indirectly to around 145 million people.

In the case of coarse cereals, the growth rate in area comprising *jowar*, *bajra*, maize, *ragi*, small millets and barley, was negative in all the three periods 1980-1981 to 1989-1990, 1990-1991 to 1999-2000 and 2000-2001 to 2011-2012. This could have been either due to relatively dry areas or shifting to other crops. However, growth in production and yield for coarse grains which was 0.40 and 1.62 per cent, respectively in the 1980s improved significantly to 3.01 and 3.85 per cent, respectively in the 2000-2001 to 2011-2012 period, majorly due to improvement in maize. Meanwhile, a sharp decline in per capita availability of pulses was observed; it fell from 69 grams in 1961 to 32 grams in 2005. The requirement was estimated to be 21.3 million tonnes by 2012. The Economic Survey 2012-2013 reports the estimated production of pulses in 2011-2012 as 17.09 million tonnes, indicating a wide gap in demand and supply. On the oilseeds front, per capita annual consumption of vegetable oil in the country at 14.10 kg is far below the global average of 23.60 kg. The production of oilseeds though has increased in recent years from 184.40 lakh tons in 2000-2001 to 297.99 lakh tons in 2011-2012, has not met with the demand for edible oils in India. A large portion of edible oil is compensated through import of palm oil from Indonesia and Malaysia. Any imbalance in the supply of palm oil from these countries will put the country in a difficult situation. A large quantity of the global production of vegetable oils is being utilized for production of bio-diesel in Europe and North America. Such non-food use of edible oils ultimately reduces their availability and pushes up their prices.

From many perspectives, agriculture in our country today is in a state of crisis. A national survey revealed that 40 per cent of farmers in India are no more having interest to be in farming. Farming is increasingly seen as an unviable activity, characterized by rising input costs and un-remunerative prices. It has to be understood that nearly 80 per cent of the land holdings in India are below 2 hectares in size. Unlike in industrialized countries where only 2 to 4 per cent of the population depends upon farming for their work and income security, agriculture is the backbone for two-third of India's population. In effect, farmers also constitute the largest proportion of consumers.

The Green Revolution had been largely confined to irrigated farming areas and to rice and wheat. The per unit area productivity of Indian agriculture today is much lower in India as compared to other major crop producing countries (Table 1). There are wide gaps in the yield among and within States. China has yield rates far ahead of India in all the three major food grain crops cultivated. As the Economic Survey of 2012-2013 has observed - "improvement in yields holds the key for India to remain self-sufficient in food grains".

Table 1: Yield of Principal Crops in different Countries in 2008 (kg/ha)

Country	*Crop*		
	Paddy	*Wheat*	*Maize*
Brazil	4229	NA	4086
China	6556	4762	5556
India	3370	2802	2324
Indonesia	4895	NA	4076
U.S.A	7672	3018	9658
Japan	6488	NA	NA
Egypt	9731	6501	7977

Hence, improving small farm production and its productivity as a single development strategy, can make the greatest contribution to eliminate hunger and poverty. Experience of countries that have succeeded in reducing hunger and malnutrition shows that growth originating in agriculture; in particular the small holder sector is at least twice as effective in benefitting the poorest as growth from non-agriculture sectors. The World Bank's World Development Report 2008 that focused on 'Agriculture for Development' had also emphasized in a similar vein, "Using agriculture as the basis for economic growth in the agriculture-based countries requires a productivity revolution in smallholder farming". As stated earlier, higher productivity requires higher investment in agriculture and agriculture research - a fact that needs to be heeded by the policy makers. India will undoubtedly remain a predominantly agricultural country during most of the 21st century, particularly with reference to livelihood opportunities. Therefore, there is a need for both vision and appropriate action in the area of shaping our agricultural destiny.

Government Initiatives in Agriculture Sector

There are milestone steps taken by government to combat the declination of agriculture sector in the country.

1. During the past one year, the government has given different perspectives to many conventional models in which India operated, including in Agriculture Sectors. Indian agriculture is typically dependent on the summer monsoon rains, making it vulnerable to the vagaries of weather. In a bid to remove persistent difficulties in Indian agriculture and revive growth, the government has approved Rs.50,000 Crore irrigation package recently under the Pradhan Mantri Krishi Sinchai Yojana (PMKSY). The scheme will also promote précised irrigation technologies, enhance recharge of aquifers and introduce sustainable water conservation practices. This scheme, when implemented, will ensure that all farm lands get water for cultivation, reducing the dependency of Indian agriculture on the monsoon and reducing the year-to-year yield fluctuations. In this context, inter-linking of large rivers of India *i.e.* Krishna and Godavari

should be done. Thus, it should also be taken up more seriously, although this would require a massive amount of political and financial capital.

2. Another hurdle facing Indian agriculture is a non-transparent agricultural marketing system. It prevents fair price to farmers, mainly because of market rigidities. It also leads to wide intra-regional price disparities and wide price fluctuations. To resolve this issue, the government also recently took the first step to create a national market for agricultural produce through an electronic platform. This will provide farmers and traders the opportunities to purchase and sale agricultural commodities in a transparent manner. It would also increase farmer's access to markets through warehouse based sales and thus the supply of commodities will increase in market.
3. The government is also recognizing its role in the food economy management. The formation of a high-level committee to redefine the roles and functions of the Food Corporation of India (FCI), the government involved in procurement, warehousing, transportation and distribution of food grains, mainly wheat and rice. As per the recommendation of the committee, according to the submitted report in January 2015, the government is also considering direct cash transfers to the Aadhaar-linked bank accounts of public distribution system (PDS) beneficiaries.

Impact of Food Distribution on Nutritional Status

The Public Distribution System (PDS) is considered as its focus on distribution of food grain in urban scarcity areas. It has elevated the situation of critical food shortage. Till 1992, PDS was a common phenomenon for all the consumers where estimated amount of food grains, edible oil and sugar were distributed through government general stores or outlets at a price lower than the market rate.

To ensure focus of PDS towards the economically backward families and in order to combat pilferage of food grains to the open market, a 'Revamped Public Distribution System' (RPDS) was launched in June 1992 in backward and remote areas. The Targeted Public Distribution System (TPDS) envisaged subsidized distribution of food grains to Below Poverty Line.

Food distribution leads to direct impact on nutritional content of food grains. Government agents and middleman play an important role to reduce the nutritional content of the food grains in PDS system. In order to create inflation in the market, food grains in bulk amount are stored in godowns. The unfavorable storage conditions lacks in maintaining food safety standards. The contamination such as water, rodents may affect the quality of food grains. It increases the chances of communicable diseases and it directly affects the human body. The nutrient content of the produce also get worsen day by day.

Challenging Nutritional Status of Country

Over the last two decades in India, there has been a progressive decline in infant and child mortality rate. Taking protein deficiency disorder into consideration, there has been a significant reduction observed in the protein energy malnutrition (PEM) especially kwashiorkor and marasmus. Body Mass Index (BMI) is considered as a measure of nutritional adequacy in adults and a better indicator of chronic energy deficiency (CED).

Women play a major role in the nutritional status of children. There is an urgent need to resolve the public health issues of under nutrition in women. The malnutrition problem and anaemic women in India is having its implications on birth outcome and under nutrition in children. The data showing state wise profile of anaemic women of age group 15-49 years in India (Table 2).

Table 2: Anaemia Prevalence in Women 15-49* Years in India-State-wise Profile

Percentage Prevalence of Anemia	*States*
< 40 per cent (5 states)	Punjab, Manipur, Mizoram, Goa, Kerala
> 40-60 per cent (16 states)	Delhi, Haryana*, Himachal Pradesh, Jammu & Kashmir, Uttaranchal, Chhattisgarh*, Madhya Pradesh*, Uttar Pradesh, Arunachal Pradesh, Meghalaya, Sikkim, Gujarat, Maharashtra*, Karnataka, Tamil Nadu
> 60 per cent (7 states)	Bihar*, Jharkhand*, Orissa*, West Bengal*, Assam*, Tripura*, Andhra Pradesh*

All India anemia prevalence is 55.3 per cent.

* Ten states have anemia prevalence over the national average of 55.3 per cent

Source: NFHS-3 (2006).

Over the past two decades, there has been a progressive decline observed in infant and child mortality rates in India. There has also been a significant reduction in the prevalence of florid nutritional deficiency disorders. It is therefore important that increasing attention is now paid to the nutritional status of the survivors. Women and child nutrition must be given the first priority.

Under Nutrition Women and Reducing Birth Rates: A Challenge

Maternal under nutrition plays a critical role in influencing maternal, neonatal and child health outcome(Mason *et al.*, 2012).With India's commitment to Millennium Development Goal (MDG 4), actions for maternal mortality reduction has received substantial attention. Several new programme directions and strategies have been introduced to improve the nutritional status of women. It has remained a low priority except for measures directed for reduction of anaemia or policy for providing supplementary food to pregnant women under the Integrated Child Development Services (ICDS) programme of Government of India. Today, the relationship of women's nutrition with birth outcomes, and stunting rates in young children is well established and the measures for improving nutritional situation in the country are high priority.

The theory of improving women's nutrition gained attention in 1998 when the conceptual framework of under nutrition in children positioned maternal nutrition as an important cause and stressed on breaking the inter-generation cycle of under nutrition (UNICEF, 1998). In recent years, the need to concentrate in the first 1000 days of life for prevention of under nutrition has been emphasized (World Bank, 2006).

In 2008, along with promotion of infant and young child feeding practices, improving nutritional status of women and adolescent girls been included in the set of selected high priority interventions proposed for reducing under nutrition rates in children in developing countries (Bhutta, *et al.*, 2008). These global guidelines were further contextualized with reference to India's nutritional epidemiology and a list of ten nutrition interventions was proposed by the Nutrition Coalition of India which includes actions for improving nutrition of adolescent girls and pregnant women (The Coalition, 2010).

The women in India increasingly getting engaged in employment in formal and non-formal sector in urban India (Imdad and Bhutta, 2011). In other developing countries, such as Sri Lanka and Vietnam, care of working mother is a high priority. Maternity leave in government and private jobs have proved effective step in reducing malnutrition. Women in India have maternity rights under the Maternity Act of 1961 and 180 days leave pattern is being followed by most of the formal sector in the states. Effective implementation of maternity leave policy is important for ensuring provision of an appropriate maternal and child care support.

In the past four-five years, a new programme for pregnant mothers "Indira Gandhi Matritva Yojna (IGMSY)" has been launched together in 52 selected districts across the country by the Department of Women and Child, as a part of the ICDS programme (IGMSY, 2011). The Scheme is designed to address socio-economic problems of pregnant women. The 12th Five Year Plan of the Government of India aims to scale up the IGMSY scheme in the entire country and is included as a component of the Food Security Bill (National Food security Act, 2013).

Steps to Combat Nutrition Deficiency

Government has started a new campaign to fortify the cereals with the essential vitamins and minerals so as to overcome the malnutrition conditions. This may leads to rise in nutritional status of the country. Iodized salt, golden rice are few examples of fortified nutrient rich foods. The launch of the Reproductive, Maternal, Newborn, Child and Adolescent Health (RMNCH+A) approach by the Ministry of Health and Family Welfare is a confirmation by Government of India to continuum of care approach in the development of state, district, block and panchayats plans of action (Table 3) (NRHM, 2013).

Future Strategies Planning

The National Food Security Mission (NFSM) launched in 2007-2008 to uplift the production of rice, wheat and pulses has been implemented across the country during the Eleventh Five Year Plan period. Extension of the Green Revolution to Eastern India comprising Assam, Bihar, Chhattisgarh, Jharkhand, Odisha, Eastern Uttar Pradesh and West Bengal under the *Rashtriya Krishi Vikas Yojana* received an allocation of rupees 400 crore in the budget for 2011-2012. It is estimated that India has the potential to cultivate oil palm in 1.03 million hectares to produce

Table 3: Continuum of Care Across Life Cycle and different Levels of Health System

Source: NRHM report (2013).

4-5 million tonnes of palm oil which would be able to mitigate the consumption requirement of 330 million people @15kg/capita/per annum. But the appropriate public policy support would be required. As the Economic Survey 2012-2013 rightly notes, "it is time to frame a price band for edible oils in a manner that harmonizes the interests of domestic farmers, processors, and consumers through imposition of import duty at an appropriate rate". The Rainfed Area Development Programme (RAPD) focused on the need for an integrated farming system strategy based on conservation agriculture. It may integrate multi-cropping, inter-cropping, mixed cropping and rotational cropping practices with allied activities like horticulture, livestock, fishery, apiculture, and agro-forestry. This approach will maximize farm returns and mitigate impact of extreme weather conditions. This will cover districts with arid, semi-arid and sub-humid agro ecosystems and less than 60 per cent of the cultivated area under irrigation. A study by the International Food Policy Research Institute (IFPRI), records the agricultural development across Asia, Africa and Latin America.

Conclusion

The primary objective of the Department of Food & Public Distribution is to ensure food security for the country through timely and efficient procurement and distribution of Food grains. This involves procurement of various Food grains, maintenance of food stocks, their storage, and time being delivery to the distributing authorities thereby monitoring production, stock and price levels of Food grains. The goal of food self-sufficiency however, can be effectively addressed only when it is ensured with the public policy support. Concluding again through the NCF reports, "Food security with home grown food grains can alone eradicate widespread rural poverty and malnutrition, since farming is the backbone of the livelihood security system in rural India. This will enable the Government to remain at the commanding height of the national food security system. Building a food security system and containing price rise with imported food grains may sometimes be a short term necessity, but will be a long term disaster to our farmers and farming. A well-defined, pro-farmer and pro-resource poor consumer Food Security Policy is an urgent necessity".

With these constraints, government has to create huge modern infrastructure for storing and transporting food grains. Creation of food-grain storage, facilities should be given infrastructure status for attracting private investments. Also for keeping the track on nutritional situation of child and women, periodical update of national and state data must be recorded through surveys along with their dietary and nutrition intake, maternal weigh gain, birth weight scenario pattern.

REFERENCES

"Agriculture for development" – World Development Report 2008. Washington DC: World Bank; (2007). World Bank.

Bhutta, Z.A., Ahmed, T., Black, R.E., Cousens, S., Dewey, K., Giugliani, E., Haider, B.A., Kirkwood, B., Morris, S.S., Sachdev, H.P.S. and Shekhar, M. (2008). What works? Intervention for maternal and child undernutrition and survival. Lancet, 371(9610), 417-440.

David, J., Spielman, Rajul Pandya-Lorch, editors. Washington DC: IFPRI; (2009). International Food Policy Research Institute (IFPRI). Millions Fed - Proven successes in agricultural development.

Economic Survey 2012-2013. New Delhi: Ministry of Finance; (2013). Government of India (GoI).

Food and Agriculture Organization (FAO) (2003). Trade Reforms and Food Security - Conceptualizing the Linkages, Corporate Document Repository, Food and Agriculture Organization of the UN, Rome. [accessed on August 24, 2013]. Available from: http: //www.fao.org/docrep/005/y4671e/y4671e06.htm.

Food and Agriculture Organization (2009). "How to Feed the World in 2050", Food and Agriculture Organization of the UN, Rome. [accessed on August 24, 2013]. Available from: http: //www.fao.org/fileadmin/templates/wsfs/docs/expert_paper/How_to_Feed_the_Worldin 2050.pdf .

Guidelines for rainfed area development programme (RADP) New Delhi: Ministry of Agriculture, Government of India; (2011). Mar, Department of Agriculture and Cooperation (DAC).

Imdad, A. and Bhutta, Z.A. (2011). Maternal Nutrition and birth outcomes: effect of balanced protein energy supplementation. BMC Public health, 11, S17.

Kharif Outlook Report, "Agricultural Outlook and Situation Analysis Reports", December (2015). National Council of Applied Economic Research, agrioutlookindia@ncaer.org.

Mason, J.B., Saldanha, L.S., Ramakrishnan, U., Lowe, A., Noznesky, E.A., Girard, A.W. and Martorell, R. (2012). Opportunities for improving maternal nutrition and birth outcomes: synthesis of country experiences. Food and Nutrition Bulletin, 33(2 Suppl 1): S104-S137.

M.S. Swaminathan and R.V. Bhavani, "Food production & availability - Essential prerequisites for sustainable food security".

National Family Health Survey (NFHS) (2006). NHFS 3, 2005-06. Mumbai: Indian Institute of Population Sciences.

National Sample Survey Organization (NSSO) *Situation assessment survey of farmers report No. 496 (59/33/3) Some Aspects of Farming NSS* 59th Round (January-December 2003) Ministry of Statistics and Programme Implementation, Government of India. 2005.

New Delhi: Ministry of Agriculture, Government of India; (2004). Dec, National Commission on Farmers (NCF). *Serving farmers and saving farming* - 1st Report.

New Delhi: Ministry of Agriculture, Government of India; (2005). Aug, National Commission on Farmers (NCF). *Serving farmers and saving farming: From Crisis to Confidence*, 2nd Report.

New Delhi: Ministry of Agriculture, Government of India; (2005). Dec, National Commission on Farmers (NCF). Serving farmers saving farming: 2006 - Year of agricultural renewal, 3rd Report.

New Delhi: Ministry of Agriculture, Government of India New Delhi; (2006). Apr, National Commission on Farmers (NCF). Serving farmers and saving farming: Jai Kisan - A draft National Policy for Farmers, 4th Report.

New Delhi: Ministry of Agriculture, Government of India; (2006). Oct, National Commission on Farmers (NCF). *Serving Farmers Saving Farming: Towards Faster and More Inclusive Growth of Farmers' Welfare*, vol, I. 5th and Final Report; pp. 187–92.

New Delhi: Ministry of Agriculture, Government of India; (2011). Mar, Department of Agriculture and Cooperation (DAC). *Guidelines for the Programme of Integrated Development of 60000 Pulses Villages in Rainfed Areas.*

NRHM (2006). ARSH (Adolescent, Reproductive and Sexual Health Programme), Implementation Guide on RCH II. Adolescent Reproductive Sexual Health Strategy, For District and Programme Manager. Government of India, 2006.

NRHM Report (2013). A strategic approach to reproductive, maternal, newborn, child and adolescent health (RMNCH+A) in India (For healthy mother and child). Ministry of Health & Family Welfare, Government of India, January, 2013.

Reddy, V., Shekar, M., Rao, P. and Gillespie, S. (1992). Nutrition in India. National Institute of Nutrition, Hyderabad,

Report on the state of food insecurity in rural India. Chennai: M.S. Swaminathan Research Foundation; (2008). Dec, M S Swaminathan Research Foundation (MSSRF).

Swaminathan, M.S. and Sinha, S.K. (1986). "Building National and Global Nutrition Security Systems". Dublin: Tycooly International Publishing Company; Global aspects of food production.

Swaminathan, M.S. (2009). The media and the farm sector. The Hindu.

The Coalition for Sustainable Nutrition Security in India, May (2010). Sustainable Nutrition Security in India: A leadership Agenda for Action.

UNICEF (2013). Improving child nutrition the achievable imperative for global progress. UNICEF.

Union Budget (2011-12). New Delhi: Ministry of Finance; 2011. Government of India (GoI)

Chapter 5

Food Distribution System and Food and Nutrition Security at National and Household Level

Tanu Jain

Introduction

Food distribution is a method of distributing or transporting food or drink from one place to another which is a very important factor in public nutrition.

FSDSs (Food Supply and Distribution System) – It is complex combinations of activities, functions and relations (production, handling, storage, transport, package, wholesale, retail, *etc.*) which enables the population to meet their food requirements. These activities are performed by different economic agents.

Two major division of Food Distribution System are:

1. Macro distribution which includes national and states economy.
2. Micro distribution includes household and household members.

Distribution Channel

Public Distribution System

Distribution of essential commodities through FPS at government controlled prices has come to be known as the public distribution system (Radhakrishna and Reddy, 2002). For this there is joint responsibility of the Central and the State Governments. Commodities which are distributed include:

Wheat, rice, sugar and kerosene to the States/UTs

Additional items to some States/UTs includes such as pulses, edible oils, iodized salt, spices.

Features of PDS

1. The aim of PDS is to provide at least a basic minimum quantity of essential items at reasonable prices to the vulnerable sections' of population and to stabilize their open market prices or at least to prevent an undue rise in such prices under conditions of shortage.
2. The prices charged from population are lower than open market prices the procurement and other costs incurred by the government.
3. Distribution of selected essential goods through the FPS which are operated by private dealers under the government's control and direction.
4. Rice, wheat and sugar occupy a predominant position. The other important items are kerosene, edible oil *etc.*,
5. The working does not hinder the functioning of the free market mechanism and could be viewed as a "dual economy" in the essential commodities.
6. The government feeds the PDS with supplies, bears the cost of subsidy, decides as to which goods to supply, at what rates, what amount to be sold per head or per family *etc.*

Revamped Public Distribution System (RPDS)

RPDS was launched in June, 1992 with the following objectives:

- To strengthen and streamline the PDS
- Improve its reach in the far-flung, hilly, remote and inaccessible areas where a substantial section of the poor live

It covered 1775 blocks where in area specific programmes such as the Drought Prone Area Programme (DPAP), Integrated Tribal Development Projects (ITDP), Desert Development Programme (DDP) and certain Designated Hill Areas (DHA)

- Food issue price to State Government – 50 paisa
- Scale of issue- 20 kg per card.

Targeted Public Distribution System (TPDS)

TPDS was launched in June 1997 and was focused on the poor. States are required to formulate and implement full proof arrangements for identification of the poor for delivery of food grains and for its distribution in a transparent and accountable manner at the FPS level. Food grains to the States/UTs were made on the basis of average consumption in the past (Kumar and Mohanty, 2012). Transitory Allocation was also provided. The end retail price is fixed by the States/UTs after taking into account margins for wholesalers/retailers, transportations charges, levies, local taxes *etc.* Under the TPDS, the States were requested to issue food-grains at a difference of not more than 50 paisa per kg over and above the CIP for BPL families. But for Antyodaya Anna Yojana where the end retail price is to be retained at Rs.2/a kg. for wheat and Rs.3/a kg. for rice.

Identification of BPL Families Under TPDS

The BPL is determined on the basis of population projections of the Registrar General of India for 1995 and the State wise poverty estimates of the Planning Commission for 1993-94. State government was advised to involve the Gram Panchayats and Nagar Palikas. The number of BPL families was increased to 652.03 lakh as against 596.23 originally estimated when TPDS was introduced in June, 1997.

Antyodaya Anna Yojana (AAY)

According to National Sample Survey - 5 per cent of the total population in the country sleeps without two square meals a day which is called as "hungry". Antyodaya Anna Yojana (AAY) was launched in December, 2000 to focused and targeted towards this category of population.

- ☆ Scale of issue – 25 kg initially, increased to 35 kg.

Expansion of AAY

A total of 50 lakh BPL households headed by widows or terminally ill persons or disabled persons or persons aged 60 years or more with no assured means of subsistence or societal support were added in 2003-2004 in the first expansion of AAY.

In 2004-05, the AAY has been further expanded by another 50 lakh BPL families by including, all households at the risk of hunger. Landless agriculture labourers, marginal farmers, rural artisans/craftsmen, such as potters, tanners, weavers, blacksmiths, carpenters, slum dwellers, and daily wagers in the informal sector. In 2005-06, the AAY has further been expanded to cover another 50 lakh BPL households.

Food Distribution in Various Welfare Schemes

Mid-Day Meal Programme

Tamil Nadu was the first to initiate a massive noon meal programme to children. Neither a child that is hungry, nor a child that is ill can be expected to learn. Realizing this need the Mid-Day Meal (MDM) Scheme was launched in primary schools during 1962-63. Mid-Day Meal improves three areas (www.mhrd.gov.in):

1. School attendance
2. Reduced dropouts
3. A beneficial impact on children's nutrition

The programme covers around 21.1 million children (beneficiaries) in class I to V standard by 1989-90 through government, local body and private-aided primary schools. Food grains at the rate of 3 Kgs. minimum per child are provided per month (300 k cal and 8-12 grams of protein per day). The Central Government supplies the full requirement of food grains for the programme free of cost. For its implementation in rural areas, Panchayats and Nagarpalikas are also involved or setting up of necessary infrastructure for preparing cooked food. For this purpose NGOs, women's group and parent-teacher councils can be utilized. The total charges for cooking, supervision and kitchen are eligible for assistance under Poverty Alleviation Programme. In several states, supplementary feeding was assisted by food supplies from Cooperation for American Relief Everywhere (CARE) and World Food Programme (WFP). There are problems of administration and quality of food that have affected the programme outcomes.

Wheat Based Nutrition Program

This program is implemented by the Ministry of Women & Child Development. The food grains allotted is utilized by the States/UTs under the Integrated Child Development Scheme (ICDS) for providing nutritious/energy food to children below 6 years of age and expectant/lactating women from disadvantaged sections.

Schemes for Supply of Food Grains to Welfare Institutions

It introduced in 2002-03 to liquidate the stocks of food grains and to meet the requirement of Hostels/Welfare Institutions *viz.* NGOs/Charitable Institutions which help the shelter less/homeless poor and other categories not covered under TPDS or under any other Welfare Schemes, an additional allocation of food grains (rice and wheat) not exceeding 5 per cent of the BPL allocation of each State/UT is made to States/UTs at BPL rates.

ST/ST/OBC Hostel

It Introduced in October, 1994 under the Ministry of Consumer Affairs, Food & Public Distribution. Ist Allocation of the scheme was made during 2001-02 to 19 States on the recommendation of Ministry of Social Justice & Empowerment.

Annapurna Scheme

This scheme was launched in 2000-01 under the Ministry of Rural Development. Indigent senior citizens of 65 years of age or above who are eligible for old age pension under the National Old Age Pension Scheme (NOAPS) but are not getting the pension are covered under this scheme and 10 Kgs. of food grains per person per month are supplied.

World Food Programme

A Country Programme Action Plan (CPAP) 2008-2012 has been signed between the Government of India and the World Food Programme (WFP). Under this programme, Department of Food & Public Distribution has committed to allocate following quantity of food grains for utilization of WFP assisted projects in India. Government of India allocates food grains at BPL issue prices for the development schemes administered by United Nations World Food Programme.

Rajiv Gandhi Scheme for Empowerment of Adolescent Girls (RGSEAG) - 'Sabla'

This scheme was launched on November, 2010 by merging two schemes namely Nutrition Programme and Adolescent Girls (NPAG) and Kishori Shakti Yojana (KSY) in to a single scheme and proposed to be implemented in 200 selected districts across the country. Nodal ministry for this scheme is the Ministry of Women and Child Development.

Scheme for Adolescent Girls (Kishori Shakti Yojna)

There was a gap in between women and child age group which was not covered by any health and social welfare programme whereas girls in these crucial groups need special attention. On one side they need appropriate nutrition, education, health education, training for adulthood, training for acquiring skills as the base for earning an independent livelihood, training for motherhood, *etc*. Similarly on the other side their potential to be a good community leader has to be realized. A scheme for adolescent girls in ICDS was launched by the Department of Women and Child Development, Ministry of Human Resource Development in 1991.

Emergency Feeding Programme

Emergency feeding programme was introduced in May, 2001. It is a food-based intervention targeted for old, infirm and destitute persons belonging to BPL households to provide them food security in their distress conditions. Food grain allocated to them was rice. Cooked food containing, inter-alia, rice- 200gms, *Dhal* (pulse) - 40 gm, vegetables- 30 gm is provided in the diet of each EFP annual beneficiary daily by the State Government.

Village Grain Bank

It is established to prevent the death of ST especially children in remote and backward areas. Allocation is made under Ministry of Tribal Affairs during 2002-2003.

Factors Affecting Food Distribution

It can be divided into two criteria:

1. External factors which include urbanization, urban food needs, the economic, political and social framework, legal and regulatory framework, institutional framework, public infrastructure, facilities and services, the regional, metropolitan and city's urbanity and spatial characteristics.
2. Internal factors include urban consumers 'food consumption and purchasing habits, flow of food products, economic players and their strategies, FSD infrastructure, facilities and services, FSD specific laws and regulations.

Problems with Macro Distribution

Infrastructure

The infrastructure lacks modern storage facilities such as silos; the grain is stored outside under plastic tarps which provide little protection from humidity and pests. It is estimated that as much as 20 per cent of the grain is wasted due to poor infrastructure facilities.

One promising model is to build modern storage infrastructure with the help of the private sector might be a solution of this problem. According to IFC Experts, public-private partnership model to build silos is economically viable if storage losses exceed 2.1 per cent annually, far below current loss estimates. However, the government seems to be hesitant to move forward and often maintains that little to no waste occurs in their facilities.

Government Purchase and Distribution Scheme

Food supply systems in India are not immune bureaucracy and corruption. Corrupt officials of government agencies and intermediaries running storage depots often rig weighing scales to indicate less grain coming in, siphoning off the excess to the gray or black markets. Officials will sometimes allow and then over-report, wastage in an effort to sell the excess supply. Similar issues arise during transport. The private sector gives better assurance that food does not go to waste, as managers typically won't be able to gain from sustained illicit transactions.

Middleman, Bargaining Power and Price Transparency

The movement of a product from a farm to a market often involves 4-5 middlemen. Transactions are handled predominantly by commission agents who have an advantage in terms of information and bargaining power. There are various incidents in which farmers often won't know the price for their product before they get to the wholesale market. Once at the market, the commission agents can dictate the price as it's not economical for the farmer to take the goods back in order to wait for a better price. We've heard that sometimes commission agents will even leave a load of rotting produce near the market as a warning to farmers who do not accept the offered prices. Commission agents have little incentive to prevent waste as they are compensated based on the total transaction value, without ever taking ownership of the product. Since they generally receive only a 2.5-6 per cent commission on sales, it makes little sense for them to invest time to find traders offering marginally higher prices – they can earn more income by completing many deals as quickly as possible. Further along the supply chain, traders also have few incentives to minimize waste. It is easier for them to deal with fewer goods at a higher price than more goods at depressed prices. As such, waste often occurs when these middlemen collude to boost prices and lower shipment quantities.

Price Volatility

This is also one of the problems at macro level distribution due to the following reasons- food spoilage/wastage, extreme or unpredictable fluctuations, farmers trying to avoid future problems and shifting of crops by farmers for economically and efficient.

Micro Level Distribution Problems

- PDS system is FSDS that encompasses micro level distribution. Major problem of micro level distribution is inefficiency in targeting beneficiaries which leads to several opportunities to manipulate the system According to Planning Commission report 2005, 57 per cent of grain is not reaching to intended poor. A large portion of the subsidized food grains and other essential commodities meant for distribution do not reach the beneficiaries.
- The difference between the open market prices and subsidized prices of these commodities under PDS determine lucrative of the leakages.
- It is estimated that a little over one-third of the food grains, supplied to PDS, do not reach actual users of the PDS-it leaks out of the programme. While some part of these may be genuine losses incurred in storage and transport, a major part is diverted to the open market.

Poor Quality of Food Grains

- Major complaints - supply of poor quality of grains.
- The lower quality of PDS wheat is evidence of inefficiencies in the operation of the public sector so find a mid-way to transfer or exchange of good quality wheat by inferior quality.
- Besides, the huge buffer stocks necessitate the storage for months and even years, thereby deteriorating the quality to a large extent.

Multiplicity of Schemes

Government is running a number of schemes, which are targeting the same section of the society

- The BPL family is entitles to 35 kg food grains under the TPDS may be beneficiary under SGRY or its Special Component or Mid Day meal.

Multiplicity of Prices

- The presence of different prices slabs under the TPDS, apart from complicating the operation of the scheme at Fair Price Shop level, also creates the problem of monitoring the Scheme.
- The presence of quite low priced food grains, as in case of the Antyodaya Anna Yojana, also increases the propensity of diversion of such food grains, as the difference between the open market rate and the subsidized food grains at FPS level is substantial.

Expensive Operation

- Most scathing criticism- Amount of annual food subsidy involved in maintaining the system. For example, the food subsidy released 2003-04,

2004-05 and 2005-06 was Rs.147664.50, 2033395.10 and 198269.50 million respectively which is 40 to 70 per cent of the total food subsidies released during these years. Thus, operations under PDS are not cost effective.

Non-Issue of Ration Card

- BPL families in Goa, Himachal Pradesh, Andaman and Nicobar Islands, Chandigarh, West Bengal, Dadra and Haveli and Lakshadweep are yet to complete.
- Only 14 states/UTs have completed identification and issue of distinctive ration cards to the beneficiaries

Ineffective Implantation of the PDS Control Order

- PDS system is unable to reach the poor effectively.
- Wrong inclusion of above poverty line households
- Exclusion of the real poor who are included in the poverty lists of villages.
- The PDS control order, 2001and 2004 provides for constant review and updating of families eligible for issue of ration card and deletion of ineligible units/households

Irregular Opening and Non-Availability of Adequate Food Grains In FPS

- The FPS doses not open daily in which some beneficiaries travel from long off distances
- Shortage of resources, delay in lifting of food grains on the part of State agencies and Fair Price Shop dealers.

Inadequate Entitlement for Households

- Under the TPDS each household is entitle to 20 kg of food grain per month. This fall is very short of individual or household need. Besides, the entitlement remains uniform for all BPL household irrespective of size. Thus gap between actual need and availability under TPDS is given more for larger family.
- In many cases the beneficiaries have to cover long distances to reach fair price shops. As a result of the Hon'ble Supreme Court's intervention in the implementation of the PDS, the commissioners to the court have reported that the opening of the ration shops has become regular and predictable.
- The consumers are generally not given the arrears of the previous month/ fortnight thereby making a room for diversion and defeating the very purpose of the scheme.

Urban Bias

- PDS has a strong urban bias resulting in undue suffering in rural areas, where most of the country's poor live.

- ☆ Studies indicate higher take off in Kerala, West Bengal, Mumbai and Delhi.
- ☆ According to Consumer survey urban bias in Tamil Nadu, Meghalaya and Goa, also marginally in Uttar Pradesh and Assam.

Multiplicity of State Agencies Handling the Scheme in a State

- ☆ Too many agencies handling the supply of food grains under TPDS in all states, which needs a lot of coordination at FCI field level.
- ☆ Most of the State agencies find difficulties in arranging necessary funds for lifting the food grains results in irregular off-take

Diversion of Food Grain

- ☆ Study (by Tata Economic Consultancy Services) found that at national level there was diversion of 36 per cent of wheat grains and 31 per cent of rice. It was also observed that the diversion is more in the Northern, Eastern and North-eastern regions.
- ☆ Uneven impact of PDS
- ☆ India Human Development Report, stated only 33 per cent of the rural households in India have reported use of the PDS, on regular basis.
- ☆ The scheme was found to be working fairly efficiently in the four southern states, 2 western states and Himachal Pradesh and on the modest level in Madhya Pradesh. It also proved that the better off states were able to corner most of the beneficiaries under the scheme

Lack of Information

- ☆ One of the major complaints regarding PDS is the supply of poor quality of grains.
- ☆ The lower qualities of PDS beneficiaries are unaware of their rights, as Citizen's Charter is seldom available in the regional language (s).
- ☆ No enough publicity and information relating to scale of issue, prices, availability of commodities to the consumers.

Lack of Training of FPS Dealers

- ☆ Lack of proper training/guidelines to the FPS owners about their duties and obligations. This, combined with absence of proper and regular inspection of the FPS, makes the situation.

System Transparency and Accountability

- ☆ The most serious flaw plaguing the system at present is the lack of transparency and accountability in its functioning.
- ☆ The system lacks transparency and accountability at all levels making monitoring the system extremely difficult.

Grievance Redressal Mechanisms

- ✩ There are numerous entities like Vigilance Committee, Anti-Hoarding Cells constituted to ensure smooth functioning of the PDS system.
- ✩ Their impact is virtually non-existent on the ground and as a result, malpractices abound to the great discomfiture of the common man.

Table 1: Per Capita Net Availability of Food Grains (Per Annum) in India (2001 to 2011)

(Kgs. per Year)							
Year	*Rice*	*Wheat*	*Other Cereals*	*Cereals*	*Gram*	*Pulses*	*Food Grains*
2001	69.5	49.6	20.5	141.0	2.9	10.9	151.9
2002	83.5	60.8	23.1	167.4	3.9	12.9	180.4
2003	66.2	65.8	17.1	149.1	3.1	10.6	159.7
2004	71.3	59.2	25.3	155.8	4.1	13.1	168.9
2005	64.7	56.3	21.7	142.7	3.9	11.5	154.2
2006	72.3	56.3	22.1	150.7	3.9	11.8	162.5
2007	70.8	57.6	20.3	148.7	4.3	12.9	161.6
2008	64.0	53.0	19.7	143.9	3.9	15.3	159.2
2009	68.8	56.5	23.3	148.6	4.7	13.5	162.1
2010	67.4	61.3	19.8	148.5	4.9	11.6	159.5
2011(P)	68.9	60.1	25.6	154.6	5.3	14.4	169.0

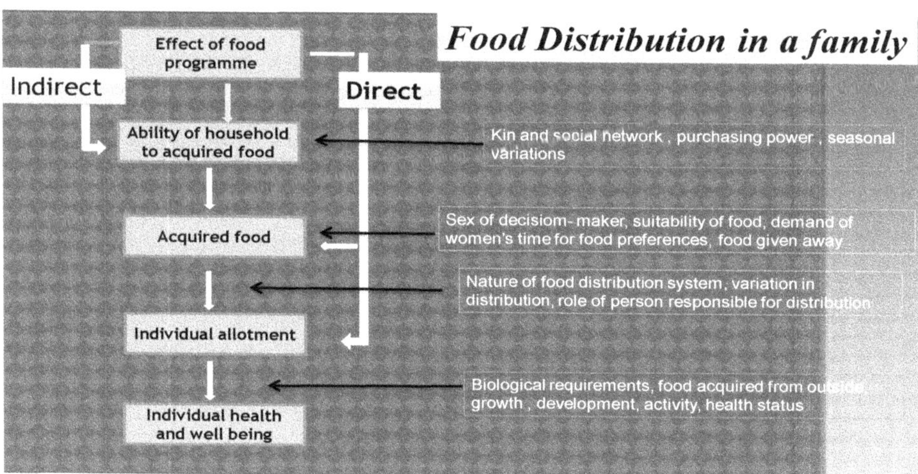

Source: www.slideshow.net

Food Security has following Dimensions

1. **Availability of food** means food production within the country, food imports and the previous year's stock stored in government godowns.

2. **Accessibility** means food is within reach of every person.
3. **Affordability** implies that an individual has enough money to buy sufficient, safe and nutritious food to meet one's dietary needs. Thus,

Food Security is Ensured in a Country Only if

1. Enough food is available for all the persons.
2. All persons have the capacity to buy food of acceptable quality.
3. There is no barrier on access to food

Overview of Food Security

The Green Revolution, launched in the late 1960s, had overwhelmingly impacted the various dimensions of food security.

- It helped India triple its food grain production between 1968 and 2000 and consequently in halving the percentages of food insecurity and poverty.
- Other livelihood indicators such as literacy rate and longevity also increased significantly.

Indian Scenario

60 million children in India are underweight and 21 per cent population is malnourished (NFHS 2015-16). Major problems are unemployment, declining wages and poor food distribution systems. World Bank Report says that a productivity loss in India is due to stunted growth, iodine deficiencies and iron deficiencies.

Nutritional security implies physical, economic and social access to balanced diet, clean drinking water, safe environment & health care for every individual

- The term malnutrition includes both under-nutrition in terms of proteins, calories, fats, vitamins and minerals and over-nutrition leading to obesity
- Though India has made great progress in food production, 50 per cent Indians particularly preschool children and women suffer from protein calorie malnutrition, micronutrient deficiency particularly Fe leading to anemia
- Every third child born in India is low birth weight.
- 30 per cent adults are under nourished, 10 per cent Indians are obese and incidence being almost 20 per cent in urban areas (NFHS 2015-16).

The malnutrition has a complex etiology and its prevention requires **Awareness & Access at Affordable** price to all the parameters of nutritional security. Countrywide diet survey shows that Indian diets are qualitatively more deficient in vitamins and minerals than proteins due to low intake of 'income-elastic' foods like vegetables, fruits, pulses and flesh foods (Suryanarayana and Silva, 2008). Malnutrition reduces immunity & infections and disease reduce appetite, impair absorption and lead to catabolic losses of precious nutrients.

Who is Food Insecure?

Although a large section of people suffer from food and nutrition insecurity in India, the worst affected groups are landless people with little or no land to depend upon, traditional artisans, and providers of traditional services, petty self-employed workers and destitute including beggars.

In the urban areas, the food insecure families are those whose working members are generally employed in ill-paid occupations and casual labour market.

Need for Food Security

- ☆ For the poor sections of the society
- ☆ Natural disasters or calamity like earthquake, drought, flood, tsunami,
- ☆ Widespread crop failure due to drought
- ☆ The poorest section of the society might be food insecure most of the times while persons above the poverty line might also be food insecure when the country faces a national disaster/calamity like earthquake, drought, flood, tsunami, widespread failure of crops causing famine, *etc.* (Figure 1).

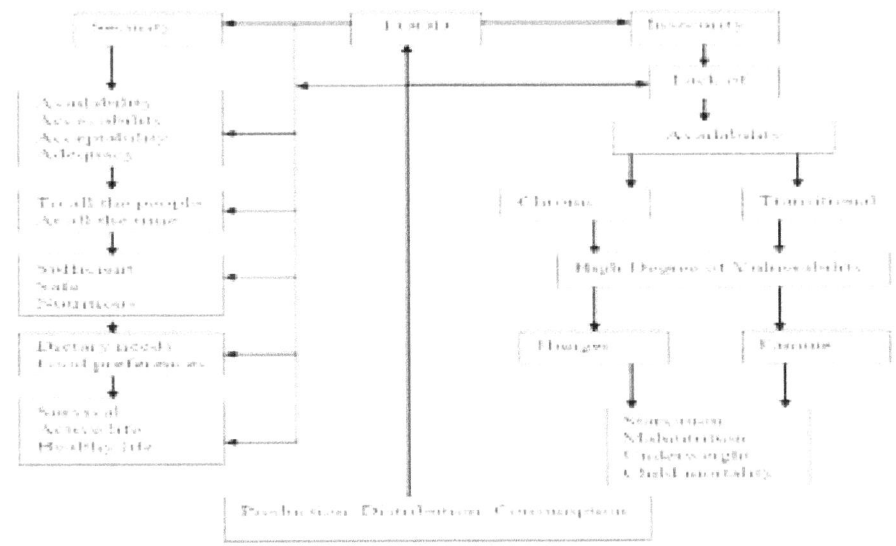

Figure 1

Chart 1. Understanding the Concept

For ensuring food security three important events have to take place, they are production, distribution and consumption which are called as a food security cycle (Figure 2).

Problem is Multi-Level–National/Household

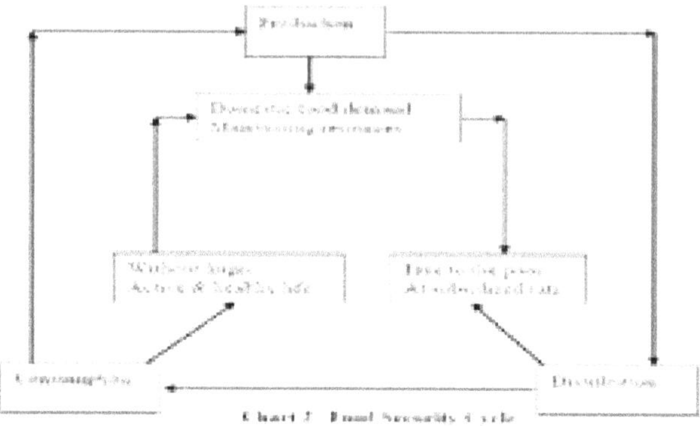

Figure 2

Food Security at Household Level: The Linkage

- ✫ Household level food insecurity
  - ❒ Low levels of household income
  - ❒ High and rising retail price of food
  - ❒ Problem mainly for poor households
  - ❒ Situation in India
  - ❒ 45 per cent suffer stunting (height for age per cent of children under 5)
  - ❒ 40 per cent underweight children (weight for age per cent of children under 5)
- ✫ Income entitlement/distribution issues
- ✫ Domestic supply issues

Distributional Issues

- ✫ 3 broad types of households
- ✫ Poor asset-less households (A)
- ✫ Poor households with low asset base (including land) (B)
- ✫ Non-poor (rich) households/large land owners (C)
- ✫ (A) and (B) are usually food insecure – low income, fully or partially dependent on markets for food
- ✫ (C) is usually food secure – sufficiently high income and/or food self-sufficient
- ✫ Taxation/re-distribution policies often ineffective
  - ❒ Income transfer programmes (cash transfer/employment programmes),
  - ❒ Consumption subsidies (public distribution)

Domestic Supply and Food Security

- ✰ 3 main features of under-developed countries
  - ❒ Restrictive government policies on investment, labour, *etc.*
  - ❒ Low asset base of the country/farmers in particular
  - ❒ Weak institutional structures (legal, property rights, security, social and economic barriers to competition in markets)
- ✰ 3 main manifestations of these features
  - ❒ Low investment levels
  - ❒ Weak infrastructure
  - ❒ Low productivity
- ✰ Outcome on food security – 3 channels
  - ❒ Low economic growth in general ➔ widespread poverty
  - ❒ Low growth in food production ➔ Insufficient food availability ➔ High food prices
  - ❒ Weak infrastructure ➔ high transport cost ➔ high retail price

Food Production vs Nutritional Security

- ✰ India's food grain (wheat & rice) production went up and kept ahead of population till mid '90s but tending to plateau.
- ✰ Pulse production has stagnated and per capita availability has declined.
- ✰ There is erosion in millet production & consumption.
- ✰ Milk, fruit & vegetable production has increased markedly, but that is not reflected in the diet of the poor due tom poor purchasing power & lack of awareness of their importance in diet.
- ✰ New technologies for bio-fortification of crops have been developed could not adopt due to inability of put in place convincing safety guidelines and measures.
- ✰ Thirty per cent of India's farm produce is wasted due to inadequate post harvest storage facilities & food processing for value addition.
- ✰ Food processing helps to prevent post harvest losses. NDRI, IVRI and several ICAR Institutes, CFTRI, DFRL *etc.* have developed useful products & storage devices that should be put into practice.
- ✰ Emphasis should be given in developing nutrient dense high shelf life fortified foods like biscuits and ready to eat mixes.
- ✰ There should be grater scientific interaction between nutrition scientists and scientists belonging to food science, agriculture, medicine, public health as well as social scientists/activists.
- ✰ Asia's food and nutrition security is under stress due to many interconnected factors that include population growth and urbanization, demographic changes, increased labour cost, high and volatile food prices, natural resource constraints and climate change.

Climate Change

Global surface temperature on an average increased by about 0.74 °C in last 100 years. Change in weather pattern, alterations in the soil moisture storage, pests and weeds may affect productivity. Climate change also decreased the freshwater availability in Asia. Also, due to melting of Himalayan glaciers, the problems of disasters like flood and drought also arose. Higher temperatures favor the growth of bacteria in food which may raise the issue of food safety

Are we Really Food Short?

India is the largest producer of wheat (15 per cent of global production), pulses (21 per cent of global production), and milk (90 million tons). India has largest irrigated land (55 million hectares) and 2nd largest arable land (184 million hectares) in the world. India is the largest producer and exporter of spices, 2nd largest producer of rice (22 per cent of global rice production) and the largest producer of world's best BASMATI RICE. Also, India is 2nd largest producer of fruits and vegetables and has largest livestock population.

The government has played a key role in ensuring food security for the poorest sections of society through various schemes such as the public distribution system, mid-day meals, food-for-work and rural employment guarantee. However, due to certain failings, a number of people still go without food.

Policies and Programmes Related to Food Security

- Food and nutrition security through government interventions in food-based programmes include the Public Distribution System (PDS), the Integrated Child Development Scheme (ICDS), the School's Mid-day Meal Scheme, Food-for-Work (FFW) and Antyodaya Anna Yojana (AAY) *etc.*
- The proposed National Food Security Act, 2009 assures that every BPL family in the country shall be entitled to 25 kg of wheat or rice per month at the rate of Rs.3/- per kg.
- Several programmes such as National Food Security Mission (NFSM) launched during 2007- 08, Bringing Green Revolution in Eastern India (BGREI) during 2010-11 are being implemented to increase the production and productivity of rice in the country.
- Ministry of Agriculture has mostly been addressing the "availability" components of food security (also contributing to the "access" component through the increased yield and income of the farmer),
- Ministry of Rural Development and Ministry of Food and Supply have been contributing mostly to the access (social safety nets and social engineering, drinking water and employment security) components.
- The Ministry of Health and Family Welfare has been addressing the absorption, fortification, child nutrition and other softer components of food security and nutritional adequacy.

- The law is also proposed to be used to bring about systemic reforms in the Public Distribution System (PDS).
- Apart from the PDS, the two major programmes such as ICDS and Mid-day Meal Scheme aimed at providing nutritional security to pregnant women and lactating mothers and young pre-school goers and school-goers, respectively.

Programmes and Activities of Different Sectors for Food Security

Food safety aims to assure a minimum amount of food consumption and/or protect households against shock to food consumption. The various ongoing social safety schemes towards enhancing household and individual level food security include Sampoorna Gramin Rozgar Yojana, Mid-Day Meal Scheme, Food for Work Programme, National Food for Work Programme, Public Distribution System, Antyodaya Anna Yojana, Annapoorna and Integrated Child Development Services Schemes. These are increasing the access to food by the low income group people.

PDS and Food Subsidy

It is now well recognized that the availability of food grains is not a sufficient condition to ensure food security to the poor. It is also necessary that the poor have sufficient means to purchase food. The capacity of the poor to purchase food can be ensured in two ways – by raising the incomes or supplying food grains at subsidized prices. While employment generation programmers attempt the first solution, the PDS is the mechanism for the second option. It stabilizes prices of food grains and makes food available at affordable prices. By supplying food from surplus regions of the country to the deficit ones, it helps in combating hunger and famine. It also provides income security to farmers in certain regions.

Under the targeted public distribution system, there are three kinds of ration cards:

- Antyodaya cards (for the poorest of the poor)
- BPL cards (for those below poverty line)
- APL cards (for all others)

Food Security Bill

- **2001:** Constitutional Court recognizes right to food in the People's Union for Civil Liberties (PUCL) case, transforming policy choices into enforceable rights (http://dfpd.gov.in)
- **2005 :** Adopts its National Rural Employment Guarantee Act and the Right to Information Act
- **2009:** India is developing a National Food Security Act
- **2011:** Food Security Bill cleared by Cabinet
- **January, 2013:** NFSB cleared by a parliamentary committee

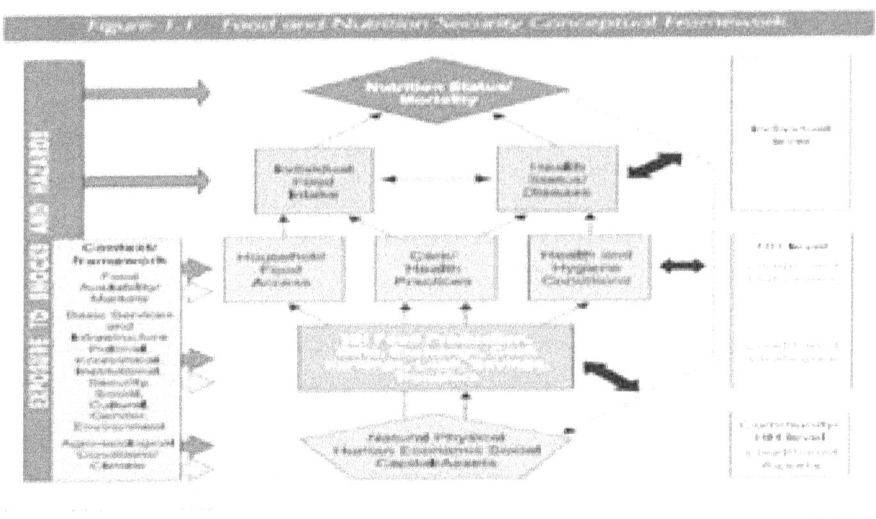

Future Vision

Focus on Freeing India of Hunger

- Our vision must be focused on free India from hunger, malnutrition and poverty. Realising that food insecurity is both a cause and consequence of poverty, particularly the increasing importance of economic access to food, income-generating of- and non-farm activities (employment) should be strongly promoted.

Food Production

- Countries in the SAARC region can benefit mutually by exchanging their technologies, research notes and learning by each other's rich experience in the field of agricultural production, post harvest management and distribution.
- Realising that under nutrition and deficiency of vital elements have adverse health consequences, especially in children and pregnant and lactating women, policies to initiate targeted nutritional interventions in forms of Integrated Child Development Services Scheme (ICDS), Mid Day Meal and Fortification have been helpful in reducing under nutrition.

Promoting Integrated Rural Employment

- Employment security in rural areas should be enhanced for improving household and individual level food security. On-farm and off-farm job opportunities are needed to promote on-farm jobs with capacity building process through training, farm credit.

Minimising Post Harvest Losses

- ☆ Priority should be given to enhance value, quality and safety along the food chain by modernizing various operations including processing, packaging, transportation, storage and marketing through the value chain system.

Recommendations for Food and Nutrition Security

- ☆ Alleviate the root cause, instead addressing only the symptoms - correct supply chain issues, which would increase availability of food while reducing costs
- ☆ FDI rules should be revised to start rapid improvement in the food supply chain due to influx in supply chain modernization techniques from foreign players like Wal-Mart
- ☆ Intervention into the food economy should be avoided by government, rather develop infrastructure for agriculture to make rural poor self-sufficient.
- ☆ It is necessary to strengthen the implementation of already existing alternate government schemes like NREGA.

Food Availability

Community Farming is one of the best options to increase food availability in India. Nutrient rich indigenous foods such as Ragi, Jowar should be emphasized by the government. Reduction in the cost of production through R&D interventions should be targeted to increase productivity. Utilization of public storage food facilities should be expanding. Compared to nations like US, India's BPL rate is very low due to which exposure of food availability benefits is limited to a very small segment of poor people, so a revision in BPL rate is much required.

Food Access

Supply Chain should be shortening, so that the beneficiaries can get the food. There is a need to start farmer-friendly marketing and processing so that farmers may own their companies. Also, there is a need strengthen the management and distribution of the food reserves. Innovative adoption of Food Coupons, Food Stamps, Food Credit Cards and Direct Cash Transfer should also be provided through Food Safety Net.

Food Utilization

Consumption of nutrient rich indigenous foods may help to increase the options for food availability and utilization. Assessment and monitoring of the nutritional status of school children and adolescents is necessary with nutrition education to create awareness regarding healthy diets. Besides it, dissemination of food and nutrition information is also important.

Creating Gender Equality

When women have an income, substantial evidence indicates that the income is more likely to be spent on food and children's needs. Analysts suggest that if women have the same access to productive resources as men, women could boost yield by 20-30 per cent, raising the overall agricultural output in developing countries by two and a half to four per cent. This gain in production could lessen the number of hungry people in the world by 12-17 per cent.

Conclusion

India has food distribution system at very large level. Many schemes are being run by the Government to provide food and ensure nutrition security of the needy and lower section of population. But there are some loop holes which are need to be rectified to ensure fair distribution. Multiplicity of schemes is one of the biggest issue through which many people get benefitted unfairly. Proper monitoring and check up are needed to stop it. Assessment and monitoring of the nutritional status at household level is also important with nutrition education to create awareness regarding healthy diets.

REFERENCES

http: //dfpd.gov.in

Kumar, B. and Mohanty, B. (2012). Public distribution system in rural India: Implications for food safety and consumer protection. Procedia – Social and Behaviour Sciences, 65: 232-38.

National Nutritional Policies (1993). Government of India, Department of Women and Child Development, Ministry of Human Resource Development, New Delhi, India.

NFHS (2015-16). National Family Health Survey 4, Ministry of Health and Family Welfare, New Delhi, India.

R. Radha Krishna & Reddy, K. V. (2002). *Food Security and Nutrition: Vision 2020.* www.planningcommision.nic.in

Singh, M. (2009, February). Food security in India leaves much to be desired.

Suryanarayana, M., & Silva, D. (2008). Poverty and Food Insecurity in India: A Disaggregated Regional Profile. Indira Gandhi Institute of Development Research, Mumbai.

www.fao.org

www.mhrd.gov.in/middaymeal.

Chapter 6

Poverty and Vicious Cycle of Low Food Productivity

Amita Yadav

Introduction

Widespread poverty resulting in chronic and persistent hunger is the single biggest scourge of the developing world today. The physical expression of this continuously re-enacted tragedy is the condition of under-nutrition which manifests itself among large sections of the poor, particularly amongst the women and children (National Nutriton Policy, GOI). The United Nations Food and Agriculture Organization say that around 1.02 billion people in the world are malnourished. This means that almost one-sixth of all humanity is suffering from hunger.

Poverty in India is widespread, and a variety of methods have been proposed to measure it. India's biggest challenge still remains ensuring food and nutritional security to its masses. It is the right of every person to have regular access to sufficient, nutritionally adequate and culturally acceptable food for an active and healthy life. Indian agriculture is ailing due to many factors and among these under investment, low productivity, irrigation and post harvest losses are the main factors. India was at 63rd position in the global hunger index (GHI) in 2015. The 55th position in 2014 was better than Pakistan (ranked 57) and Bangladesh (57), but trails behind Nepal (44) and Sri Lanka (39).

Cycle of Poverty

In economics, the cycle of poverty is the "set of factors or events by which poverty, once started, is likely to continue unless there is outside intervention".

The cycle of poverty has been defined as a phenomenon where poor families become impoverished for at least three generations, *i.e.* for enough time that the family includes no surviving ancestors who possess and can transmit the intellectual, social, and cultural capital necessary to stay out of or change their impoverished condition. In calculations of expected generation length and ancestor lifespan, the lower median age of parents in these families is offset by the shorter life spans in many of these groups.

The followings are causes of poverty:

- Low productivity
- Low salary
- Poor infrastructure and governance
- Business failure
- Ignorance, lack of skills and technology
- Unhealthiness or diseases
- Disaster
- Inability to access to resources such as land, finance, information, technical assistance
- No ongoing education
- Lack of Education and knowledge

Vicious Cycle of Poverty

Different economists have different opinions about the vicious cycle of poverty. According to Prof. Nurkse, "The main reason of vicious circle of poverty is the lack of capital information."

Similarly, Kindleberger opined that vicious cycle of poverty takes place due to the small size of the market."

Under nutrition is a condition resulting from inadequate intake of food or more essential nutrient(s) resulting in deterioration of physical growth and health. The inadequacy is relative to the food & nutrients needed to maintain good health, provide for growth and allow a choice of physical activity levels, including work levels that are socially necessary. This 'condition of under-nutrition, therefore, reduces work capacity and productivity amongst adults and enhances mortality and morbidity amongst children. Such reduced productivity translates into reduced earning capacity, leading to further poverty, and the vicious cycle goes on (figure 1 below). The nutritional status of a population is therefore critical to the development- and well being of a nation.

Vicious Cycle of Poverty

However the reasons of vicious cycle of poverty can be classified into three groups:

1. Supply side of vicious cycle.

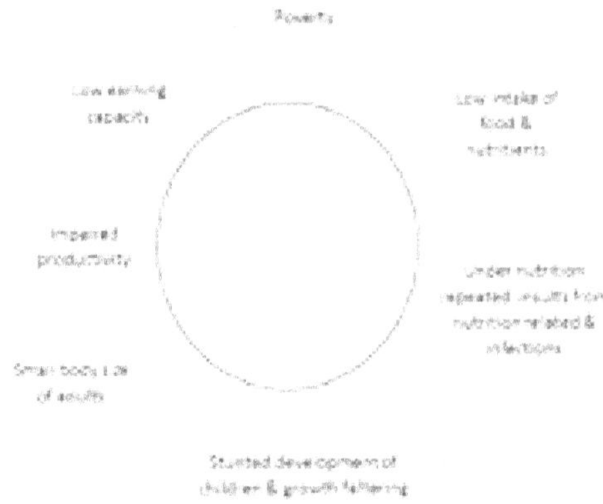

Figure 1

2. Demand side of vicious cycle.
3. Vicious cycle of market imperfections.

Supply Side of Vicious Cycle

Supply side of vicious cycle indicates that in under developed countries, productivity is so low that it is not enough for capital formation. In the words of Prof. Nurkse on the supply side there is a small capacity to save resulting from low level of national income. The low real income is a reflection of low productivity, which in turn is due largely to the lack of capital. The lack of capital is a result of the small capacity to save and so the circle is complete (Figure 2).

Figure 2

Demand Side of Vicious Cycle

According to Prof. Nurkse, "On the demand side the inducement of invest may be low because of the small purchasing power of the people, which is due to the real income, which is again due to low productivity. The level of productivity however, is the result of small amount of capital used in production which intern may be caused or at least partly caused by small inducement to invest (Figure 3).

Figure 3

Vicious Cycle of Market Imperfections

Meier and Baldwin have described a third vicious based on capital deficiency due to market imperfections. In underdeveloped countries, resources are underdeveloped and people are economically backward. Existence of market imperfections prevents optimum allocation and utilization of natural resources ant the result is underdevelopment and this, intern, leads to economic backwardness (Figure 4).

The development of natural resources depends upon the character of human resources. But due to lack of skill and low level of knowledge, natural resources will remain unutilized, underutilized or misutilized.

The vicious cycle of poverty is a result of the various vicious cycles which were on the sides of supply of the demand and demand for capital. As a result capital formation remains low productivity and low real incomes. Thus, the country is caught in vicious cycle of poverty which mutually aggravating and is very difficult to break them.

Need for Nutritional Policy within the Development Context

The need for a National Nutrition Policy is implicit in both the paramountcy of nutrition in development as well as in the complexity of the problem. This general problem of under-nutrition should be seen as a part of a larger set of processes that produces and consumes agricultural commodities on farms, transforms them into food in the marketing sector and sells the food, to customers to satisfy nutritional, aesthetic and social needs. Within this set, there are three sub-sets of issues, within the broad sectors of agriculture, food and nutrition, with various linkages among them. In fact, the third subset, *viz.* Nutrition, is the net- result of the other two subsets.

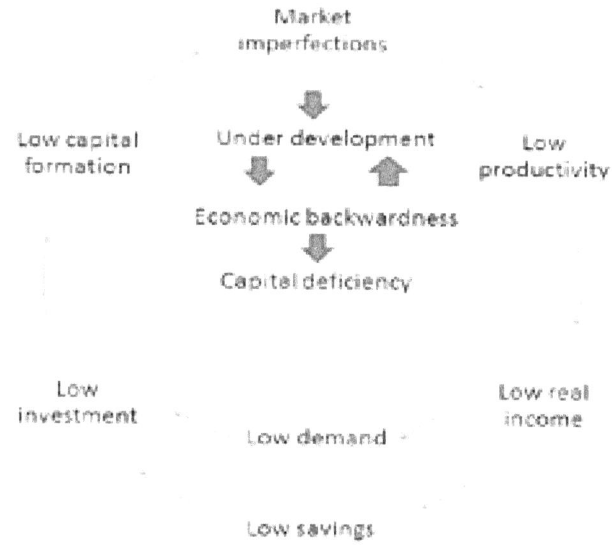

Figure 4

Low Productivity

Low productivity of the crops is also the worrying factor. European Union produces about 202 million tones of cereals by using only 37,040 thousand ha of land, whereas India produces 242 million tones of cereals on 142.7 million ha of land. China had similar pressures on its food security as it started its economic reforms in late 1970s. China used hybrid rice to raise yields. Now, more than 60 per cent of the area under rice in China is hybrid. As a result, Chinese yields of rice are almost double those of India, which has less than 3 per cent area under hybrids. China produces almost 200 million tons of paddies from 29 million ha compared to India's 150 million tons coming from 44 million ha.

Thus, our immediate attention and action is needed to improve the productivity of wheat, rice, pulses and oilseeds in the Indo-Gangetic plains and eastern India, particularly in Bihar, Jharkhand, Chhattisgarh, Orissa, Eastern Uttar Pradesh, West Bengal and Assam. Central Government has put an ambitious plan by focusing on these unrealized potential areas. The Central Government has launched various schemes like National Food Security Mission, Rashtriya Krishi Vikas Yojana, National Horticulture Mission, and National Agriculture Extension System to increase public investment in agriculture and rapid dissemination of technology. All these efforts are targeted to maximize returns to the farmers and to boost the food grain production. There is also urgent need to focus on other key areas which are irrigation, biotechnology, protected cultivation, post harvest technology and crop specific missions (Hazra, 2012).

As Indian agriculture is mostly rain fed, there is need for augmenting the irrigation facilities and this can be done by rainwater harvesting and by ambitious plan of interlinking of the rivers. There is also urgent need for application of biotechnology in agriculture. The biotechnological tools should be used to produce high yielding, pest-resistant and drought and flood resistant varieties of major food grains, pulses and oilseed crops. There is need to create more technical human resource and more technological interventions in the form of high yielding crop varieties and better agricultural practices to accelerate the growth of the agriculture to 4 per cent. Emphasis is also required on better post harvest management practices to minimize the 10-15 per cent losses in food grains and 30-35 per cent post harvest losses in fruits and vegetables (Gautam and Kumar, 2012)

Conclusion

Tackling hunger, poverty low food production and food insecurity in rural India will require multi-sectoral efforts; elected local bodies, together with the concerned departments of the government *i.e.* health, education, women and child welfare, rural and tribal development, *etc.* should prepare micro level action plans for food security initiatives. On the one hand, renewed attention to food and nutrition should prompt action for designing and implementing sound food and nutrition policies and at the same time, it is also necessary to empower the small scale food producers through cooperatives, self-help groups and other socially viable methods of group endeavor both at the production and post harvest phases of farming. The neglect of agriculture has to be stopped and greater attention should be paid to financing agriculture for food security. So, ultimately combating food crisis in rural India will require -

- More food aid for the rural poor
- More attention towards proper storage and distribution of food
- Establishment of good governance at the national/state/regional level administration as well as within the Panchayati Raj system for proper implementation of food security initiatives at the village-level
- Much more investments in agriculture, especially in small farm sectors
- Universal access to safe water and proper sanitation for the rural poor
- More investment in social security schemes designed especially for the health and educational betterment of the vulnerable and backward section of the society
- More comprehensive framework of monitoring mechanism at every level of service delivery
- System for the food and nutrition security initiative.

REFERENCES

Gautam, Harender Raj and Kumar, Er. Rohitashw (2012). Right to food. Kurukshetra March 2012.

Hazra, Anupam (2012). Food Security in Rural India: Poverty in the Land of Plenty, Kurukshetra March 2012.

Kindleberger Charles. Causes of vicious cycle of poverty. Article accessed online at http: //www.econmicsdiscussion.net/poverty/3_major_causes_of_vicious_circle_of_poverty_with_diagram/4592. Accessed on 08/06/2016.

Meier and Baldwin. Causes of vicious cycle of poverty. Article accessed online at http: //www.econmicsdiscussion.net/poverty/3_major_causes_of_vicious_circle_of_poverty_with_diagram/4592. Accessed on 08/06/2016.

National Nutrition Policy. (1993). Government of India Department of Women & Child Development, Ministry of Human Resource Development, New Delhi.

Nurkse Ragnar. Obstacles to economic development. In: The Economics of Development and Planning, 40th edn. By M.L. Jhingan. Vrinda Publication (P) Ltd.

Chapter 7

Food Price Control and Consumer Subsidy for Sustainable Development

Shripad Bhatt and Purushottam

Introduction

Food is any substance consumed to provide nutritional support for the body. It is usually of plant or animal origin, and contains essential nutrients such as carbohydrates, fats, proteins, vitamins or minerals. In ordinary usage, price is the quantity of payment or compensation given by one party to another in return for goods or services. The price controls are government-mandated legal minimum or maximum prices set for specified goods, usually implemented as a means of direct economic intervention to manage the affordability of certain goods. Governments most commonly implement price controls on staples, essential items such as food or energy products. Inflation in food items affects the food and nutritional security of consumers. On the other hand, a lower farm price upsets the livelihood of farmers. To safeguard the interests of both farmers and consumers, government tries to keep the prices both affordable for consumers and at the same time, remunerative for the farmers.

Need for Food Price Regulation

In India, malnutrition levels are higher. According to the Food and Agriculture Organization (FAO, IFAD and WFP, 2015), India has the second highest number of undernourished persons in the world with 19.46 crore people amounting to 15.2 per cent of national population. India ranks 114th of 132 countries in stunting with 38.7 per cent incidence level and ranks 170th out of 185 nations in anaemia prevalence among women of reproductive ages with 48.1 per cent incidence level. India ranked

97th of the 118 nations in 2016 according to the Global Hunger Index calculated by the International Food Policy Research Institute (IFPRI). Further, the share of total expenditure on food is also higher in case of poor. Because of all these reasons, it is essential to have a food price regulation mechanism and a public distribution system to fight hunger and to ensure nutritional security in India.

Policy Options for Food Price Regulation

There are several ways and means through which government can intervene and regulate the foodgrains market in India. Government monitors the prices of essential food items on regular basis. Government uses different policy tools such Minimum Support Prices (MSP), buffer stock operations, execution of Essential Commodities Act, 1955 and Prevention of Black-marketing and Maintenance of Supplies of Essential Commodities Act, 1980, Public Distribution System, usage of Price Stabilization Fund (PSF) and through trade policies such as import-export restrictions and minimum export prices and import prices to regulate the prices.

Minimum Support Prices (MSP)

Minimum Support Prices are announced to safeguard farmers' remuneration. Commission for Agricultural Costs and Prices (CACP) recommends the minimum support prices for 23 crops. Based this recommendation, central government fixes the MSP and announces it before the start of season. CACP decides on the prices based on many factors, among which major are the cost of production, supply and demand conditions, inter-crop price parity and implications of MSP on the consumers among other factors. Government undertakes the procurement at the MSP majorly for paddy and wheat though in the last two years pulses were also procured.

Objectives of MSP

1. To protect farmers' interest through remunerative prices and improve the foodgrain production
2. To prevent farm price collapse during distress sale
8. To develop a production pattern based on the needs of the economy through price cues

Buffer Stock

Food Corporation of India (FCI) maintains the buffer stock of foodgrains. Rice and wheat procured from farmers are stored and then distributed through Targeted Public Distribution System (TPDS) and Other Welfare Schemes (OWS) to consumers. Besides this, buffer stocks also help to stabilize the food prices (through Open Market Sale Scheme) and help to overcome emergency situations like drought and war. From time to time, the central government fixes the foodgrain stocking/ buffer norms (the level of stock to be maintained in the central pool at a given time). At present, the government has fixed the stocking norms taking into consideration the operation stocks and food security reserves needed.

Objectives of Buffer Stock

1. To supply foodgrains to PDS and other welfare schemes
2. To ensure food security during emergency situations
3. To stabilize the food prices in the economy

Public Distribution System

The public distribution system (PDS) in India aims at providing foodgrains to weaker sections of the society at cheaper price. This food management policy has helped the poor and weaker sections of the population and insulated them from the inflationary pressures in food grains. If the PDS is effectively, it has the potential to ensure nutritional security (Kattumuri, 2011).

Milestones in Food Management Policies in India

1. The PDS was initiated during war period (1940's) and was continued up to 1992 as general entitlement scheme
2. The Revamped Public Distribution System (RPDS) was launched during 1992 with an intention to cover drought prone, tribal and hill areas.
3. The Targeted Public Distribution System TPDS was launched in 1997 to focus on "poor in all areas". The families above poverty line were grouped under APL. From 2000, beneficiaries have been classified as AAY families (poorest among BPL families); BPL and APL and accordingly Central Issue Prices (CIP) are fixed.
4. The Antyodaya Anna Yojana (AAY) was started during 2000 to provide food grains at highly subsidized prices to the poorest of poor. The programme started with one crore beneficiaries families which were later increased to 2.5 crore.
5. The National Food Security Act (NFSA), 2013 has the objective of providing subsidized food grains to two-thirds of the population-covering upto 75 per cent of rural and upto 50 per cent of urban population. As per this act, eligible persons would get 5 kg of food grains per month at a price of Rs. 3/kg of rice, Rs. 2/kg of wheat and Re. 1/kg of coarse grains. As on April, 2016, 33 states/UT's have enacted this act providing food grains to 72.45 crore beneficiaries (14.8 crore families) and other states are in the advanced stage of implementation.

Besides these schemes, there are Other Welfare Schemes like Annapoorna Scheme, Mid-Day Meal Scheme, Integrated Child Development Services Scheme, and Rajiv Gandhi Scheme for Empowerment of Adolescent Girls, Village Grain Bank Scheme, Relief and Defence schemes *etc.* are operating in India[1].

Responsibility of handling PDS lies with both Central and State governments (Table 1). With the help of FCI, the Central Government provides foodgrains to

---

1 Visit http://fci.gov.in/sales.php?view=47 for detailed descriptions about various schemes.

State Governments/Union Territories (UT's) at the central issue price. The State Governments identifies the beneficiaries and distributes the grains through Fair Price Shops.

Table 1: Role of Central and State Governments in Operating PDS in India

	Central Government	State Government
Responsibility	Procurement of food grains from farmers Storage Transportation and allocation to state governments	Identification of beneficiary consumers and issuing ration cards Allocation of food grains to eligible families through Fair Price shops
Price	Procurement at MSP Allocation at Central Issue price	After 2001, state governments/UTs were given the power to fix the retail price except for AAY and NFSA, 2013

Procurement and off take of rice and wheat are shown in the Figure 1. The total procurement and off take are moving together over years, and any shortfall/ surplus demand is met with the help of stocks. The present procurement is to the tune of 60 million tonnes while the off take is about 56 million tonnes (2014-15).

Figure 1: Procurement and off Take of Rice and Wheat in PDS (*Source*: Economic Survey 2015-16).

Food Subsidy

The word 'subsidy' originated from a Latin word *subsidium* meaning assistance. In India, there are different subsidies provided like the food subsidy, petroleum subsidy, fertilizer subsidies are the major ones.

Share of different costs incurred by the FCI in handling rice and wheat have been depicted in Figures 2 and 3. Figures show that, in both the foodgrains, share of pooled cost of grains account for two-thirds of the economic cost with procurement incidentals and distribution cost accounting for the rest. Pooled costs include the MSP and bonus, and this cost is on the rise due to higher cost of production. Procurement incidentals are the costs incurred on APMC cess, commission charges and other administrative charges. The difference between economic cost (present economic cost for rice Rs. 3266.70/q in case of rice and Rs. 2344.80/q in wheat) and the issue price is borne by the government in the form of food subsidy.

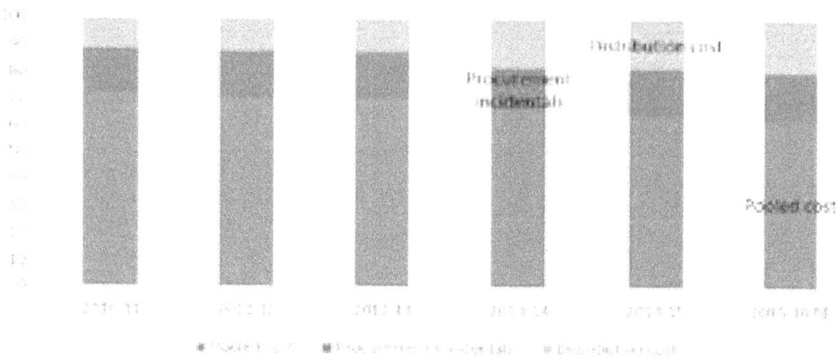

Figure 2: Share of Different Costs in Total Economic Cost of Rice (per cent).

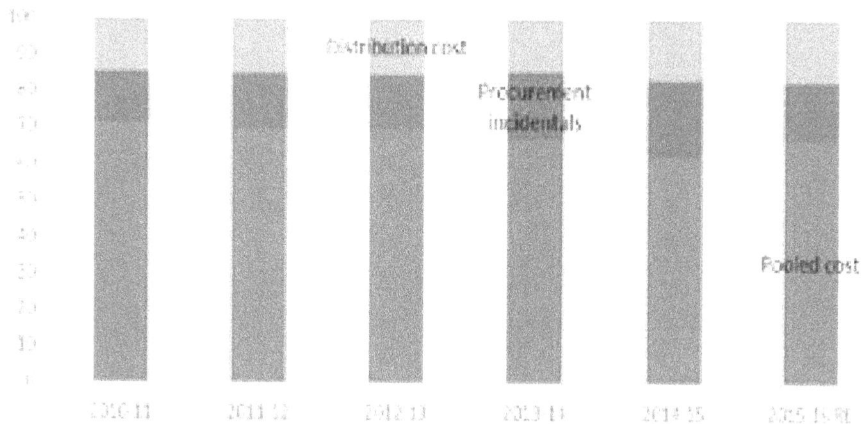

Figure 3: Share of Different Costs in Total Economic Cost of Wheat (per cent)

Central Issue Prices (CIP) of wheat and rice are fixed by the Government of India which are uniform across the nation. From 2002, the CIP has not been increased. Table 2 presents the CIP across different schemes.

Table 2: Central Issue Prices (CIP) of Wheat and Rice (Rate: Rs./Quintal.)

Commodity	*APL*	*BPL*	*AAY*	*NFSA*	*Other than NFSA*
Wheat	610	415	200	200	610
Rice	830	565	300	300	830

Figure 4 shows the food subsidy over the years. Due to increasing costs of procuring, handling and storing the food grains, there is an exponential growth in the food subsidies provided in India. Present food subsidy has surpassed Rs. one lakh crore.

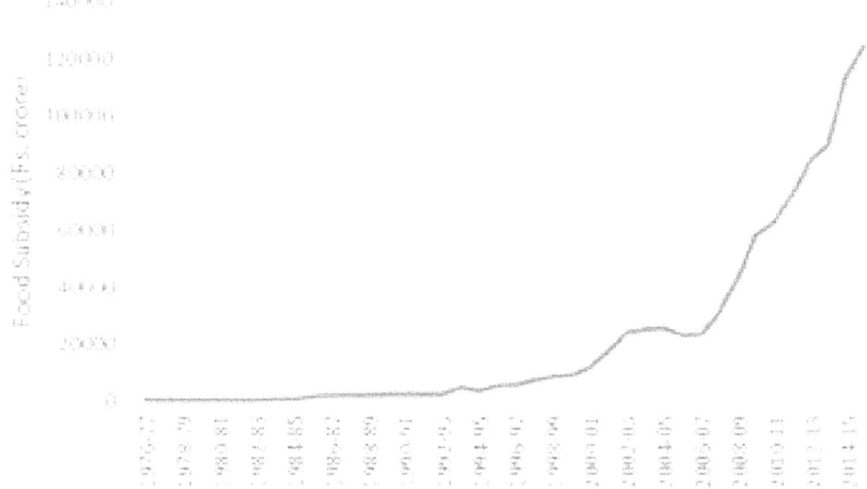

Figure 4: Food Subsidies in India.
[*Source*: Department of Food and Public Distribution and Economic Survey (Various years)].

However there is some criticism against food subsides in India. Many studies (Radha Krishna and Subbarao, 1997) have pointed out that there are many leakages in the present PDS and the benefits the end consumer gets is very small compared to the total economics costs. Further there are WTO rules which stipulate that subsidies should be within the prescribed limits as subsidies distort the market. Some studies have shown that, contrary to the public perception, PDS has been more or less an effective food policy system (Khera, 2011) while some studies advocating modification in PDS (Kochar, 2015).

New Initiatives

To reduce the leakage/diversion, government has digitised all the ration card details to bring in transparency. Around 56 per cent of the ration cards have been linked with Aadhaar cards and point of sale devices (June, 2016). Supply chain has been computerised; Grievance Redressal Mechanism and Transparency Portal have been created for accountability. Further, on pilot basis, Direct Cash Transfer of food subsidy to the bank accounts of the beneficiaries, who can buy the foodgrains of his choice from market, has been started in Chandigarh and Puducherry in September, 2015. End-to-end digitisation and automation would help in increasing effectiveness of PDS programmes.

REFERENCES

FAO, IFAD and WFP, (2015). The State of Food Insecurity in the World 2015.

Kattumuri, R. (2011). Food security and the targeted public distribution system in India. *Asia research centre working paper 38.*

Khera, R. (2011). Revival of the public distribution system: evidence and explanations. *Economic and Political Weekly, 46*(44-45), 36-50.

Kochar, A. (2005). Can targeted food programs improve nutrition? An empirical analysis of India's public distribution system. *Economic development and cultural change, 54*(1), 203-235.

Radha Krishna, R., & Subbarao, K. (Eds.). (1997). *India's public distribution system: a national and international perspective* (Vol. 380). World Bank Publications.

Chapter 8

National Nutrition Policy for Health and Food Safety

Uttara Singh and Sadhna Singh

Introduction

India's concern for Nutrition is as old as her civilization. Its holy books and other ancient scriptures contain guiding principles for nutrition and health. In the post independent India there has been an unequivocal commitment to the cause of nutrition through Constitutional provisions.

The Constitution of India states explicitly in Article 47 that the "State shall regard the raising of the level of the nutrition and the standard of living of its people and the improvement of public health among its primary duties and, in particular, the State shall endeavour to bring about prohibition of the consumption except for medicinal purposes of intoxicating drinks and of drugs which are injurious to health." The adoption of the National Nutrition Policy (NNP) by the Government in 1993 and the International Conference on Nutrition (ICN), held in December, 1992 at Rome are the important landmarks in the field of nutrition. The institution of National Nutrition Policy under the auspices of the Department of Women and Child Development, Ministry of Human Resource Development with specific objective of Operationalizing the multi-sectoral strategy enshrined in the Policy through the development of National Plan of Action on Nutrition reflects the vision and abiding interest of the planners as well as Nutrition Scientists in nutritional welfare of the country. The National Plan of Action, therefore, would serve the dual purpose of having a more or less set pattern for implementing the National Nutrition Policy as well as meeting the commitment made at the International Conference on Nutrition.

Present Nutrition Situation

The concept of health, as defined by the World Health Organization is the "state of complete physical, social and mental well-being and not merely absence of disease or infirmity". Nutrition and health are not synonymous, but without good nutrition, health cannot be at its best. Food has always played an extraordinarily vital role in the rise and growth or the fall and decline of a nation because of its effect on the health efficiency of its populace. Despite spectacular increase in the food grain production in recent years the problem of chronic malnutrition continues to exist extensively; especially among children and women, because they are caught in sequence of ignorance, poverty, inadequate food intake, disease and early death. Today, 'Malnutrition' is no longer considered an outcome of food deficiency or a health problem but as a multi-dimensional problem interfacing all efforts of developing human resources. The nutritional status of a nation is closely related to food adequacy and its distribution, levels of poverty, status of women, rate of population growth and access of its population to health, education, safe drinking water, environmental sanitation, hygiene and other social services. The nutritional status is, thus, as outcome of complex and inter-related set of factors. The nutrition problems of major importance can be classified in three broad groups:

1. Protein Energy Malnutrition
2. Micro-Nutrient Deficiencies, and
3. Prevalence of Chronic Diseases

The extent of Protein Energy Malnutrition (PEM) is reflected by the growth indicated by height and weight of children and adults in comparison to the established standards. This is often a direct outcome of maternal nutritional status resulting in foetal intra-uterine malnutrition and the low-birth weight of infants.

Micronutrient Deficiencies relate to poor intake of vitamins and minerals and can be identified by specific overt clinical signs. Among micronutrient deficiencies those of vitamin 'A', iron and iodine in particular are widely prevalent and have no relevance to the calorie and protein levels in the individual diets. Both PEM and the micronutrient deficiencies can lead to specific impaired functions of human performance like the decreased (work) productivity in adults and poor cognitive development and schooling in children.

Prevalence of Chronic Diseases

Obesity, hypertension, cardiovascular diseases and diabetes mellitus are closely linked to some metabolic disorders or to inappropriate diets often characterized by excessive intake of energy and fat, particularly the saturated fats, and low fiber intake. The trend in nutritional improvement in India during the last 15 years or so has been positive, though modest. The major gains of sound Government policies on food security and management of health and nutrition sectors have been complete elimination of famines in the last two decades, rare appearances of classic nutritional deficiency syndromes of cardiac beriberi, peripheral neuritis, pellagra *etc.*, of 50s and early 60s, steep decline in Kwashiorkor of the classical kind and almost disappearance of keratomalacia which was a major cause of nutritional blindness amongst children during the 60s.

The levels of food grain production and the per capita food availability are the important factors for assessing the nutrition situation in the country. While rapid strides in agricultural production in recent years has helped India in achieving national food security, there has not been a significant impact on the overall nutritional status of its population since the per capita availability of food, which is a key index to the food situation, has not changed appreciably as brought out by the Food Consumption Surveys. However, the energy intake of landless agricultural labour registered an average increase of 130 kcal/CU/day, yet remaining below the recommended level. The availability of food does not, in- itself, ensure its consumption by the population. Food prices and economic status of the people are the important determining factors.

India is known for its population size and the manner in which it is increasing. Population of this size will indeed have a tremendous impact on the benefit of development measures. Food, Agriculture, Health, Education and Rural Development are some of the important governmental sectors whose development measures are strongly dependent on population size. India's population has been one of the major factors that have mitigated dramatic improvement in the nutritional status of India's population so far.

Need for an Integrated and Inter-Sectoral Approach for Policy Implementation

Since India's independence in 1947, the Government has been striving continuously towards a better standard of life for its population. This was inherent in the principal objective of making the newly independent India progressive, self-reliant and fully developed in all possible means. Nutrition had been mentioned frequently even in the First Five Year Plan as an important determinant of health. The analysis of various phases of nutrition- programmes in India indicates that nutrition programmes started appearing in India within the provincial health sectors. The four distinct phases of nutrition programmes from the start to the present stage can be classified as:

1. Medical/Clinical Phase
2. Food Production and Technology Phase
3. Community Development Phase
4. Multi-Sectoral Phase

The medical/clinical phase was supported by bio-chemists and laboratory specialists and was pre-dominated by clinical studies on various types of malnutrition to understand the causes, course and methods for diagnosis of nutritional deficiencies. This phase provided an insight that the malnutrition was a community based problem and that an individual approach of diagnosis and treatment will not touch even the fringe of the problems. It also helped in realization that causes of malnutrition need not always be within the purview of medical sector and that measures even outside the health sector were necessary. A number of food processing and food fortification measures including the iodization of common salt, were initiated during the phase. The benefits of increased food production and technological advancement, however, did not touch the poorest segments of the population who remained where they were. It was being increasingly understood that the solution to the problem of malnutrition could not be restricted to any one sector and any nutrition programme designed to have an impact on the population must be directed towards those in the lowest economic level.

During the mid-60s Community Development Phase prevailed on the nutrition scene. The Applied Nutrition Programme, the very first integrated programme dominated the nutrition scene for about 10 years. The programme, though on paper, a multi-sectoral coordinated programme, in actual practice emphasized predominantly on production aspect only through school gardens, kitchen gardens, backyard poultry *etc.* The positive outcome of this phase was the increasing realization that combating malnutrition needed community participation and involvement and that at the Governmental level there was a need for a coordinated approach. During early 70s, diverse nutrition programme directed towards specific nutritional deficiencies were initiated such as Crash Feeding Programme, the Supplementary Nutrition Programme for pre-school children in rural, urban and tribal areas for bridging the calorie and protein gap, Prophylaxis Programme against nutritional anemia in women and prophylaxis programme against nutritional blindness in children *etc.* The national level Integrated Child Development Services (ICDS) culminated from this realization which took birth on 2nd October, 1975 marking the start of a multi-sectoral phase.

The ICDS programme is probably the most rational approach towards combating malnutrition in children and mothers. This phase is remarkable in that it has increasingly become clear that malnutrition and socio-economic deprivations are virtually the cause and consequence of each other. This realization has led to the formulation of National Nutrition Policy and the National Plan of Action on Nutrition within the framework of national development.

National Nutrition Policy

It advocates a comprehensive, integrated and inter-sectoral strategy for alleviating the multi-faceted problem of malnutrition and achieving the optimal state of nutrition for the people. The strategy highlights the importance of direct nutrition interventions for vulnerable groups as short term measures as well as various development policy instruments as long-term institutional and structural changes which will create conditions for improved nutrition. The Policy has identified Key areas for action in various spheres like food production, food supply, education, information, health care, rural development, women and child development, people with special needs, and monitoring and surveillance. The Policy requires a series of action to be undertaken by various concerned Departments/Ministries of the Government. Some of the direct interventions enshrined in the Policy are:

1. Expanding the nutrition intervention net (ICDS, UIP, ORT)
2. Empowering mothers with nutrition and health education

3. Reaching the adolescent girls
4. Ensuring better coverage of expectant women
5. Controlling micronutrient deficiencies, and
6. Fortifying essential foods with nutrients.

The indirect policy instruments or the long-term institutional changes cover the areas of:

1. Food security
2. Improving dietary pattern
3. Purchasing power
4. Public distribution system
5. Land reforms
6. Health and family welfare
7. Nutrition and health education
8. Education and literacy
9. Nutrition surveillance
10. Information and communication
11. Community participation

The Government has initiated several measures in the spheres of women and child development, health and family welfare, education, rural development *etc.*, in the recent years which match the strategies outlined in the National Nutrition Policy. The major task in the war against malnutrition, therefore, is to ensure that nutritional objectives are not only articulated in various development/sectoral plans and policies, but are also matched by plans of action and there is close coordination between these sectors to achieve the Goals set in the National Nutrition Policy.

National Nutrition Goals

The ultimate goal of the Government is to develop and implement a comprehensive, integrated and multi-sectoral strategy based on an intersectoral approach. While each sector would have its own goal for maximizing its contribution towards nutrition, the National Nutrition Goals to be as under:

1. Reduction in moderate and severe malnutrition among pre-school children
2. Reduction in chronic under nutrition and stunted growth in children
3. Reduction in incidence of low birth weight to less than 10 percent
4. Elimination of blindness due to vitamin 'A' deficiency
5. Reduction in iron deficiency anaemia among pregnant women
6. Universal iodization of salt for reduction of iodine deficiency disorders
7. Improving household food security through poverty alleviation programmes
8. Promoting appropriate diets and healthy lifestyles

Sectoral Plans of Action

National development policies and strategies can have an important bearing, on the nutritional status of the population. Since many of the basic causes of malnutrition lie outside the immediate field of nutrition, the most effective governmental strategies to reduce malnutrition on a national scale can be those that focus on incorporating nutritional considerations in all development policies and programmes including even the non-nutritional interventions. The Sectoral Plans, thus, represent a starting point in their process of analysis, articulation, planning, reinforcing and monitoring each sector's contributions for reducing malnutrition in the country. The sectoral plans define the general objectives, specific objectives and activities for the sectors which can contribute to nutrition improvement through their respective programmes: These sectors are Agriculture, Civil Supplies, Education, Food, Mother & Child Health, Women and Child Development, Health, Environment and Forest, Rural Development and Information & Broadcasting.

Sectoral Plan-Agriculture

General Objective

To ensure national level food security including adequate buffer stocks and nutritional considerations in Agriculture Policy.

Specific Objectives

1. To ensure increase production of various foods with a view to match the nutritional requirements of the population.
2. To enhance food security by minimizing losses during harvest in the field.
3. To link Food, Nutrition and Population issues in formal and nonformal agricultural training.
4. To strengthen the machinery for agriculture extension for promoting nutrition oriented horticultural activities, home gardening *etc.*
5. To give due emphasis to the development of Horticulture and promote the production of Vitamin 'A' (β-carotene) and Iron rich foods and increase awareness to improve consumption.

Sectoral Plan-Civil Supplies and Public Distribution

General Objective

Ensuring food security at the household level by making the essential foods available through the Public Distribution System to the people, particularly to the disadvantaged sections.

Specific Objectives

1. To expand the network of Fair Price Shops with emphasis on remote, far-flung and inaccessible areas.

2. To ensure effective and speedy implementation of Revamped Public Distribution System in identified areas like drought-prone areas, desert areas, tribal areas and far-flung and hilly areas.
3. To provide a better package to the identified priority areas by including additional commodities such as pulses, edible oil, coarse grains, iodized salt *etc.*
4. To make additional allocations to the State/UTs to augment the supply to the Revamped Public Distribution System areas for help during lean periods.
5. To take care of infrastructure requirements such as additional fair price shops and additional storage capacity in the identified areas within a specific time-frame.
6. To include O.R.S. (Oral Rehydration Salts) packets in Public Distribution System and ensure its availability in remote areas as to prevent deaths in children due to diarrhoea.
7. To create adequate buffer stocks to meet the requirements of the seasonally "at risk" population during unforeseen natural calamities and other contingencies.
8. To take effective steps against hoarding and black marketing of essential food articles to ensure easy availability at reasonable prices to the public.

Sectoral Plan-Education

General Objective

To provide convergent services under Education sector to enhance the nutritional and health status of the community with special emphasis on girls education and improved status of women.

Specific Objectives

1. To achieve universalisation of education of children, particularly girls and disadvantaged groups.
2. To incorporate basic health and nutrition education in school curriculum.
3. To organize family life education for both boys and girls and equip them, especially the adolescent girls, with basic knowledge of mother and child care, basic nutrition and health.
4. To impart functional literacy for improving economic status and nutritional well being of the community.
5. To create nutritional awareness through various formal and non formal education programmes and adult education programme.

Sectoral Plan-Forestry

General Objective

Popularizing the growing of plants/trees supplying foods/fruits with special emphasis on β-carotene (vitamin 'A') rich species in the Social Forestry Programmes with a view to create nutritional awareness and promote the consumption of nutritious foods.

Specific Objectives

1. To increase the, percentage of the species supplying food/fruits in the Social Forestry Programmes.
2. To integrate the production, harvesting, processing, transportation and marketing in the Forestry sector with a view to promote the consumption of nutritious foods/fruits.
3. To ensure remunerative prices for forest produce.
4. To establish linkages with on-going development programmes.
5. To organize Education, Training and Information for Forestry sector at State, District and Village levels for creating awareness about nutrition, plants/trees, supplying nutritious foods and their use.
6. To identify forest species rich in different nutrients with special emphasis on β-carotene for different geographical regions and adopt in Social/Farm Forestry Programme.

Sectoral Plan-Maternal and Child Health (Family Welfare)

General Objective

Improving the nutritional status of women and children through nutrition prophylaxis programmes, health and nutrition education and public health measures, besides achieving a small family norm.

Specific Objectives

1. Elimination of blindness due to vitamin 'A' deficiency and reduction of Bitot spots in pre-school children.
2. Reduction in iron deficiency anemia among pregnant women.
3. Universal coverage under Child Survival and Safe Motherhood (CSSM) Programme.
4. Reduction in incidence of low birth weight babies.
5. Promotion of exclusive breast feeding upto 4 months and sound complementary feeding practices of children after 4-6 months of age.
6. Universal use of O.R.S. (Oral Rehydration Salts) by 1997 with a view to effectively manage persistent diarrhoea and reduce deaths due to diarrhoea in children under 5 by 50 per cent.
7. Access by all couples to birth control information and services.

Sectoral Plan-Food

General Objective

Ensuring Food Security in the country.

Specific Objectives

1. To introduce nutritional considerations in Food Management and Procurement Policy.
2. To provide subsidy in order to ensure distribution of foodgrains to the common people at affordable prices.
3. To maintain national grain reserves/buffer stocks to meet the demands during lean periods.
4. To create larger and better storage structures to meet the growing demands in future.
5. To promote scientific storage of food grains.
6. To ensure remunerative price to the farmers for their produce.
7. To procure foodgrains of adequate nutritional quality.
8. To allocate adequate foodgrains for supplementary nutrition programmes for vulnerable groups.

Sectoral Plan-Food Processing Industries

General Objective

Meeting the nutritional needs of the people by giving nutrition orientation to the projects in food processing sector.

Specific Objectives

1. To set up Nutrition Cell in the Ministry of Food Processing Industries.
2. To develop and encourage milling technology based on nutritional considerations.
3. To promote agro-based food processing units with adequate linkages for marketing.
4. To develop and produce low cost instant health foods particularly for children.
5. To promote nutritious beverages like vegetable protein (groundnut cake) based milk beverages.
6. To undertake fortification and enrichment of common foods with vital nutrients like vitamin 'A', iron and iodine.
7. To prepare special nutritious supplements like ARF (Amylase Rich Flour) for therapeutic purposes.
8. To ensure training of personnel in Food Processing Industry on basic nutrition concepts.

Sectoral Plan-Health

General Objective

Achieving health for all through prevention and control of various forms of malnutrition, diseases related to inappropriate diets, creating health awareness among the people and ensuring adequate primary health care for all.

Specific Objectives

1. Ensuring adequate primary health care for r all with special emphasis on vulnerable groups.
2. Reducing IDD-Goitre prevalence.
3. Protecting consumers through improved food quality and safety.
4. Promoting appropriate diets and healthy life styles.
5. Creating health awareness among the people.
6. Managing health of people in natural calamities.

Sectoral Plan-Information & Broadcasting

General Objective

Creating a climate of awareness in the country about the importance of nutrition for the well-being of the people and ways and means of preventing various forms of malnutrition though its different units.

Specific Objectives

1. According high priority to awareness generation programmes concerning nutrition on Doordarshan, AIR as well as through Field Publicity Units.
2. Allocating free time for communicating nutrition messages during the prime time on Doordarshan.
3. Screening all commercial advertisements having a bearing on nutrition and health of people with a view to check misinformation reaching the people.
4. Screening of all commercial advertisements with a view to ensuring that they are in conformity with Infant Milk Substitutes Act, 1992.

Sectoral Plan-Labor

General Objective

Protecting and promoting the nutrition of various types of labor agricultural, construction, industrial *etc.*, with special emphasis on children and women at work, through formulation and enforcement of appropriate labor laws.

Specific Objectives

1. To ensure optimum nutrition besides safety, health and welfare of labor.
2. To review policy relating to special target groups such as women and child labor with a view to incorporate nutrition components in the same.

3. To give due emphasis to the problem of child labor as part of the overall task of securing human development.

Sectoral Plan-Rural Development

General Objective

Improving purchasing power of the people in rural areas through employment generation and poverty alleviation programmes with a view to improve food security at the household level.

Specific Objectives

1. To strengthen rural poverty alleviation programmes focusing on employment & income generation.
2. To make a forceful dent on the purchasing power of the economically disadvantaged sections of the population with special focus on landless;
3. To improve women's socio-economic status through DWACRA;
4. To upgrade the skills of rural youth with special focus on adolescent girls through TRYSEM;
5. To accelerate developmental programmes in difficult areas, such as Drought Prone and Desert areas;
6. To strengthen Rural Sanitation Programme and provide universal access to safe drinking water for preventing water borne diseases

Sectoral Plan-Urban Development

General Objective

Ensuring access to social services relating to Health Care, Nutrition, Women & Child Development, Pre-school and Non-Formal Education and Physical amenities like potable water supply, sanitation, sewerage, drainage etc.; with a view to improve the nutrition level of the urban poor.

Specific Objectives

1. To strengthen Urban Poverty Alleviation Programme.
2. To extend the Urban Basic Services for the poor both in area and content with increased emphasis on improving the nutrition of the vulnerable groups like infants, preschool children, expectant and lactating mothers *etc.*
3. To revitalize the scheme of Environmental Improvement of Urban Slums.
4. To ensure access to safe drinking water to not less than three fourth of urban population.
5. To contribute towards the achievement of Mid-decade and NPA Goals with specific reference to improving the nutritional status of groups at risk including women; adolescent girls, children 0-2 years and pre-school

children among the urban poor with emphasis in Class I and II size cities/ towns which comprises over 70 per cent of urban population.

Sectoral Plan-Welfare

General Objective

To promote nutrition of the disadvantaged sections of society by ensuring nutritional components in various welfare programmes.

Specific Objectives

1. To set up a Nutrition Cell in the Ministry of Welfare to facilitate incorporation of nutrition objectives and components in various welfare programmes.
2. To give due emphasis to nutritional needs_ and care of children in orphanages, street children and aged people.
3. To incorporate nutrition components in the form of nutrition education and training in tribal development programmes, education programmes for tribal girls, welfare of the disabled, drug addiction *etc.*
4. To prevent disabilities due to nutritional deficiencies.

Sectoral Plan-Women and Child Development

General Objective

Ensuring appropriate development of human resources both through direct nutrition interventions for especially vulnerable groups as well as through various development policy instruments for improved nutrition as laid down in National Nutrition Policy. Improving nutrition and health of women and children through strengthening and expansion of ICDS programme and setting up of appropriate systems for monitoring the follow-up actions under National Plan of Action for Children.

Specific Objectives

1. To achieve mid-decade and decade goals set in the National Plan of Action on Children (NPA).
2. To ensure inter-sectoral coordination among various sectors involved in achieving nutrition goals.
3. Monitoring the nutrition situation in the country by establishing a Nutrition Surveillance System and Data Base in the Department.
4. To develop district level diet and nutrition profiles of various States/UTs with a view to enable area specific programme and nutrition education interventions.
5. To undertake direct nutrition interventions for specially vulnerable groups through
   (a) Expanding Integrated Child Development Services so as to cover all the community development blocks in the country and 50 per cent of the urban slums.

(a) Extending services under ICDS especially to adolescent girls from poor families in all CD blocks and 50 per cent of the urban slums of the country.

(c) To reduce incidence of low birth weight babies to 10 per cent by various strategies such as care of pregnant and nursing mothers and improving capabilities of young girls and women to look after themselves and children.

6. To promote breast feeding and appropriate complementary feeding to check growth faltering and malnutrition among children.
7. To create nutritional awareness among the people through interpersonal as well as mass media communication using the infrastructure of FNB, ICDS, Health care system as well as Home Science Colleges and NGOs working in this field.
8. To undertake fortification of foods with essential nutrients like iron, Vit. 'A' *etc.*
9. To popularize production of low cost nutritious foods from locally available food materials by involving women's groups, NGOs *etc.* for meeting the needs of supplementary feeding programmes.
10. To control support of micronutrient deficiencies by sensitizing policy makers, professional groups, programme personnel, extension workers and beneficiaries.
11. To identify areas of nutrition related research to be carried out in various sectors.
12. To develop communication strategy for improving knowledge, attitude and practices related to nutrition.
13. To promote convergence of services of related sectors through effective coordination mechanism at all levels and extending the outreach of these programme with a view to improving the access to social services.

Implementation Strategy

The National Plan of Action on Nutrition is a framework to translate the National Nutrition Policy into forceful, viable and realistic action programmes. The measures enumerated in the sectoral plans would need to be administered by various governmental and non-governmental organizations. The role of the Department of Women and Child Development, being the nodal Department, is to coordinate, monitor and wherever required regulate the implementation of the National Plan of Action on Nutrition.

The National Nutrition Council, headed by the Prime Minister will be the highest body for overseeing the implementation of the National Nutrition Policy and the Plan of Action. It will issue policy guidelines for improving and strengthening the various sectoral programmes. The

Council will also be the national forum for policy coordination, review and direction at national level. Special Working Groups will be constituted in all concerned Departments to analyze the nutritional relevance of sectoral proposals and incorporate nutritional considerations in the light of the Nutrition Policy wherever necessary.

Nutrition Surveillance System is an important instrument of the Nutrition Policy. Recognizing the need for a nation-wide nutrition monitoring system, a task force for developing a National Nutrition Surveillance System, involving experts, NIN, NNMB, ICMR, NSSO *etc.* has been set up in the Department. The mechanism for inter-sectoral planning and coordination at the State level would be similar to that of the Central Government. An apex State-level Nutrition Council will be constituted under the chairmanship of the Chief Minister and would comprise concerned Ministers of the State Government, experts and representatives of professional bodies and leading NGOs. There would be an Inter-Departmental Coordination Committee under the chairmanship of the Chief Secretary which will coordinate, oversee and monitor the implementation of the National Nutrition Policy. The Committee would also focus on the state level targets for various nutrition related indicators set under the National Plan of Action of Nutrition. The Secretary of the Department dealing with women and children would be the convener of this Committee. Special Working Groups will be set up in the concerned sectors of the State Government which will be responsible for vetting the various sectoral schemes from the point of view of nutrition before they are finalized.

Mobilizing resources for nutrition interventions would be a major area of concern for the State Governments and local bodies (including municipal and panchayat bodies). A concerted effort will be made towards building effective community support and ultimately community participation in these schemes. In order to create a strong data-base for nutrition surveillance, it is imperative to have such Cells at the State Level also for collecting requisite information from various sources. The present day soft-ware technology can help analyze such data within a short time if adequate set up and linkages exist.

The State Governments may consider constituting similar bodies like Coordination Committees, Nutrition Council *etc.*, at their district level also. The mechanism and structures set up for coordinating the development, review and monitoring of the National Plan of Action on Nutrition will enable a flexible, sensitive and responsive process of continuous assessment, analysis and implementation of Nutrition relevant actions for reducing malnutrition in the country.

Conclusion

All the states in India though at different pace area in this transition phase. Factors pertaining to food, production, utilization, food safety, micronutrient deficiencies and nutrition related chronic morbidities are some of the newer challenges. These issues must receive political attention in order to be cornered effectively. The National Nutrition Policy recommends actions needed to sustain this position and innovative multi-sector and sector specific objective that will help to assure food and nutrition improvement. This policy recognizes and focuses on the national resolve to substantially reduce the prevalence of stunting and to improve household food security particularly among the most vulnerable families.

REFERENCES

WHO, (2010). A Review of Nutrition Policies. Draft Global Nutrition Policy Review: World Health Organization, 2010.

Food and Nutrition Policy, (2016). Public Health Knowledge of World.http: // ocw.jhsph.edu/index.cfm/go/viewCourse/course/FoodNutritionPolicy/ coursePage/index/. Retrieved on 06th June 2016.

Kanjilal, B., Mazumdar, P.G., Mukherjee, M. and Rahman, M.H. (2010). Nutritional status of children in India: household socio-economic condition as the contextual determinant.Int J Equity Health. 9: 19.

National Plan of Action on Nutrition. (2001). Food and Nutrition Board Ministry of Women & Child Development Government of India,

National Nutrition Policy. (1993). Government of India. Department of Women & Child Development, Ministry of Human Resource Development, Nutrition. UNICEF India, http: //unicef.in/Story/1124/Nutrition. Retrieved on 25th May 2015.

Chapter 9

Contribution of National and International Organization for Agricultural Development

Vishakha Sharma

Introduction

Food is essential for life. We depend on agricultural outputs for our food requirements. India produces large quantity of food grains such as millets, cereals, pulses, *etc.* A major portion of the food-stuffs produced is consumed within the country. Our farmers' works day and night to feed our population that counts over 1.21 billion.

Besides agriculture with a commercial bias, subsistence agriculture with its emphasis on the production of food for the cultivator's family is widespread. Traditionally, Agriculture is followed as the simplest method of obtaining food for the family. Agriculture in India is more a 'way of life' then a 'mode of business'.

Nearly 800 million people across the globe will go to bed hungry tonight, most of them smallholder farmers who depend on agriculture to make a living and feed their families. Despite an explosion in the growth of urban slums over the last decade, nearly 75 per cent of poor people in developing countries live in rural areas. Agriculture is the science and practice of activities relating to production, processing, marketing, distribution, utilization, and trade of food, feed and fiber. This definition implies that agricultural development strategy must address not only farmers but also those in marketing, trade, processing, and agri-business. Agriculture plays a vital role in India's economy. Over 58 per cent of the rural households depend on agriculture as their principal means of livelihood. Agriculture, along with fisheries and forestry, is one of the largest contributors to the Gross Domestic Product (GDP). The Department of Agriculture and Cooperation under the Ministry of Agriculture is responsible for the development of the agriculture sector in India. It manages several other bodies, such as the National Dairy Development Board (NDDB), to develop other allied agricultural sectors.

National Organization for Agricultural Development

Centrally Sponsored Missions

1. National Food Security Mission (NFSM)
2. National Mission on Sustainable Agriculture (NMSA)
3. National Mission on Oilseeds and Oil Palm (NMOOP)
4. National Mission on Agricultural Extension and Technology (NMAET)
5. Mission of Integrated Development of Horticulture (MIDH)

Central Sector Schemes

1. National Crop Insurance Programme (NCIP)
2. Integrated Scheme on Agriculture Cooperation (ISAC)
3. Integrated Scheme for Agriculture Marketing (ISAM)
4. Integrated Scheme on Agriculture Census, Economics & Statistics (ISACE&S)
5. Secretariat Economic Service (SES)

State Plan Scheme

1 Rashtriya Krishi Vikas Yojna (RKVY)

Centrally Sponsored Missions

National Food Security Mission (NFSM)

NFSM aims to increase the production of rice, wheat, pulses and Coarse Cereals through area expansion and productivity enhancement; restoring soil fertility and productivity; creating employment opportunities; and enhancing farm level economy. The basic strategy of the Mission is to promote and extend improved technologies, *i.e.*, seed, micronutrients, soil amendments, integrated pest management, farm machinery and resource conservation technologies along with capacity building of farmers.

National Mission on Sustainable Agriculture (NMSA)

NMSA has been formulated to make agriculture more productive, sustainable, remunerative and climate resilient by promoting location specific integrated/ Composite Farming Systems; conserve natural resources through appropriate soil and moisture conservation measures; adopt comprehensive soil health management practices; optimize utilization of water resources through efficient water management to expand coverage for achieving 'more crop per drop; develop capacity of farmers & stakeholders, in conjunction with other on-going Missions and pilot models in select blocks for improving productivity of rainfed farming by mainstreaming rainfed technologies.

National Mission on Oil Seeds and oil Palm (NMOOP)

The Mission aims to expand area under oilseeds, harness the potential in the area/districts of low productivity, strengthening inputs delivery mechanism, strengthening of post harvest services besides a focus on tribal areas for tree bourn oilseeds.

National Mission on Agricultural Extension & Technology (NMAET)

The Mission has four components *viz.*: Sub Mission on Agriculture Extension, (SMAE) Sub Mission on Seed and Planting Material (SMSP), (iii) Sub Mission on Agricultural Mechanization (SMAM) and (IV) Sub Mission on Plant Protection and Plant Quarantine (SMPP).The Mission aims to disseminate information and knowledge to the farming community in local language/dialect in respect of agricultural schemes.

Mission of Integrated Development of Horticulture (MIDH)

The Missions aims to promote holistic growth of horticulture sector, including bamboo and coconut through area based regionally differentiated strategies, which includes research, technology promotion, and extension, post harvest management, processing and marketing, in consonance with comparative advantage of each State/region and its diverse agro-climatic features; encourage aggregation of farmers into farmer groups like FIGs/FPOs and FPCs to bring economy of scale and scope; enhance horticulture production, augment farmers, income and strengthen nutritional security and improve productivity by way of quality germplasm, planting material and water use efficiency through Micro Irrigation.

Central Sector Schemes

National Crop Insurance Scheme (NCIP)

The Scheme aims to provide insurance coverage and financial support to the farmers in the event of crops failure as a result of natural calamities, pests and diseases as also to encourage farmers to adopt progressive farming practices, high value inputs and higher technology in agriculture.

Integrated Scheme on Agriculture Cooperation (ISAC)

The objective of the scheme is to provide financial assistance for the activities of cooperatives like agro-processing, marketing of food grains, input supply, development of weaker section cooperatives, computerization of co-operatives *etc.* as also to develop cooperative awareness amongst the people and to cater to the education and training requirements of cooperative personnel and State Government officials.

Integrated Scheme on Agriculture Marketing (ISAM)

The Scheme aims to to promote creation of agricultural marketing infrastructure by providing backend subsidy support to State, cooperative and private sector investments; to promote creation of scientific storage capacity and to promote pledge financing to increase farmers' income; to promote Integrated Value Chains (confined up to the stage of primary processing only) to provide vertical integration of farmers with primary processors; to use ICT as a vehicle of extension to sensitize and orient farmers to respond to new challenges in agricultural marketing; to establish a nation-wide information network system for speedy collection and dissemination of market information and data on arrivals and prices for its efficient and timely utilization by farmers and other stake holders; to support framing of grade standards and quality certification of agricultural commodities to help farmers get better and remunerative prices for their graded produce; to catalyze private investment in setting up of agribusiness projects and thereby provide assured market to producers and strengthen backward linkages of agri-business projects with producers and their groups; and to undertake and promote training, research, education, extension and consultancy in the agri marketing sector.

Integrated Scheme on Agriculture Census, Economics and Statistics (ISACE&S)

The Scheme aims to collect/compile data of operational holdings in the country to provide aggregates for basic Agricultural Characteristics for use as the benchmark for inter-census estimates.

Secretariat Economic Service (SES)

The Scheme aims to provide support and servic☐es to the employees/officers of the Department of Agriculture & Cooperation including provision of office equipments, furniture, office accommodation, renovation of rooms, transport services, newspaper, Magazines, Publicity and Advertisement expenditure, *etc.*

State Plan Scheme

Rashtriya Krishi Vikas Yojna

Rashtriya Krishi Vikas Yojana (RKVY) is a special Additional Central Assistance Scheme which was launched in August 2007 to orient agricultural development strategies, to reaffirm its commitment to achieve 4 per cent annual growth in the agricultural sector during the 11th plan. The scheme was launched to incenti*viz*.e the States to provide additional resources in their State Plans over and above their baseline expenditure to bridge critical gaps.

Areas of Focus Under the Rkvy

- ☆ Integrated Development of Food crops, including coarse cereals, minor millets and pulses
- ☆ Agriculture Mechanization
- ☆ Soil Health and Productivity
- ☆ Development of Rainfed Farming Systems
- ☆ Integrated Pest Management
- ☆ Promoting extension services

- ☆ Horticulture
- ☆ Animal Husbandry, Dairying & Fisheries
- ☆ Sericulture
- ☆ Study tours of farmers
- ☆ Organic and Bio-fertilizers
- ☆ Innovative Schemes

National Bank for Agriculture and Rural Development (NABARD)

At the instance of Government of India Reserve Bank of India (RBI), constituted a committee to review the arrangements for institutional credit for agriculture and rural development on 30 March 1979, under the Chairmanship of Shri B.Sivaraman, former member of Planning Commission, Government of India to review the arrangements for institutional credit for agriculture and rural development. The Committee, in its interim report, submitted on 28 November 1979, felt the need for a new organizational device for providing undivided attention, forceful direction and pointed focus to the credit problems arising out of integrated rural development and recommended the formation of National Bank for Agriculture and Rural Development (NABARD). The Parliament, through Act, 61 of 1981, approved the setting up of NABARD. The bank came into existence on 12 July 1982 by transferring the agricultural credit functions of RBI and refinance functions of the then Agricultural Refinance and Development Corporation (ARDC). NABARD was dedicated to the service of the nation by the late Prime Minister Smt. Indira Gandhi on 05 November 1982.

NABARD was set up with an initial capital of 100 Crore. Consequent to the revision in the composition of share capital between Government of India and RBI, the paid up capital as on 31 March 2015, stood at 5000 Crore with Government of India holding 4,980 Crore (99.60 per cent) and Reserve Bank of India 20.00 Crore (0.40 per cent).

Mission

- ☆ Promote sustainable and equitable agriculture and rural prosperity through effective credit support, related services, institution development and other innovative initiatives.
- ☆ Providing refinance to lending institutions in rural areas
- ☆ Bringing about or promoting institutional development and
- ☆ Evaluating, monitoring and inspecting the client banks

NABARD is set up as an apex Development Bank with a mandate for facilitating credit flow for promotion and development of agriculture, small-scale industries, cottage and village industries, handicrafts and other rural crafts. It also has the mandate to support all other allied economic activities in rural areas, promote integrated and sustainable rural development and secure prosperity of rural areas. In discharging its role as a facilitator for rural prosperity NABARD is entrusted with.

Minimum Support Price Scheme

The Minimum Support Prices were announced by the Government of India for the first time in 1966-67 for Wheat in the wake of the Green Revolution and extended harvest, to save the farmers from depleting profits. Since then, the MSP regime has been expanded to many crops. Minimum Support Price is the price at which government purchases crops from the farmers, whatever may be the price for the crops. The MSP is announced by the Government of India for 25 crops currently at the beginning of each season *viz.* Rabi and Kharif. Following are the 25 crops covered by MSP:

- Cereals (7) - paddy, wheat, barley, jowar, bajra, maize and ragi
- Pulses (5) - gram, arhar/tur, moong, urad and lentil
- Oilseeds (8) - groundnut, rapeseed/mustard, toria, soyabean, sunflower seed, sesamum, safflower seed and nigerseed
- Copra
- Raw cotton
- Raw jute
- Sugarcane (Fair and remunerative price)
- Virginia flu cured (VFC) tobacco

Pradhan Mantri Gram Sinchayee Yojana

Har Khet ko Pani "Prime Minister Krishi Sinchayee Yojana"

Government of India is committed to accord high priority to water conservation and its management. To this effect Pradhan Mantri Krishi Sinchayee Yojana (PMKSY) has been formulated with the vision of extending the coverage of irrigation **'Har Khet ko pani'** and improving water use efficiency **'More crop per drop'** in a focused manner with end to end solution on source creation, distribution, management, field application and extension activities. The Cabinet Committee on Economic Affairs chaired by Hon'ble Prime Minister has accorded approval of Pradhan Mantri Krishi Sinchayee Yojana (PMKSY) in its meeting held on 1st July, 2015.

PMKSY has been formulated amalgamating ongoing schemes *viz.* Accelerated Irrigation Benefit Programme (AIBP) of the Ministry of Water Resources, River Development & Ganga Rejuvenation (MoWR, RD&GR), Integrated Watershed Management Programme (IWMP) of Department of Land Resources (DoLR) and the On Farm Water Management (OFWM) of Department of Agriculture and Cooperation (DAC). PMKSY has been approved for implementation across the country with an outlay of Rs. 50,000 crore in five years. For 2015-16, an outlay of Rs.5300 crore has been made which includes Rs. 1800 crore for DAC; Rs. 1500 crore for DoLR; Rs. 2000 crore for MoWR (Rs. 1000 crore for AIBP; Rs. 1000 crores for PMKSY).

International Organization for Agricultural Development

The World Bank

The World Bank Group's Partnership Strategy for India (2013-2017) will help India lay the foundations for achieving "faster, sustainable, and more inclusive growth" as outlined in the government's 12th five year plan. The World Bank Group will support India with an integrated package of financing, advisory services, and knowledge. During the World Bank financial year (July 2013-June 2014), funding for India was $5.2 billion $2.0 billion in International Bank for Reconstruction and Development (IBRD), $3.1 billion in International Development Association and $0.1 billion in CTF or Clean Technology Fund) across 16 projects.

1. Enhancing agricultural productivity, competitiveness, and rural growth
2. Poverty alleviation and community actions
3. Sustaining the environment and future agricultural productivity

The Bank's Agricultural and Rural Development portfolio is clustered across three broad themes with each project, generally, showing a significant integration of these themes.

- ☆ Agriculture, watershed and natural resources management
- ☆ Water & irrigated agriculture
- ☆ Rural livelihood development

International Fund for Agriculture Fund for Development (IFAD)

Since it was created in 1977, IFAD has focused exclusively on rural poverty reduction, working with poor rural populations in developing countries to eliminate poverty, hunger and malnutrition; raise their productivity and incomes; and improve the quality of their lives. The Fund has designed and implemented projects in very different natural, socio-economic and cultural environments. Many IFAD-supported projects and programmes have been in remote areas, and have targeted some of the poorest and most deprived segments of the rural population. IFAD has recognized that vulnerable groups can and do contribute to economic growth. These groups have shown that they can join the mainstream of social and economic development; provided the causes of their poverty are understood and enabling conditions for development are created. IFAD has been working in India for more than 30 years. Today, India is IFAD's largest borrower and one of its main contributors. OVER the years, the Fund and the Government of India have achieved significant results in the following areas:

- ☆ Commercialization of small holder agriculture
- ☆ Grass-roots institution building
- ☆ Women's empowerment
- ☆ Tribal development

The World Food Programme (WFP)

It is the world's largest humanitarian agency fighting hunger worldwide. Born in 1961, WFP pursues a vision of the world in which every man, woman and child has access at all times to the food needed for an active and healthy life. It works towards that vision with our sister UN agencies in Rome – the Food and Agriculture Organization (FAO) and the International Fund for Agricultural Development (IFAD) as well as other government, UN and NGO partners. On average, WFP reaches more than 80 million people with food assistance in 82 countries each year. 11,367 people work for the organization, most of them in remote areas, directly serving the hungry poor.

WFP's strategic plan lays out four objectives and all our work is geared towards achieving them. They are:

1. Save lives and protect livelihoods in emergencies
2. Support food security and nutrition and rebuild livelihoods in fragile settings and following emergencies
3. Reduce risk and enable people, communities and countries to meet their own food and nutrition needs
4. Reduce under nutrition and break the intergenerational cycle of hunger.

FAO (Food and Agriculture Organization)

Achieving food security for all is at the heart of FAO's efforts – to make sure people have regular access to enough high-quality food to lead active, healthy lives. FAO works with three main goals they are:

- The eradication of hunger, food insecurity and malnutrition;
- The elimination of poverty and the driving forward of economic and social progress for all; and
- The sustainable management and utilization of natural resources, including land, water, air, climate and genetic resources for the benefit of present and future generations.

Conclusion

Every nation's economic environment is made up of a complex aggregation of individual laws and regulations. All governments, including those in the industrialized West, do dumb things–sometimes out of ignorance, sometimes in response to interest group pressure, and sometimes in an attempt to achieve noneconomic ends. The basic question is whether economic stupidity is the exception or the rule–whether, in essence, the government acts as if its role is to manipulate the economy.

India is mainly an agricultural country. Agriculture is the most important occupation for most of the Indian families. In India, agriculture contributes about sixteen per cent (16 per cent) of total GDP and ten per cent (10 per cent) of total exports. Over 60 per cent of India's land area is arable making it the second largest country in terms of total arable land. Agricultural products of significant economic value include rice, wheat, potato, tomato, onion, mangoes, sugar-cane, beans, cotton, *etc.*

The agriculture sector in India is expected to generate better momentum in the next few years due to increased investments in agricultural infrastructure such as irrigation facilities, warehousing and cold storage. Factors such as reduced transaction costs and time, improved port gate management and better fiscal incentives would contribute to the sector's growth. Furthermore, the growing use of genetically modified crops will likely improve the yield for Indian farmers.

REFERENCES

Bose, A. (2015). Importance of agriculture in Indian economy. Important India. Retrieved from-http: //www.importantindia.com/4587/importance-of-agriculture-in-indian-economy/on 25 July, 2016.

Department of Agriculture & Cooperation Ministry of Agriculture Government of India. (2007). Guidelines for National Agriculture Development Programme (NADP) Rashtriya Krishi Vikas Yojana (RKVY).

Retrieved from: http: //www.fao.org/about/en/on 25 July, 2016.

Retrieved from https: //www.wfp.org/about on 28 July, 2016.

Retrieved from http: //pmksy.gov.in/on 28 July, 2016.

Schemes for Development of Agriculture and Farmers' Welfare. (2014). Press Information Bureau Government of India Ministry of Agriculture. Retrieved from: http: //pib.nic.in/newsite/PrintRelease.aspx?relid=106668 on 25 July, 2016.

Chapter 10

Nutritional Impact of Agricultural Programmes

Tanu Jain

Nutrition Security is the strength of any nation. If a nation is nutritionally secured, it may overcome any and every kind of problem. To be nutritionally secured should be the most important duty of nation. The backbone of any nutritionally secured nation is agriculture because it can only be achieved with the sufficient production and consumption of food.

To make agricultural sector strong enough, kinds of programmes are run which directly and indirectly linked with nutrition and its impact on population. Joint programme was also held on "Food Security and Nutrition" to discuss how to improve policies, programmes and interventions for making agriculture and food systems more responsive to nutrition (www.wfp.org).

Improving nutrition must begin with food and agriculture because the poor and most nutritionally vulnerable depend in large part upon agriculture for their livelihoods. It is the food system as a whole *i.e.* the post-production sector beyond agriculture including processing, storage, trade, marketing and consumption that contributes significantly more to the eradication of malnutrition.

"Nutrition-enhancing" is an approach that addresses the underlying determinants or basic causes of malnutrition. Nutrition-enhancing agriculture and food systems are those that effectively and explicitly incorporate nutrition objectives, concerns and considerations, improve diets and raise levels of food and nutrition security (OIWPS, 2010).

Food-based approaches recognize the central role of food, agriculture and diets in improving nutrition. Agriculture and food-based strategies focus on food as the primary tool for improving the quality of the diet and for addressing and preventing malnutrition and nutritional deficiencies. The approach stresses the multiple benefits derived from enjoying a variety of foods, recognizing the nutritional value of food for good nutrition, and the importance and social significance of the agricultural and food sector for supporting rural livelihoods (www.gainhealth.org).

In particular, there has been increased attention to an interest in improving the synergies between agriculture and nutrition. The agricultural sector has a key role to play in improving the livelihoods of the vulnerable through provision of food and income

From the programmatic perspective, the interest in leveraging agriculture to maximize nutrition impact has prompted development institutions to consider how best to implement the linkages in programmes. Several guidance documents have been produced for linking agriculture and nutrition, and there is emerging consensus on the best practices to achieve nutrition-sensitive agriculture. Programme designs incorporated best practices for agriculture programming for nutrition.

The Synthesis of Guiding Principles on Agriculture Programming for Nutrition was published by FAO in February 2013. This work involved a focused literature review and consultative process to elicit the best practices of agriculture programming for nutrition. The comprehensive review included guidance, institutional strategies and other publications released by international development institutions and United Nations agencies.

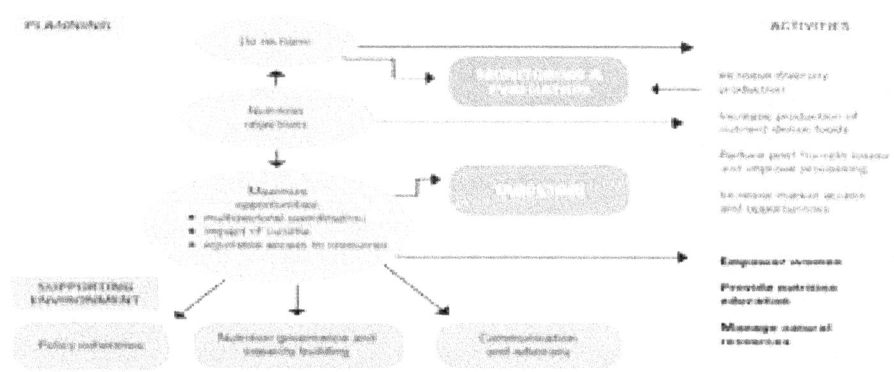

Figure 1: (Adapted from FAO, 2013).

Agricultural projects have a profound effect on household food security and nutritional wellbeing. There are multiple pathways through which agriculture can influence the nutritional outcomes. Agricultural programmes can maximize nutritional impact by following ways:

1. Agriculture programme may help to integrate nutrition counseling through agricultural extension which is highly useful especially when women are involved.

2. Home gardening is one of the effective way of improving diet and nutrition. Women are the caretaker of home. If the knowledge of home gardening would be provided to women, the healthy habits may be developed for household food consumption decisions.
3. Introduce micro-nutrient-rich crop varieties. Agricultural projects that utilize micro nutrient-rich plant varieties have major potential for ensuring needed nutrients and improving nutritional wellbeing. The agricultural programmes are working in this direction.

The agricultural sector presents many opportunities for improving nutrition and health. Agriculture sector can become strong with the intervention of new technologies in research laboratories and on farm experiments. Some technological advancement has been adopted through which the crop production has been increased unexpectedly.

Plant Breeding

Plant breeding is a science of optimizing a plant's genetic makeup to produce desired characteristics. It can be accomplished through a number of techniques. Through these techniques, higher yielding crops cab is produced which are better in quality, tolerant to environmental pressures, resistant to pests and diseases and tolerant to insecticides and herbicides.

Hybridization

Hybridization is a tool that farmers have used to develop high-yielding seeds since the early 1900s. It involves crossing of two or more crop lines to produce hybrid crops with more favorable traits, resulting from combining genes from the selected parents. It provides more tolerance to diseases, pests and environmental stresses such as hybrid corn, sorghum, canola, sunflower, and rice.

Molecular Marker-Assisted Selection (MAS)

Molecular markers are small sequence differences between various lines in a plant breeding population that can be used, when physically linked to traits, as a surrogate for the presence or absence of a desired trait without having to field test for the attributes of that trait. Molecular markers are detected through DNA sequencing methods using DNA derived from plant samples. The practice of molecular marker-assisted selection enables plant breeders to combine desirable plant traits rapidly and in large numbers. MAS is being used for breeding tomato to make it pathogen resistance

Nuclear Techniques: Nuclear techniques are used extensively in agriculture to make food crops more resistant to disease, boost crop yields and combat pests and animal diseases, manage soils and water management and identify causes of ocean acidification, largely responsible for the depletion of fish stocks in our oceans. Using nuclear techniques in conservation agriculture, crop yield can be increased, soil quality can be improved and land carbon sequestration can be enhanced. These techniques are being applied in extensive soybean cultivation in Brazil Another important contribution of agriculture towards nutrition and health is increased rural income, allowing people to improve their diets. The poor are overwhelmingly located in rural areas and derive a significant share of their income from agricultural activities.

International and National Research Organizations

International and national research organizations are working in the direction to make the world food secured and safe. Some of them are:

Asia-Pacific Association of Agricultural Research Institutions (APAARI)

Agricultural Research for Development (ARD) in the Asia-Pacific region is effectively promoted and facilitated through novel partnerships among NARS and other related organizations so that it contributes to sustainable improvements in the productivity of agricultural systems and to the quality of the natural resource base that underpins agriculture, thereby enhancing food and nutrition security, economic and social well being of communities and the integrity of the environment and services it provides. APAARI's mission is to promote the development of NARS (national agriculture research system) in the Asia-Pacific region through inter-regional and inter-institutional cooperation.

Consultative Group on International Agricultural Research (CGIAR)

(Formerly the Consultative Group on International Agricultural Research). An international organization which funds and co-ordinates research into agricultural crop breeding (http://www.cgiar.org). It has a network of 15 research centers known as the CGIAR (Consortium of International Agricultural Research Centers). The CGIAR's vision is supported by four strategic objectives:

- Reducing rural poverty
- Improving food security
- Improving nutrition and health
- Sustainably managing natural resources

Indian Council of Agriculture Research (ICAR)

The Indian Council of Agricultural Research (ICAR) is an autonomous organization under the Department of Agricultural Research and Education, Ministry of Agriculture, government of India. It is formerly known as the Imperial Council of Agricultural Research. It is established in 1929 as a registered society under the Societies Registration Act, 1860 in pursuance of the report of the Royal Commission on Agriculture (http://icar.org.in). The Council is the apex body for coordinating, guiding and managing research and education in agriculture including horticulture, fisheries and animal sciences in the country. It has 101 ICAR institutes and 73 agricultural universities spread across the country. The ICAR has played a role in enabling the country to increase the production of foodgrains by four, horticultural crops by six, fish by nine (marine five and inland 17, milk six times and eggs 27 times since 1950.

Programmes Working Towards Nutrition Security

To rid the world of extreme poverty and hunger continued investment and the application of research and development in agriculture and food production is essential. For this purpose many international and national level programmes are being run.

International Atomic Energy Agency (IAEA)

The IAEA, in partnership with the Food and Agriculture Organization of the United Nations (FAO), helps Member States to produce better and safer food using nuclear technology while promoting the sustainable use of agricultural resources. Eight Millennium Development Goals (MDGs) have been adopted by the international community as a foundation for global development activities. These goals aim to make significant steps to combat poverty, hunger, disease, illiteracy, environmental degradation and discrimination against women. The International Atomic Energy Agency's Statute states that the IAEA shall seek to accelerate and enlarge the contribution of atomic energy for peace, health and prosperity throughout the world (http://www.iaea.org). The IAEA contributes in many ways like Better crops and improved land and water management (Enhancing crop varieties, Reducing soil erosion, Optimizing fertilizer and water use), Controlling and eradicating insect pests, Increasing livestock productivity (Diagnosing and preventing disease), Food safety (Controlling food contaminants, Food irradiation), Monitoring and improving nutrition and Improving reproduction.

Adaptation for Smallholder's Agriculture Programme (ASAP)

It is launched by the International Fund for Agricultural Development (IFAD) in 2012 to make climate and environmental finance work for smallholder farmers. ASAP is driving a major scaling up of successful 'multiple-benefit' approaches to smallholder agriculture which improves production while reducing and diversifying climate-related risks. It increases capacity of at least 8 million smallholder farmers to expand their livelihood options in an uncertain and rapidly changing environment (http://www.ifad.org).

Direction of Work

1. Improve land management and promote gender- sensitive, climate-resilient agricultural practices and technologies
2. Increase availability and efficient use of water for smallholder agriculture production and processing
3. Increase capacity to manage short- and long-term climate risks and reduce losses from weather-related disasters
4. Document and disseminate knowledge on climate-smart smallholder agriculture
5. Increase climate resilience of rural infrastructure.

Conclusion

Nutrition security is the utmost requirement of any nation. It can be only achieved when a nation is food secured *i.e.* there is sufficient production, availability and affordability of food. Agriculture sector contribute in nutrition security through various national and international programmes. IAEA, ASAP, are some programmes which are being run in this direction. APAARI, ICAR are agricultural organization which works to make nation nutrition secured.

REFERENCES

FAO (2013). Food and Agriculture Organization of the United Nations June.

http: //www.gainhealth.org/programs/gain-national-food-fortification-program

http: //www.wfp.org/nutrition

Kadiyala, Suneetha (2011). Strengthening the Role of Agriculture for a Nutrition Secure India December.

OIWS (2010). Food Security in India: Performance, Challenges and Policies September OIWPS – VII.

www.apaari.org

www.cgiar.org

www.iaea.org

www.icar.org

www.ifad.org

Part B

Food Processing and Technology

Chapter 11

Impact of Agricultural Marketing System and Post Harvest Processing of Foods on Nutrition Security in India

Ragini Ranawat

Introduction

The relationship of agriculture with food and nutritional security per se is as old as time itself. This straightforward link is but a complicated one. By and large, India was immensely benefitted from Green Revolution. The country which was dependent on imported food grains to meet the day-to-day food requirement of the common people was transformed into a food exporter. Whereas the food crop production this decade increased by 22.74 per cent from the last one, the per capita food availability increased only by 4.8 per cent. The difference clearly points to factors like failure of supply chains and losses of food post – harvest. Despite the fact that a large percentage of the world's population is still undernourished, a huge amount of food is lost or wasted within the food system. This loss occurs throughout the system - from the production stage to household consumption. Majority of the rural poor rely on short Food Supply Chains (FSCs) with limited post harvest infrastructure and technologies. Loss occurs at every stage of the supply chain. Following harvest, about 60-70 per cent of food grain is stored on farms for variable periods, normally in traditional structures and at dangerously high moisture levels. This makes them particularly vulnerable to infestations of pests and micro-organisms. Significant losses also occur during processing, where the number of mills is insufficient to meet demand, and most processing units are small and use outdated technologies. The highest rates of loss are in perishable fruits and vegetables, where there is a lack of proper storage facilities, absence of proper handling, transportation, pre- and postharvest treatment and processing.

Interventions within agricultural marketing systems focusing on training and upgrading technical capacity to reduce losses, increase efficiency and reduce labour intensity of the technologies employed and improving infrastructural facilities on and off farm can enhance availability of balanced and healthy food to the population thus ensuring nutritional security.

Agricultural Marketing System and Nutrition Security

Agricultural marketing is the study of all the activities, agencies and policies involved in the procurement of farm inputs by the farmers and the movement of agricultural products from the farms to the consumers. The agricultural marketing system is a link between the farm and the non-farm sectors.

In earlier days when the village economy was more or less self-sufficient the marketing of agricultural products presented no difficulty as the farmer sold his produce to the consumer on a cash or barter basis.

Today's agricultural marketing has to undergo a series of exchanges or transfers from one person to another before it reaches the consumer. There are three marketing functions involved in this, *i.e.*, assembling, preparation for consumption and distribution. Selling of any agricultural produce depends on some couple of factors like the demand of the product at that time, availability of storage *etc.* The products may be sold directly in the market or it may be stored locally for the time being. Moreover, it may be sold as it is gathered from the field or it may be cleaned, graded and processed by the farmer or the merchant of the village.

Most of the agricultural products in India are sold by farmers in the private sector to moneylenders (to whom the farmer may be indebted) or to village traders. Products are sold in various ways. For example, it might be sold at a weekly village market in the farmer's village or in a neighboring village. If these outlets are not available, then produce might be sold at irregularly held markets in a nearby village or town, or in the mandi.

The past two decades have seen a major transformation and consolidation of food retail, and an expansion of the food processing and food service industries, especially in middle income countries and urban areas of low income countries. These trends are being driven by such factors as rising household incomes which induce increased demand for convenience, quality, and food safety, urbanization, improvements in transportation, life-style changes (that see more women working outside the home) and liberalized foreign investment regulations. Food processing, food retail, and food services are increasingly influencing the basket of food items available for consumption as more and more people work farther from home and are more likely to purchase meals prepared outside the house. The food services industry is also influencing what is being produced by farmers around the world through its very extensive networks of suppliers. Advertising techniques also encourage changes in food consumption patterns throughout the developing world.

When we talk about the three major aspects of food and nutrition security (availability, access and absorption), the second one *i.e.* physical and economic access of all people to nutritionally balanced food the factors like food distribution and price along with purchasing power comes into play. Physical access requires efficient marketing, transport, and storage system to carry the food within an easy reach or at a reasonable distance from human settlements (villages). Economic access of every household to food depends on its purchasing power and prices of food at which it is available.

There has been a considerable improvement in physical access of households to food in different parts of the country, which was contributed by several factors. First, the share of rice, which is more geographically dispersed, in total staple food, continues to be quite high at around 45 per cent. Second, the expansion of network of public distribution system helped in reaching cereals to deficit and geographically difficult regions (hilly and desert areas). And third, expansion of road networks, creation of primary market yards, and building-up of storage facilities in the rural areas increased physical access of rural households to food even in otherwise deficit areas.

Cereals being staple food in India, supply of these at affordable prices have been an essential component of food-security policy. Chronic food insecurity is being addressed through subsidized food distribution, food for work, and employment generation and guarantee programmes. Transitory food-insecurity is addressed through short-term relief programmes. And nutritional insecurity of women and children is addressed through supplementary nutrition and mid-day meals programme in schools. As per the assessment of World Food Programme, food assistance programmes in India have moved from 'food for the nation' to 'food for the people' and recently to 'food security for the vulnerable' (Acharya, 2009).

Food Distribution Systems Impact Over Food Security

Public Distribution System (PDS) provides rationed amount of basic food items and other non-food items at subsidized prices to consumers through a network of "fair price shops". The coverage and functioning of PDS underwent several changes overtime but it essentially remained an instrument to augment food security.

Food Corporation of India (FCI) along with Central Warehousing Corporation (CWC) and respective State Warehousing Corporations (SWC) are the primary agencies entrusted with the task of grain storage.

According to a World Bank study in 2011, postharvest loss occurs in 7- 10 per cent of Indian food grains from farm to market, and 4-5 per cent at distribution levels. This accounts for a loss of 11-15 million metric tons (mmt) of food grains annually – 3-4 mmt of wheat and 5-7 mmt of rice inter alia. For fruits and vegetables, postharvest loss is estimated to be about 30 per cent. Other estimates vary from the official GOI figure of 0.87 per cent grain loss to industry estimates as high as 50 per cent.

The Food Corporation of India which is responsible for implementing the Public Distribution System Scheme faces immense problems regarding its storage facilities.

The Problem Can be Divided into two Parts

1. One of the reasons for the wastage of food is that grains are not moved out of the warehouses in time and distributed. Due to this inefficiency in transporting grains out of the warehouses and to the ration shops, massive quantities of grains pertaining to years 2008-09 were found in the FCI godowns in March 2012. This storage of old crops has led to considerable depreciation in the quality of food grains. Due to this problem, 1.06 Lac tonnes of wheat under the custody of State Government Agencies in Haryana and Punjab was damaged. If we assume a family needs 30Kg of wheat for a month, this quantity of wheat would have fed 1 lakh families for 3 years.
2. The Food Corporation of India owns a storage capacity of about 156 lac tonnes including the storage space hired by it. The total storage capacity comes to about 336.04 lac tonnes. However, the Stock of Food grains procured in the year 2011-12 was 667.89 lac tonnes, leading to severe strain on the FCI storage facilities. This led to the use of State Government Agencies for the storage of grains and the use of extremely poor storage facilities- such as storing grains in the open, susceptible to damage due to rain, floods and rat infestation.

Another problem that can and does occur in PDS/FCI is that if the quality of grain stored/distributed is not up to the mark, it may lead to infections and prove to be nutritional drainages rather than sources.

Food Processing and Preservation–Critical Aspect of Nutritional Security

Increasing agricultural productivity is critical for ensuring global food security, but this may not be sufficient. Food production is currently being challenged by limited land, water and increased weather variability due to climate change. To sustainably achieve the goals of food security, food availability needs to be also increased through reductions in the post harvest losses at farm, retail and consumer levels.

Post harvest Food Loss (PHL) is defined as measurable qualitative and quantitative food loss along the supply chain, starting at the time of harvest till its consumption or other end uses (Buzby and Hyman, 2012). PHLs can occur either due to food waste or due to inadvertent losses along the way. Food loss on the other hand, is the inadvertent loss in food quantity because of infrastructure and management limitations of a given food value chain. Food losses can either be the result of a direct quantitative loss or arise indirectly due to qualitative loss. Food loss and food waste add to contribute to post harvest food losses

While the nation is debating on achieving food and nutritional security, in its backyard millions of tons of fruits and vegetables are being dumped into dustbins. India is the country which affords to lose 30 per cent of fresh fruits and vegetables. The report of The Associated Chambers of Commerce & Industry of India (ASSOCHAM) mentions that postharvest loss touched in India Rs 2.13 lakh

Crore in 2011-12 and may cross Rs 2.50 lac Crore in 2013-14. Processing of fruits and vegetables is as low as 2 per cent, 35 per cent in milk, 21 per cent in meat, and 6 per cent in poultry products. India's population is expected to reach 1.69 billion by 2050, adding almost 29 per cent more mouths to be fed by agricultural sector. The total food grain demand of the nation is projected to be about 377 Mmt by 2050 (Upali and Amarasinghe, 2007).

Post harvest losses signify loss or non-conformance to quality and involve economic and environmental costs during the various stages of production. When food losses are incurred in the supply chain, nutritious, safe and affordable food commodities become scarce in the markets. Further the prices of these essential food commodities create a perfect storm in the budgets of small and middle-class households. Food security for the poor stems from the issue of access to nutritious and safe food at affordable prices. The food supply chain provides employment for millions of people and food losses results in lost opportunities for these stakeholders to increase their incomes thus endangering their access to basic amenities including healthy food. Farmers and end-consumers are most affected sections of the society in terms of economic losses suffered between farm and fork stages of the food supply chain.

As per the specifications of National Institute of Nutrition at least 300 g of fruits and vegetables are to be consumed by an individual for a balanced diet. Thus when population is one billion, the minimum requirement is 110 million tons to meet our basic nutritional requirement. It is evident that one way to meet the demand is minimization of post harvest losses.

Reduction in food losses would increase the amount of food available for human consumption and enhance food security. A reduction in food loss also improves food security by increasing the real income for all the consumers. Given the significant role food loss reductions could have toward sustainably contributing to global food security, it is important to analyze the causes, consequences and policy strategies needed to minimize the food wastages.

Processing technologies are improving the nutritional content of household consumption through fortification; at the same time some processes involve additives that raise health concerns. The share of processed food in total consumption has also been rising over time. In India, recent evidence points to a notable shift towards processed foods over time. A demand model developed at the International Food Policy Research Institute (IFPRI) projects that Indian expenditures on processed foods will rise faster than expenditures on both total food and non-food goods (Ravi and Roy 2006). Food processing also increases the availability of food by reducing post harvest losses and extending the shelf life of food products. Improved food preservation is enabling developing countries to expand their food exports.

With reference to fruit and vegetables in particular, adoption of small-scale technologies could produce major benefits. Low temperature drying coupled with other preservation techniques could enhance storage capacity, while higher-grade containers and packaging materials could lengthen shelf-life. Most postharvest technologies are capital-intensive, but India pioneers in development of low-cost postharvest handling methodologies, which include the CoolBot cold storages, low cost Controlled Atmosphere storage, insulated trucks for 'cool transport' and other no-cost methods for handling product.

Conclusion

With our technological advances in leaps and bounds, we have come a long way, but many have been left behind, with some 925 million people still counted as undernourished as of 2010 (FAO 2010; IFRC 2011). Looking at increasing the production manifolds is just one aspect of the solution.

Strengthening the agriculture marketing systems with focus on quality and enrichment can be a big step in achieving nutritional security. Proper grading and storage with the help of sound on-farm infrastructural facilities can aid in increasing the availability of good quality healthy food under nominal prices for the population.

Post harvest losses, specifically, food loss and waste are a global endemic challenge that affects all nations, rich and poor. Food losses and waste are seen during the food supply chain, from initial agricultural production stages to the end household consumption stage. In developing nations food losses are borne during the post harvest and processing stages, while developed and high income countries register food waste during the retail and consumption stages of the end consumer.

Each country has a unique set of drivers that are responsible for causing food losses and food waste. Country specific factors such as selection of local crops and production trends, available infrastructure, presence and alignment of marketing chains and channels to facilitate distribution and consumption of agricultural commodities determine food losses. Poor and developing nations have a common set of factors in terms of poor roads, infrequent supply of electricity at high tariffs, absence of dry and cold storage facilities, and fragmented logistics which tend to hamper the efforts of these nations in reducing food losses and waste from farm to fork. The quantity of post harvest losses may represent a small percentage of the total agricultural output in India, yet the monetary loss to the national exchequer is significant and the opportunity cost of lost agricultural exports will run into millions of dollars. India's efforts to address the challenges of combating hunger, malnutrition and poverty will bear fruit when it realizes self sufficiency in protecting its surplus production of agricultural commodities during the post harvest cycles.

REFERENCES

Acharya, S.S. (2009). Food Security and Indian Agriculture: Policies, Production Performance and Marketing Environment. Agricultural Economics Research Review. Vol. 22 January-June. pp 1-19

Buzby, J.C. and J. Hyman. (2012). "Total and per capita value of food loss in the United States." Food Policy, 37(5), 561-570.

FAO. (2010). Global hunger declining, but still unacceptably high. Food and Agriculture Organization of the United Nations (FAO). http: / /www.fao.org/ docrep/012/al390e/al390e00.pdf

IFRC. (2011). World disasters report - focus on hunger and malnutrition. International Federation of Red Cross and Red Crescent Societies (IFRC). http: //www.ifrc.org/publications-and-reports/world-disasters-report/wdr2011/

Ravi, C. and Roy, D. (2006). Consumption patterns and food demand projections: A regional analysis. Paper Presented at the Workshop Plate to Plough: Agricultural Diversification and Its Implications for the Smallholders, organized by the International Food Policy Research Institute, Asia office, and the Institute of Economic Growth, at New Delhi, September 20-21.

Upali A. Amarasinghe. (2007). India's water future to 2025-2050: business-as-usual scenario and deviations.

Chapter 12

Post Harvest Losses, Food Spoilage and its Causes

Anurag Singh, Ankur Ojha and Ashutosh Upadhyay

Post Harvest Loss of food is defined as the qualitative and quantitative loss of the food products during the supply chain. It means the losses occurring between the production and actual consumption of the food product. Postharvest activities include harvesting, handling, storage, processing, packaging, transportation and marketing of the food commodities. The losses occur at various stages after harvesting and a number of factors are responsible for the loss. The food products, those are classified as *perishable* and *semi-perishable* in nature, are more susceptible to undergo the post harvest losses. For example, horticultural crops such as fruits and vegetables face more post harvest loss than the cereal-grains (Parfitt *et al.*, 2010). Worldwide post harvest fruit and vegetables losses are as high as 30 to 40 per cent and even much higher in some developing countries. Post harvest loss not only reduces the food for human consumption but is also a loss of resources utilized for the production of the food. The energy, water and nutrients required to grow a food commodity just go waste due to post harvest losses. Moreover the post harvest loss create additional burden of waste management on the society. Infrastructural and managerial limitations of a given food value chain are the main cause for such post harvest losses. To combat the prevailing situation of hunger and malnourishment globally as well as to feed the growing population, the reduction in the food waste is essential. It is estimated that the world population will grow from 5.7 billion inhabitants in 1995 to 8.3 billion in 2025. To feed this population not only the increased production but the reduced food loss is important. Then only we can ensure that sufficient food, both in quantity and in quality is available to every inhabitant in our planet.

Post Harvest Losses in Developed Countries vs Developing Countries

Food losses during the food value chain are low in developed countries as compared to the developing ones. This is due to the use of more-efficient farming systems, better transport facilities, better management, availability of storage and processing facilities. All these ensure that the maximum portion of harvested produce reaches to the market (Hodges *et al.*, 2011). If the perishable agricultural produce is to be stored for longer time, the availability of cold chain system is required. The extensive and effective cold chain systems available in developed countries also help to prolong the shelf-life of food products. On the other hand, in low-income/ developing countries, the major post harvest loss occurs in the early and middle stages of the food supply chains. Many factors *viz.* premature harvesting, poor storage facilities, lack of infrastructure, lack of processing facilities, and inadequate market facilities are the main reasons for high food losses along the entire Food Supply Chain. According to FAO post harvest losses in developing countries can range from 15 per cent up to 50 per cent. Food and Agriculture Organization of U.N. predicts that about 1.3 billion tons of food are globally wasted or lost per year (Gustavasson, *et al.* 2011).

Scenario of Post Harvest Loss in India

As per the study conducted by Central Institute if Post Harvest Engineering and Technology (CIPHET)-Ludhiana in 2015 the annual value of harvest and post harvest losses of major agricultural produces in India is of Rs. 92,651 Crore. As per the study, the post harvest losses of cereals were low in the range of 4.65–5.99 per cent while that of pulses was between 6.36–8.41 per cent and oil seeds 3.08–9.96 per cent. As per the report, the maximum losses were in fruits and vegetables (4.58–15.88 per cent) and fisheries–marine (10.52 per cent).

Categorization of Post Harvest Losses

Post harvest losses can be categorized as:

1. Quantitative loss
2. Qualitative loss

Quantitative food loss can be defined as reduction in weight of edible grain or food available for human consumption. The quantitative loss is caused by the reduction in weight due to factors such as spillage, consumption by pest, chemical changes and also due to reduction in moisture content due to changes in temperature (FAO, 1980).

The *Qualitative loss* is due to unwanted changes to the food product quality characteristics like taste, color, texture, flavor *etc.* Qualitative loss of the food products can happen due to incidence of insect pest, mites, rodents and birds, or from handling, physical changes or chemical changes in fat, carbohydrates and protein, and by contamination of mycotoxins, pesticide residues, insect fragments, or excreta of rodents and birds and their dead bodies. Qualitative deterioration makes food unfit for human consumption and leads to food loss (Buzby and Hyman, 2012).

Remedies to Reduce Post Harvest Losses

Post harvest losses can be minimized by establishing an efficient food supply chain (Shukla & Jharkharia, 2013). This includes improving and upgrading of all operations and activities in the supply chain like harvesting, handling, storage, processing, packaging, transportation and marketing. The focus for supply chains in developing countries is to improve and upgrade processes, quality and distribution. There is a dire requirement of knowledge and implementation of improved postharvest technologies in many developing countries (Kitinoja *et al.*, 2011).

Food Spoilage

Food spoilage can be defined as any metabolic process that makes foods unfit for human consumption due to changes in sensory characteristics. Spoiled foods are not always harmful if consumed. Spoiled food may be safe to eat, means they may not cause illness because there are no pathogens or a toxin present, but due to the changes in texture, smell, taste, or appearance, they are rejected by the consumer (Rawat, 2015).

Factors Affecting Food Spoilage

There are various factors affecting the types of spoilage of a particular food item such as:

The Composition of Food

The composition of food affects the susceptibility of a food to spoilage. Food products having high moisture and sugar content are more prone to microbial growth.

Structure of the Food Item

Whole healthy tissues of food from inside are sterile or low in microbial content. Skin, rind or shell on food works as its protective covering from spoilage microorganisms.

Types of Microorganisms Involved

The types of microorganisms present in food depend on its composition of food. If these are spoilage causing microorganisms who can survive at room temperature, the product will spoil fast.

Conditions of Storage of the Food

Conditions of storage of food affect the growth of microorganisms. Even if the proper storage of food is done, the food loses its freshness and nutritive value if it is stored for too long.

Causes of Food Spoilage

The causes responsible for food spoilage can broadly be classified as *physical causes, chemical causes* and *biological causes*. Physical causes consist of temperature and physical abuse. Chemical causes include reaction of food constituents with oxygen and light as well as chemical reactions within food constituents. Biological causes comprise of growth and activity of microorganisms such as bacteria, yeast and moulds; activity of food enzymes and damage due to pests, insects and rodents *etc.*

Microbial Spoilage

Spoilage causing microorganisms are widely available in nature like in soil, water and air, on the skin of cattle, fruits and vegetables, on the feathers of poultry, on the hulls of grains, and shells of nuts, on the clothing and skin of handling personnel, on processing equipment and within the intestines and body cavities of animal and human body. These microorganisms, if get the suitable environment, they grow and spoil the food product. There are three types of microorganisms that cause food spoilage - yeasts, molds and bacteria (Rawat, 2015).

1. **Yeasts** are a subset of a large group of organisms called fungi that also includes molds and mushrooms. Yeasts growth causes fermentation which is the result of yeast metabolism. There are two types of yeasts *true* yeast and *false* yeast. *True yeast* metabolizes sugar producing alcohol and carbon dioxide gas. This is known as fermentation. *False yeast* grows as a dry film on a food surface, such as on pickle brine. False yeast occurs in foods that have a high sugar or high acid environment.
2. **Molds** grow in filaments forming a tough mass which is visible as `mold growth'. Molds form spores which, when dry, float through the air to find suitable conditions where they can start the growth cycle again. Mold can cause illness, especially if the person is allergic to molds. Usually though, the main symptoms from eating moldy food will be nausea or vomiting from the bad taste and smell of the moldy food.
3. Both yeasts and molds can thrive in high acid foods like fruit, tomatoes, jams, jellies and pickles. Both are easily destroyed by heat. Processing high acid foods at a temperature of 100°C (212°F) in a boiling water canner for the appropriate length of time destroys yeasts and moulds.
4. **Bacteria** are round, rod or spiral shaped microorganisms. Bacteria may grow under a wide variety of conditions. There are many types of bacteria that cause spoilage. They can be divided into: *spore-forming* and *nonspore-forming*. Bacteria generally prefer low acid foods like vegetables and meat. In order to destroy bacteria spores in a relatively short period of time, low acid foods must be processed for the appropriate length of time at 116°C (240°F) in a pressure canner (retort/autoclave).

Action of Native Enzymes

Food products have a variety of enzymes available naturally. These enzymes are working as catalyst for various reactions. The activity of these endogenous enzymes in plant and animal foods is often intensified after harvest/slaughter due to lack of control mechanisms in the harvested plant food/slaughtered animal. The activity of these native enzymes, if not controlled, may lead to different types of changes (spoilage) in food products *e.g. oxidative browning* in apple or yam when they are cut or bruised and exposed to air. *Ripening* of fruits and vegetables leading to senescence stage, if not controlled. The native enzyme may be inactivated by heat, radiation/by the use of specific chemicals.

Insects, Parasites and Rodent

Insects, parasites and rodents destroy cereal grains, fruits and vegetables by not only consuming the food but contaminating the food also with their excreta, droppings, urine and filth. These contaminations cause the microbial spoilage of the food products. Parasites enter the human body mostly through poultry which have been improperly cooked.

Chemical Reactions

Every food product has various food constituents in different proportions. When these chemical constituents react to each other or to the environment, the food deteriorates. For example, Oxidation of unsaturated fatty acids leading to oxidative rancidity (off flavor development), putrefaction of protein, hydrolysis of fats releasing free fatty acids, loss of vitamins due to oxidations *etc.*

Environmental Factors

Air and oxygen can have detrimental effects on vitamin A, C, food colour, flavour and other constituents. These compounds get oxidized due to the presence of oxygen in the air. Light destroys riboflavin, vitamin A, Vitamin C and also promotes light induced oxidation reactions affecting flavour and colour of food.

Time

The quality of food remains at its peak for some time soon after its harvest/ slaughter and thereafter as time progress, the deterioration in the quality of the food also progress. Simple option for maintaining the food quality is through temperature maintenance and creating hygienic conditions. The harvested/slaughtered food must be cleaned and cooled immediately. This delays the onset of deterioration of food quality but does not prevent it.

Physiological

Food commodities like fruits and vegetables, which are living commodities, undergo natural respiratory losses. As a result of respiration, weight loss of the commodities and heat generation occurs. Changes occurring during ripening and senescence may increase the susceptibility of the commodities to mechanical damage and microbial spoilage. All these physiological changes lead to a lowered nutritional level and reduced consumer acceptance.

Mechanical

Improper handling of food commodities may cause damage. Particularly in case of horticultural crops, bruising, cutting, excessive pressure during stacking *etc.* can damage the crop and this will lead to intensified enzymatic action or the entry of the microbes that will spoil the product.

Physical

Improper maintenance of physical parameters like temperature and humidity may also lead to food spoilage. Excessive or insufficient heat or cold can spoil foods. Improper atmosphere in closely confined storage at times causes losses.

Measures to Avoid Food Spoilage

Food spoilage can be prevented by following good post harvest practices and preservation methods. These includes various operations like storage of food products at low temperature, storage under controlled/modified atmosphere conditions, proper handling after harvesting to avoid mechanical damage, use of chemical preservatives within prescribed limits, Thermal processing, use of natural preservatives *e.g.* sugar, salt, acids, proper packaging *etc.* Depending on the type of the food product, the suitable preservation method may be used to void spoilage.

Conclusion

From the above discussion this can be concluded that the loss of the food products after harvesting till it reaches to the consumer (post harvest losses) is mainly due to the improper food value chain management. There is also a need for creating awareness among the farmers for proper post harvest management of food commodities so that the loss at farm level may be minimized. Food deterioration or spoilage is affected by several factors including biological, physical and chemical. If these factors are governed properly, the spoilage of food may be avoided and the shelf life may be extended.

REFERENCES

Parfitt, J., Barthel, M. & Macnaughton, S., (2010). Review Food waste within food supply chains: quantification and potential for change to 2050. Philosophical Transactions of the Royal Society B: Biological Sciences, 365, 1554: 3065-3081

Hodges, R.J., Buzby, J.C. & Bennett, B., (2011). Postharvest losses and waste in developed and less developed countries: opportunities to improve resource use. The Journal of Agricultural Science, 149, S1: 37-45.

Gustavsson, J., Cederberg, C., Sonesson, U., van Otterdijk, R., and Meybeck, A. (2011)."Global Food Losses and Food Waste: Extent Causes and Prevention." Rome, Food and Agriculture Organization (FAO) of the United Nations.

Food and Agriculture Organization. (1980). Assessment and Collection of Data on Post harvest Food Grain Losses, FAO Economic and Social Development Paper 13. Rome.

Buzby, J.C., and J. Hyman. (2012). "Total and per capita value of food loss in the United States." Food Policy, 37(5), 561-570.

Shukla, M., & Jharkharia, S., (2013). Agri-fresh produce supply chain management: a state-of-the-art literature review. International Journal of Operations & Production Management, 33, 2: 114-158.

Kitinoja, L., Saran, S., Royb, S.K. & Kader, A.A., (2011). Postharvest technology for developing countries: challenges and opportunities in research, outreach and advocacy. Journal of the Science of Food and Agriculture, 91, 4: 597-603.

Rawat S., (2015). Food Spoilage: Microorganisms and their prevention, Asian Journal of Plant Science and Research, 5(4): 47-56.

Chapter 13

Effect of Food Processing Techniques on Nutritional Value, Food Packaging and Labeling

Manali Khatri and Anurag Singh

Introduction

Food processing techniques includes a set of physical, chemical or microbiological methods that are used to convert raw ingredients into commercial food products of various forms that can be easily prepared and served by/to the consumer. Food processing is a generic term which includes various activities like macerating, mincing, emulsification, cooking, pickling, pasteurization, canning and various other preservation techniques. Primary processing methods such as slicing, dicing, cutting, drying, freezing which subsequently results in variety of secondary products are also included under the umbrella of food processing (Dandsena and Barik, 2016).

The basic concept of food processing revolves around preservation and toxin removal thus rendering processed food less susceptible to early or fast spoilage as compared to fresh foods and therefore is best suited for long distance transportation from its source. The act of processing significantly improves the sensory aspects of food along with its consistency and ease in marketing and distribution tasks (Ani *et al.*, 2012).

Despite of having number of benefits, there are certain noticeable drawbacks that also run parallel with the food processing. One major issue is nutritional density and availability. All the processing method or technique decreases food's nutritional density and amount of nutrition loss depends on food and processing method and its duration. In particular, processes that expose food to high levels of light, heat, and/or oxygen cause the greatest nutrient loss. Beside this, to maintain the hygiene and safety of food during the entire processing length is also an issue of concern.

This chapter provides an overview of the impact of thermal, non thermal and other emerging technologies on the nutritional value of the food systems.

Effect of Food Processing on the Proteins

Proteins present in food are responsible for various functions. Protein is having the ability to form gels, networks and develop films, absorb fat, hold water, foam, emulsify and dissolve under various pH conditions. These protein functions can be affected both by intrinsic and extrinsic factors of the test food matrix.

Processing the food alters its protein solubility by increasing the net charge, exposes surface amino acids to new environments and by altering the isoelectric pH. Solubility can be increased or decreased depending upon the processing treatment given. High temperature processing denatures the protein and induces protein-starch cross linking and causes agglomeration and therefore the insolubility. On the other hand when the net charge becomes negative above the isoelectric pH, the protein solubility enhances.

Heat induced protein denaturation although decreases the protein solubility but the ability to trap water typically increases thus increasing the overall water holding capacity. Fat absorbtion capacity is also affected greatly by processing method. Cleavage of disulphide bond, aggregation, unfolding, formation of dimmers, large oligomers of proteins are also the result of thermal treatment and these alterations or modifications can be desirable or detrimental. Heating proteins induces maillard reactions thereby several chemosensory properties such as color and flavors are created. Thermal processing may also modify the antigenicity of proteins by inducing the covalent modifications in protein structure (Aryee *et al.*, 2018).

Recently adopted non thermal and emerging food process technologies like pulse electric field, High pressure processing use low energy and short treatment durations to cast the similar effects as of thermal processing without compromising its nutrients and overall sensory appeal. However the magnitude of effects depends on several parameters.

Effect of Food Processing on the Carbohydrates

Carbohydrates perform various functions in living organisms. The complex carbohydrates (polysaccharides) serve as storage of energy or as structural components and can also supply important vitamins and proteins. Like all other biomolecules the nutritional quality of food carbohydrates can also be altered by processing in a number of ways.

Firstly, there is a considerable loss of low molecular weight carbohydrates in the wet heat treatment process in the processing water decreasing its carbohydrate content. Secondly, the two possible major changes that occur when carbohydrates are subjected to high temperature treatment are caramelization (Non enzymatic browning) and starch gelatinization. Caramelization is a result of non enzymatic browning in carbohydrate rich foods at significantly high temperature. In protein rich foods the carbohydrate protein reactions results in another form of browning *i.e.* Maillard browning (Non enzymatic browning). The reaction is nutritionally important to consider as it decreases the bioavailability of amino acid especially lysine thus decreasing the overall protein value (Danago, 2009).

Gelatinization is an irreversible loss of crystalline regions in starch granules in the presence of water at high temperatures. The nutritional importance of the process is that it dramatically increases the availability of starch for digestion by the amylolytic enzymes. Studies showed that the gelatinized starch is thermodynamically instable thus it re-associates on ageing and may reduce the digestibility of starch again; process is known as retrogradation.

The Effect of Food Processing on Nutritional Qualities of Vitamins and Minerals

Vitamins are very important for human health because they carry out various biological functions inside human body. It is recommended that humans must consume vitamins in their daily diet. Nevertheless, the benefits are numerous; but processing can proved to be detrimental, affecting the nutritional quality of foods. Blanching, for example, results in leaching losses of most of the water soluble vitamins and some of the minerals. Also, milling and extrusion can cause the physical removal of minerals during processing (Reddy and Love, 1999). The nutritional quality of minerals in food depends on their quantity as well as their bioavailability. The time and temperature of processing, product composition and storage are the factors that substantially impact the vitamin status of our foods. Water-soluble vitamins (B-group and C) are highly unstable than fat-soluble vitamins (K, A, D and E) during food processing. In milling the husk part of grains are removed which contains most of the B-group vitamins, phytochemicals and some minerals and thereby decreasing the overall vitamin mineral content from cereals. Processes like canning and pasteurization also results in heavy losses of vitamins. Researchers also state that processes which do not restrict contamination of transition metal ions from equipment or water supplies used in processing, and storage conditions which do not block light and limit exposure to air (by packaging under nitrogen) promote significant vitamin losses due to the catalytic effects of these elements. Non thermal processes like high pressure processing which focuses on the application of high pressure to kill the microorganisms and not on heat, impacts less on the vitamin content thereby retaining majority of vitamins in the products.

Minerals too have diverse functionalities and potentials in the body's metabolism and deficiency of these bioactive constituents can result in probable incidences of common disorders and disease symptoms. The most dominant minerals to fortify various food preparations are iron, calcium, zinc and iodine as these mineral deficiencies are popular among human population. Modern processing techniques (*e.g.*, high pressure and sonication) compared with the conventional processes have lower negative impacts on the content of micro and macro minerals. Accumulation of mineral elements in the edible tissues of crops using agro-biotechnological techniques (*e.g.*, gene over expression and activati control) and their direct fortification in the form of premixes, into formulation of processed foods along with nano-encapsulation could enhance the concentration and bio-accessibility of these bioactive ingredients (Gharibzahadi and Jafri, 2017). One of the most important issue in the processing of mineral rich foods is the formation of strong complexes between bivalence elements (*e.g.*, Ca2+, Mg2+, Zn2+, Fe2+, Cu2+), and, phytate, fiber, and compounds of tannin and lectin. De-phytization is a process which is introduced to break-down the phytates of infant cereals in order to improve the uptake efficacy of minerals; transport level and release rate of bivalent minerals such as Fe, Zn and even Ca.

Effect of Novel Technologies on Polyphenols

There is a fast growing acceptance across the globe that polyphenols which are phenolic secondary metabolites in nature and present majorly in plant derived foods exhibits various beneficial effects in the prevention of various degenerative diseases because of their anti-inflammatory, antioxidant, anti-mutagenic and anti-tumor activities (Pandey and Rizvi, 2009). These polyphenols are very vulnerable to heat, light and other physical and chemical treatments so today's researcher's growing interest is towards on limiting these losses during food processing. The fast emerging novel food processing technologies divided into thermal and non-thermal categories have great potential to produce safe food products with high quality in terms of the nutrient retention and safety. In general, these processing techniques can minimize the adverse effects of conventional processing on the levels of bioactive compounds and their retention in vegetables, fruits and their products. There is also growing interest in utilization of these novel food processing techniques to meet the increasing consumer demand for nutritious foods, in terms of bioactive compound retention and at the same time ensuring its food safety. During the last few years, several interesting applications of these novel technologies have emerged to improve the bioactivity and technological properties of food therefore their use is well known.

Although novel food processing techniques have known to improve the nutritional quality of processed foods but better understanding of the complex physicochemical mechanism of action of these processing technologies and their effects on the technological, sensorial and functional properties of fruits, vegetables and their products, would also contribute to reinforce the presence of their applications.

Food Processing and Food Packaging

Packaging of food is required to prevent product deterioration by retaining the beneficial effects of processing along with extending its shelf-life and thereby maintaining or increasing the overall quality and safety of food. While doing this, packaging provides protection from chemical, biological, and physical deterioration (Abdullahi, 2014).

As the name indicates the biological protection provides a barrier from biological agents *i.e.* microorganisms (pathogens and spoiling agents), insects, rodents, and other animals, thereby preventing disease and spoilage. In addition, biological barriers also maintain conditions which control the metabolic activities such as ripening and aging thereby retarding their senescence and ultimately increasing the shelf life. Such barriers act via a multiplicity of mechanisms which includes their preventing access to the product, preventing odor transmission, and maintaining the set/target internal environment of the package.

Chemical protection is devised/designed to limit/reduce the compositional changes initiated by environmental influences such as exposure to gases (typically oxygen), moisture (gain or loss), or light (visible, infrared, or ultraviolet). There are many different packaging materials which provide a barrier to such type of chemical deterioration. Usually, Glass and metals provide a nearly absolute barrier to chemical and other environmental agents but closure devices used along with these may contain materials that allow minimal levels of permeability. For example, plastic caps have some permeability to gases and water vapors, similar is the case with gasket materials used in caps to facilitate closure and in metal can lids to allow sealing after filling. Thus, Packaging made of plastics (Polymers) in itself is a broad area of study which offers a wide range of barrier properties but is generally more permeable than glass or metal.

Physical protection is just to protect food from mechanical damage which includes cushioning against the shock and vibration that a product must encounter during its distribution and transportation and even during handling. These are typically developed from paperboard and corrugated fibre board materials. Physical barriers resist impacts, abrasions, and crushing damage, so they are widely and continuously used as shipping containers and as packaging for delicate foods such as eggs and fresh fruits. Appropriate physical packaging also protects consumers from different hazards. For example, child-resistant closures hinder access to potentially dangerous products. In addition to this the utilization of plastic packaging for products ranging from sauces to carbonated beverages bottles has reduced the danger from broken glass containers.

In recent years, where new food processing techniques are attracting a lot of attention as compared to conventional methods, there are techniques which require the processing of foods inside their package, it is therefore very important and required to understand the interaction between the package, food and the process itself. New processing methods such as the use of high intensity pulsed light, high pressure, irradiation, and high intensity electric fields, are emerging in food industry so as to meet the health conscious consumer demand for safer and good quality food products. Researchers are still investigating to prove the soundness of their effects on microorganisms and overall acceptability of foods. Processes like high pressure, irradiation, microwave cooking and thawing; ozone treatment might require the treatment of foods in the packages. The success of most of these preservation methods depends on how well the processed food is protected from adverse environmental conditions and this protection is majorly accomplished by packaging.

For Example, packages used in microwave processing requires a special attention on migration studies due to high temperatures attained during cooking. Laminates made up of PET/polypropylene (PP) plus a barrier layer made of ethylene-vinyl alcohol (EVOH), polyvinylidene chloride (PVdC), polyethylene (PE) and other aiding materials are the commonly used materials/polymers for microwable packaging of foods (Ozen and Floros, 2001). Similarly, PP, PE, nylon 6, cellophane, and rubber hydrochloride were found to be the apt polymers for the packaging of marine foods during radiation. Water vapor and oxygen permeability of several laminated plastic films (PET/Al/CPP, PP/EvOH/PP, OPP/PVOH/PE,KOP/CPP,) were not affected from high pressures between 400 and 600 MPa therefore justify their suitability and usage in high pressure processing technology.

Food Package Labeling

Nutrition information on food lebels is a cost effective method of communicating nutrition information to consumers. The information appears to the consumer at the point of purchase for most packaged foods (Miller and Cassady, 2015). The lebeling of food items must adhere to regulatory standards. Such claims therefore provide good information about the contents of a product and could influence and satisfy consumers' attitudes, preferences and choices. Manufacturers choose which claims, if any, to make, and presumably this selection process is purposeful, not random. Thus a manufacturer's choice of claims might directly influence consumer judgments. For most foods, no single claim would provide a complete characterization of the product as a whole. For example, knowing that a food is a good source of vitamins and iron does not tell consumers whether it is also high or low in saturated fat, sodium, sugar, or calories. Thus it is required that while evaluating consumer effects of package claims, it is important to consider not just claim-specific outcomes, but also how claims may affect their broader judgments about a product.

Figure 1: What all to Look for in a Food Package Label. (*Source*: http://claytowne.com).

Food labeling in a nut shell can provide consumers with the information they need and desire to make food choices. On a whole a food label is designed in such a way that it may tell consumers about: the qualities of a product, the appropriate use of the product, and the benefits of the product, possible risks from the product which include presence of allergens, trans fats, saturated fats *etc.*, and last not the least where and how a product is produced (Figure 1).

Conclusion

Food processing affects the nutritive value of food by affecting the available nutrients. Thermal processing has more adverse effects as compared to the novel technologies. Food Packaging not only protects the products but also tells the consumer about the product. Hence, the processing should be done in such a way that the safety of food can be ensured without comprising with the quality. Moreover packaging should be done keeping all the regulatory requirements for the packaging.

REFERENCES

Abdullahi, N. (2014). Hazard chemicals in some food packaging materials: A review. *Annals Food Science and Technology*, 15(1): 115-120.

Ani, Babiyans Roy P., Hameedunissa, Begum A. and Selvakumar, K. (2012). A changing scenario of food processing technology in India. *Journal of Biological and Information Sciences*, 1(3): 17-20.

Aryee, A.N.A., Agyei, D., and Udenigwe, C.C. (2018). Impact of processing on the chemistry and functionality of food proteins. Proteins in Food Processing (Second Edition) A volume in Woodhead Publishing Series in Food Science, Technology and Nutrition, pp. 27-45.

Dandago, M.A. (2009). Changes in Nutrients during Storage and Processing of Foods: A Review. *Techno Science Africana Journal*, 3(1): 24-27.

Dandsena, N. and Banik, A. (2016). Processing and value addition of the underutilized agriculture crops and indigenous fruits of Bastar region of Chhattisgarh. *International Journal of Multidisciplinary Research and Development*, 3(3): 214-223.

Gharibzahedi, S.M.T. and Jafari, S.M. (2017). The importance of minerals in human nutrition: Bioavailability, food fortication, processing effects and nano encapsulation. *Trends in Food Science & Technology*, 62: 119-132.

Miller, L.M.S. and Cassady, D.L. (2015). The effects of nutrition knowledge on food label use: A review of the literature. *Appetite*, 92: 207-216.

Ozen, B.F. and Floros, J.D. (2001). Effects of food processing techniques of packaging materials. *Trends in Food Science & Technology*, 12: 60-67.

Pandey, K.B. and Rizvi, S.I. (2009). Plant polyphenols as dietary antioxidants in human health and disease. *Oxid Med Cell Longev*. 2(5): 270-278.

Reddy, M.B. and Love, M. (1999). The Impact of Food Processing on the Nutritional Quality of Vitamins and Minerals. In: Jackson L.S., Knize M.G., Morgan J.N. (eds) Impact of Processing on Food Safety. *Advances in Experimental Medicine and Biology*, Vol. 459. Springer, Boston, MA, p. 99-106.

Chapter 14

Qualitative and Quantitative Changes during Post Harvest Handling and Processing of Food

Preeti B. Dixit

Introduction

As soon as a crop is removed from the ground, or separated from its parent plant, it begins to deteriorate. Postharvest treatment largely determines final quality, whether a crop is sold for fresh consumption, or used as an ingredient in a processed food product. The most important goals of post harvest handling are keeping the product cool, to avoid moisture loss and slow down undesirable chemical changes, and avoiding physical damage such as bruising, to delay spoilage. Sanitation is also an important factor, to reduce the possibility of pathogens that could be carried by fresh produce. After the field, post harvest processing is usually continued in a packing house. This can be a simple shed, providing shade and running water, or a large scale, sophisticated, mechanized facility, with conveyor belts, automated sorting and packing stations, walk-in coolers and the like. In mechanized harvesting, processing may also begin as part of the actual harvest process, with initial cleaning and sorting performed by the harvesting machinery. Initial post harvest storage conditions are critical to maintaining quality. Each crop has an optimum range for storage temperature and humidity. Also, certain crops cannot be effectively stored together, as unwanted chemical interactions can result. Various methods of high-speed cooling, and sophisticated refrigerated and atmosphere controlled environments, are employed to prolong freshness, particularly in large-scale operations. Regardless of the scale of harvest, from domestic garden to industrialized farm, the basic principles of post harvest handling for most crops are the san handle with care to avoid damage (cutting, crushing, and bruising), cool immediately and maintain in cool conditions, and cull (remove damaged items).

Postharvest Shelf Life

As the crop is harvested from the field, numerous biochemical processes initiate continuous change in the original composition of the crop until it becomes unmarketable. The period during which consumption is considered acceptable is defined as the time of "postharvest shelf life." Postharvest shelf life is typically determined by objective methods that determine the overall appearance, taste, flavor, and texture of the commodity. These methods usually include a combination of sensorial, biochemical, mechanical, and colorimetric (optical) measurements.

Importance of Post harvest technology lies in the fact that it has capability to meet food requirement of growing population by eliminating avoidable losses making more nutritive food items from low grade raw commodity by proper processing and fortification, diverting portion of food material being fed to cattle by way of processing and fortifying low grade food and organic wastes and by-products into nutritive animal feed. Post harvest technology has potential to create rural industries. In India, where 80 per cent of people live in the villages and 70 per cent depend on agriculture have experienced that the process of industrialization has shifted the food, feed and fibre industries to urban areas. The purpose of post harvest processing is to maintain or enhance quality of the products and make it readily marketable.

The Post Harvest Industry Includes the Following Main Components

Harvesting and Threshing	*Drying and Storage*
Processing (conservation and/or trans-formation of the produce)	Utilization by consumer including home processing.
Transportation and distribution.	Marketing
Grading and quality control	Pest control.
Packaging	Communication among all concerned
Information, demonstration and advisory systems	Price stabilization Management and integration of the total system

Main Elements of the Post harvest System

Harvesting

The time of harvesting is determined by the degree of maturity. With cereals and pulses, a distinction should be made between maturity of stalks (straw), ears or seedpods and seeds, for all that affects successive operations, particularly storage and preservation.

Pre-Harvest Drying

It is mainly intended for cereals and pulses. Extended pre-harvest field drying ensures good preservation but also heightens the risk of loss due to attack (birds, rodents, insects) and moulds encouraged by weather conditions, not to mention theft. On the other hand, harvesting before maturity entails the risk of loss through moulds and the decay of some of the seeds.

Transport

Much care is needed in transporting a really mature harvest, in order to prevent detached grain from falling on the road before reaching the storage or threshing place. Collection and initial transport of the harvest thus depend on the place and conditions where it is to be stored, especially with a view to threshing.

Post harvest Drying

The length of time needed for full drying of ears and grains depends considerably on weather and atmospheric conditions. In structures for lengthy drying such as cribs, or even unroofed threshing floors or terraces, the harvest is exposed to wandering livestock and the depredations of birds, rodents or small ruminants. Apart from the actual wastage, the droppings left by these marauders often result in higher losses than what they actually eat. On the other hand, if grain is not dry enough, it is vulnerable to mould and can rot during storage. Moreover, if grain is too dry it becomes brittle and can crack after threshing, during hulling or milling. This applies especially to rice if milling takes place a long time (two to three months) after the grain has matured, when it can cause heavy losses. During winnowing, broken grain can be removed with the husks and is more susceptible to certain insects (*e.g.* flour beetles and weevils). Lastly, if grain is too dry, this means a loss of weight and hence a loss of money at the time of sale.

Threshing

If a harvest is threshed before it is dry enough, this operation will most probably be incomplete. Furthermore, if grain is threshed when it is too damp and then immediately heaped up or stored (in a granary or bags), it will be much more susceptible to attack from micro-organisms, thus limiting its preservation.

Storage

Facilities, hygiene and monitoring must all be adequate for effective, long-term storage. In closed structures (granaries, warehouses, hermetic bins), control of cleanliness, temperature and humidity is particularly important. Damage caused by pests (insects, rodents) and moulds can lead to deterioration of facilities (*e.g.* mites in wooden posts) and result in losses in quality and food value as well as quantity.

Processing

Excessive hulling or threshing can also result in grain losses, particularly in the case of rice (hulling) which can suffer cracks and lesions. The grain is then not only worth less, but also becomes vulnerable to insects such as the rice moth (Corcyra cephalonica).

Marketing

Marketing is the final and decisive element in the post harvest system, although it can occur at various points in the agro-food chain, particularly at some stage in processing. Moreover, it cannot be separated from transport, which is an essential link in the system.

Changes in Food due to Various Processing Operations

Changes in Sensory Characteristics of Foods

There are number of definitions of quality of foods which are discussed by Cardello (1995). To the consumer most important quality attributes of a food are its sensory characteristics (texture, flavor, aroma, shape and color). These determine an individual's preferences for specific products, and small differences between brands of similar products can have a substantial influence on acceptability.

Textural Changes

The moisture and fat contents, and the types and amount of structural carbohydrates (cellulose, starches, and pectic materials) and proteins that are present mostly determine the texture of food. Changes in texture occur due to loss of moisture, or fat, formation or breakdown of emulsion and gels, hydrolysis of polymeric carbohydrates, and coagulation or hydrolysis of proteins.

Taste, Flavor and Aroma

Taste attributes consist of saltiness, sweetness, bitterness and acidity and some of these attributes can be detected in very low thresholds in foods. The taste of foods is largely determined by the formulations used for particular food and is mostly unaffected by processing. Exception to this includes increased sweetness due to respiratory changes in fresh foods and changes in acidity or sweetness during food fermentations. Fresh foods contain complex mixtures of volatile compounds, which give characteristics flavours and aromas, some of which are detectable at extremely low concentrations. These compounds may be lost during processing, which reduced the intensity of flavors or reveal other flavors compounds. Volatile aroma compounds are also produced by the action heat, ionizing, radiation, oxidation or enzyme activity on proteins, fats and carbohydrates. Includes the maillard reaction between amino acid and reducing sugars or carbonyl groups and the products of lipid degradation, or hydrolysis of lipids to fatty acids and subsequent conversion to aldehydes, esters and alcohols.

Color

Many naturally occurring pigments are destroyed by heat processing, chemically altered by change in pH or oxidized during storage. As a result the processed food may lose its characteristic color and hence its value. Synthetic pigments are more stable to heat, light and changes in pH and they are therefore added to retain the color of processed foods. Maillard browning is an important cause of both desirable changes in food color and in the development of off colors. Blanching causes unintentional separation of water-soluble vitamins and sugars.

The heat processing causes major changes on nutritional quality of a food, which involves heating the food at high temperature. Longer the duration, the nutritional losses will be higher. Heating food leads to losses in heat sensitive vitamins reduces the biological value of proteins (owing to destruction of amino acids or maillard browning reactions) and promotes lipid oxidation. Heat processing is also done in foods to modify some desirable properties like gelatinization of starches and coagulation of proteins improve their digestibility, and anti nutritional components are destroyed (trypsin inhibitor in legumes).

Oxidation is one of the most important factors affecting nutritional value of a food. Oxidation may occur due to exposure to air, too much heat and action of oxidative enzymes (peroxidase or lipoxygenase). Oxidation leads to degeneration of lipids to hydroperoxides and subsequent reaction to form a wide variety of carbonyl compounds, hydroxyl compounds and short chain fatty acids for and in frying oil to toxin compounds. Oxidation also leads to destruction of oxygen sensitive vitamins.

Effect of Heat

The preservative effect of heat processing is due to the denaturation of proteins, which destroys enzyme activity and enzyme controlled metabolism in microorganisms the rate of destruction is a first order reaction; that is when food is heated to a temperature that is high enough to destroy contaminating micro-organisms, the same percentage die in a given time interval regardless of the numbers present initially. This is known as the logarithmatic order of death and is described by a death rate curve.

Effect of Heat on Nutritional and Sensory Characteristics

The destruction of vitamin and many aromatic compounds and pigments by heat following by heat follows a similar first order reaction to microbial destruction. Nutritional and sensory properties are better retained by the use of higher temperatures and shorter time during heat processing.

Changes Due to Size Reduction

Size reduction is one of the food processing operations which are used to control the textural or rheological properties of food and to improve its overall efficiency in mixing and heat transfer. Changes due to size reduction include changes in aroma and flavor of foods. The disruption of cells and resulting increase in surface area promotes oxidative deterioration and higher rates of microbiological and enzymatic activity. Size reduction therefore has little or no preservative effect. Dry foods like grains and nuts have a sufficiency of low a_w to permit storage for long after milling without substantial loss in nutritional value or eating quality. However, moist food deteriorates rapidly if other preservatives measures are not taken in to consideration.

There are small but largely unreported changes in the color, flavor and aroma of dry foods during size reduction. Oxidation of carotene bleaches flour and causes losses in nutritional value. There is loss of volatile constituents from spices and some nuts, which may accelerated if the temperature is allowed to rise during the milling. In moist food disruption cells allows enzymes and substrates to become more intimately mixed, which causes accelerated deterioration of flavor, aroma and color. Additionally the release of cellular materials provides a suitable substrate for microbiological growth and this can also result in the development of off flavor and aromas.

The increase in surface area of foods during size reduction causes loss of nutritional value due to oxidation of fatty acids and carotenes. Loss of vitamin C and thiamin in chopped or sliced fruits and vegetables are substantial. Losses during storage depend on the temperature and moisture content of the food and on the concentration of oxygen in the storage atmosphere. In dry foods the main loss in nutritional value results from separation of the product components after size reduction (*e.g.* Separation of bran from rice, wheat or maize)

Changes Due to Homogenization

To achieve desired mouth feel in many liquid, semi liquid drinks/foods emulsification, stabilization and homogenization is done. Homogenization has an effect on the colour of some foods (for example milk) because the larger number of globules causes greater reflectance and scattering of light. Flavour and aroma are improved in many emulsified foods because volatile components are dispersed throughout the food and hence have greater contact with taste buds when eaten. Homogenization in milk reduces the average size of fat globules from 4 μm to 1 μm, creating a more creamy texture in milk. Viscosity of liquid also increases due to higher number of globules and absorption of casein on to the surface of these globules.

Lager number of globules produced due to homogenization causes greater reflectance and scattering of light thus affecting the color of foods.

Flavor and aroma are improved in many emulsified food because volatile components are dispersed throughout the food and hence have greater contact with taste buds when eaten. Separation of components changes the nutritional value of emulsified foods leading to a more digestible form of fat and proteins due to the reduction in particle size (Fellows, 2009).

Changes Due to Fermentation and Enzymes

Fermentation leads to complex changes in structures of protein and carbohydrates leading to a softer texture of fermented foods. Flavor changes includes changes in reduction in sweetness and increase in acidity due to fermentation of sugars and organic acids, an increase in saltiness in some foods like pickle, soy sauce, fish, meat *etc.* due to salt addition and reduction in bitterness in some foods due to the action of deliberating enzymes. The aroma of fermented products is due to the large number of volatile compounds (*e.g.* amines, fatty acids, aldehydes, ketones and alcohol) and products from interaction of these compounds during fermentation and maturation. In bread and cocoa subsequent unit operation of baking and roasting produce the characteristics aromas. The color of fermented food is retained owing to the minimal heat treatment and/or a suitable pH range for pigment stability. Changes in color may also occur owing to the formation of brown pigments by proteolytic activity, degradation of chlorophyll and enzymic browning.

Complex changes in composition of protein, fats and carbohydrates along with improved availability of some vitamins due to the microbial growth in fermented foods causes drastic changes in nutritional value of fermented foods. Microorganism utilizes fatty acids, sugars amino acids and vitamins from the food for their life processes. Hydrolysis of polymeric compounds to produce substrate for cell growth, improves the bioavailability and digestibility of proteins and polysaccharides.

Changes Due to Ionization

At commercial dose levels, ionizing radiation has little to no effect on the digestibility or the composition of essential amino acids (Josephson *et al.*, 1975) but at higher dose levels, cleavage of sulphydryl group from sulphur amino acids in protein causes changes in aroma and taste of foods. Carbohydrates are hydrolyzed and oxidized to simpler compound ad depending on the dose received may become deploymerized and more susceptible to enzymic hydrolysis. However there is no changes in the degree of utilization of the carbohydrate and hence no losses in nutritional value. The effect on lipid is similar to that of auto oxidation, to produce hydroperoxides and the resulting unacceptable changes in aroma and flavor. The effect is reduced by irradiating foods while frozen, but foods that have high concentration of lipids are generally unsuitable for irradiation.

Changes Due to Blanching

The heat received by a food during blanching inevitably causes some changes to sensory changes and nutritional qualities. However, the heat treatment is less severe than for example for heat sterilization. In addition, the resulting changes in food quality are less pronounced. In general, the time temperature combination used for blanching is a compromise, which ensures adequate enzyme inactivation but prevents excessive softening and loss of flavor in food.

Some minerals, water-soluble vitamins and other water-soluble constituents leached out during blanching. Losses of such components are mostly because of leaching out than thermal destruction and to a lesser extent oxidation. Maturity, variety of food and method adopted for preparation of food like extent of cutting, slicing or dicing are the determining factors for the extent of vitamin losses during blanching. Time and temperature combination of blanching is also important, as the losses of vitamins are less at higher temperature for shorter duration. Losses of ascorbic acid used as an indicator of food quality and therefore the severity of blanching.

Blanching brightens the color of some foods by removing air and dust on the surface and thus altering the wavelength of reflected light (Selman, 1987). Sodium carbonate (0.125 w/w) or calcium oxides are often added to blancher water to protect chlorophyll and to retain the color of vegetables, although the increase in pH may increase losses of ascorbic acid. Enzymatic browning of cut apples and potatoes is prevented by holding the food in dilute (2 per cent w/w) brine prior to blanching. Under blanching can lead to development of off flavors during storage of dried or frozen foods.

Effect of blanching on texture results softening of the vegetables to facilitate easy filling into containers prior to canning. However, when used for freezing or drying, the time and temperature conditions needed to achieve enzyme inactivation causes an excessive loss of texture in some type of foods (potatoes) and in large pieces of food. Calcium chloride (1-2 per cent) is therefore added to blancher water to form insoluble calcium pectate complexes and thus to maintain firmness in the tissues.

Changes due to Pasteurization

Pasteurization is relatively mild heat treatment and even when combined with other unit operations like irradiation and chilling there are only minor changes to the nutritional and sensory characteristics of most foods. However, the shelf life of pasteurized foods is usually only extended by a few days or weeks compared with many months with the more severe heat sterilization.

In fruit juices the main causes of color deterioration is enzymatic browning by polyphenoloxidase. This is accelerated by the presence of oxygen and fruit juices are therefore routinely deaerated prior to pasteurization. Losses in vitamin C and carotenes are minimized by deareation. Changes in milk are confined to a 5 per cent loss of serum protein and small changes to the vitamin content (Allen and Joseph, 1985). Thiamin and vitamin C losses are highest among all vitamins due to higher temperature and shorter time pasteurization.

Heat Sterilization

The purpose of heat sterilization is to extend shelf life of foods for storage at ambient temperatures, while minimizing the changes in nutritional value and eating quality. The time temperature combination used in canning have substantial effect on most naturally occurring pigments in food. In meat, the red myooxyglobin is converted into a brown pigment called metmyoglobin and purplish myoglobin is converted in to red-brown myohaemichromogen. Maillard browning and caramelization also contribute to the color changes in sterilized meat. However, this is a desirable change in cooked meats. Sodium nitrite and sodium nitrate used often in some meat products to cease the growth of *C. botulinum*, the resulting red-pink coloration is due to nitric oxide myoglobin and metmyoglobin nitrite.

In vegetables and fruits chlorophyll changes in to pheophytin, carotenoids are isomerizes from 5, 6- epoxides to less intensely colored *8-epoxides and anthocyanins* are degraded to brown pigments. Discoloration of canned food often is associated with the reaction of iron and tin with anthocyanins to form a purple pigment, and leucoanthocyanins forms pink anthocyanin complexes complexes in quinces and pears.

Caramelization, maillard reaction, and changes in the reflectivity of casein micelles may be changed in sterilized milk.

Changes due to Evaporation

Compounds specifically aroma compounds which are highly volatile in nature in comparison to water are lost during evaporation leading to losses in sensory characteristics of the most concentrates. Loss of flavor in fruit juices and concentrates often results due to evaporation. Sometimes loss of unplea sant volatiles compounds improves the overall product quality. To recover and retain volatiles vapour condensation and fractional distillation is often used. Some volatiles can be recovered by stripping volatiles from the feed liquor with inert gas and adding them back after evaporation.

Evaporation darkens the colors of foods, partly because of the increase in concentration of solids, but also because the reduction in water activity promotes chemical changes (*e.g.* maillard reaction). As the time and temperature dependent, short residence times and low boiling temperatures produce concentrates which have a good retention of sensory and nutritional qualities.

Distillation in food processing is mostly confined to the production of alcoholic spirits and separation of volatile flavors and aroma compounds. When a food that contains components having different degrees of volatility is heated, those that have a higher vapour pressure (mole volatile components) are separated first. These are termed the 'distillate' and components that have a lower volatility are termed 'bottoms' or residues.

Changes due to Extrusion

Extrusion is a process, which combines several unit operations including mixing, cooking, kneading, shearing, shaping and forming. Extruders are classified according to the method of operation (cold extruders or extruder cookers) and the method of construction (single-or twin-screw extruders).the principles of operation is similar in all types: raw materials are fed into the extruder barrel and the screw (s) then convey the food along it. Further, down the barrel, smaller flights restrict the volume and increase the resistance to movement of the food. As a result, it fills the barrel and the spaces between the screw flights and becomes compressed. As it moves further along the barrel, the screw kneads the material into a semi-solid, plasticized mass. If the food is heated above 100°C the process is known as extrusion cooking (or *hot extrusion*).

Cold extrusion, in which the temperature of the food remains at ambient is used to mix and shape foods such as pasta and meat products. Low pressure extrusion, at temperatures below 100°C, is used to produce, for eg, liquorice, fish pastes, surimi and pet foods.

Extrusion cooking is a high temperature short-time (HTST) process which reduces microbial contamination and inactivates enzymes. However, the main method of preservation of both hot and cold extruded foods is by the low water activity of the products (0.1-0.4) and for semi moist products in particular, by the packaging materials that are used.

Gelatinization of starch usually causes an increase in viscosity, but in extrusion cookers the intense shearing action can also break macro molecules down to smaller units, resulting in a reduction in viscosity.

Sensory characteristics: production of characteristics texture is one of the important features of extrusion technology. The extent of changes to starch determined by the operating conditions and feed materials, produces the wide range of product textures that can be achieved. The HTST condition in extrusion cooking has only minor effects on the natural color and flavors of food. However, in many foods the color of the product is determined by the synthetic pigments added to the feed material as water-or–oil soluble powders, emulsion. Fading of color due to product expansion, excessive heat or reactions with proteins, reducing sugars or metals ion may be a problem in some extruded foods. Added flavors are mixed with ingredients before cold extrusion, but this is largely unsuccessful in extrusion cooking as the flavors are more suitable but expensive. Flavors are therefore more often applied to the surface of extruded foods in the form of sprayed emulsions or viscous slurries. However, this may cause stickiness in some products and hence require additional drying.

Vitamin losses in extruded foods vary according to the type of food, the moisture content, the temperature of processing and holding time. Generally, losses are minimal in cold extrusion. The HTST condition in extrusion and the rapid cooling as the products emerges from the die, cause relatively small losses of most vitamins and essential amino acids. For example at an extruder temperature of 154°C there is 95 per cent retention of thiamin and little loss of riboflavin, pyridoxine, niacin or folic acid in cereals. However, losses of ascorbic acid and β-carotene are up to 50 per cent, depending on the time that the food is held at the elevated temperatures (Harpe,1979), and loss of lysine, cystine and methionine in rice product varies between 50 and 90 per cent depending on processing conditions (Seiler,1984). In soy flour the changes to proteins depend on the formulation and processing conditions. High temperatures and the presence of sugars cause Maillard browning and a reduction in protein quality. Lower temperatures and low concentrations of sugar result in an increase in protein digestibility, owing to rearrangement of the protein structure. Destruction of anti-nutritional components in soya products improves the nutritive value of texturised vegetable protein.

Changes Due to Dehydration

All products undergo changes during drying and storage that reduces their quality compared to the fresh materials aim of improved drying technologies is to minimize these changes while maximizing process efficiency. The main changes due to drying are textural, loss of flavor or aroma, and in some foods color and nutritional losses also occurs as a result of dehydration.

Changes in texture of solid foods are an important cause of quality deterioration. The nature and extent of pretreatment like addition of calcium chloride in blanching water, the type and extent of size reduction and peeling *etc.* affects the texture of rehydrated fruits and vegetables. Gelatinization of starch, crystallization of cellulose and localized variation in the moisture content set up internal stresses are causes of the loss of texture in food products. These rupture, crack, compress and permanently distort the relatively rigid cells which cause shrunken shriveled appearance in dried food. Though on rehydration products absorb water but at relatively slower rate and does not allow food to regain the firm texture of the fresh material.

In general, rapid drying and high temperatures causes greater changes to the texture of foods than do moderate rate of drying and lower temperatures. As water is removed during drying, solutes move from the interior of the food to the surfaces.

The mechanism and rate of movement are specific for each solute and depend on the type of food and the drying condition used. Evaporation of water causes concentration of solutes at the surface. A high air temperature (particularly with fruits, fish and meats) causes complex physical and chemical changes to the solutes at the surface, and in the formation of a hard impermeable skin. This is termed *case hardening* and it reduces the rate of drying to produce a food with a dry surface and a moist interior. It is minimized by controlling the drying conditions to prevent excessively high moisture gradients between the interior and the surface of the food.

Heat not only vaporizes water during drying but also causes loss of volatile components from the food and as a result most dried foods have fewer flavors than the original material.

The extent of volatile loss depends on the temperature and moisture content of the food and on the vapour pressure of the volatiles and their solubility in water vapour. Volatiles having high relative volatility and diffusivity are lost at an early stage in drying. Foods that have high economic value due to their characteristic flavor like herbs and spices are dried at low temperatures. Most fruits and vegetables contains only small amount of lipid, but oxidation of unsaturated fatty acids to produce hydroperoxides, which react further by polymerization, dehydration or oxidation to produce aldehydes, ketones and acids, causes rancid and objectionable odors. Some foods (for example carrot) may develop an odor of 'violets' produced by the oxidation of carotenes to -ionone (Rolls and Porter, 1973).

Drying changes the surface characteristics of a food and hence alters its reflectivity and color. In fruits and vegetables, chemical changes to carotenoids and pigments are caused by heat and oxidation during drying and residual polyphenoloxidase enzyme activity causes browning during storage. This is prevented by blanching or treatment of fruits with ascorbic acid or sulphur dioxide. For moderately sulphured fruits and vegetable the rate of darkening during storage is inversely proportional to the residual sulphur dioxide content. However sulphur dioxide bleaches anthocyanins and residual sulphur dioxide is also linked with health concerns. The rate of darkening increases markedly at high drying temperatures, when the moisture content of the product exceeds 4-5 per cent and at storage temperature above 38°C (Lea, 1958).

Drying affects the nutritional value specially water soluble vitamins. Vitamin C is sensitive to heat and oxidation therefore losses are high at higher temperature, higher moisture content and higher oxygen level during storage. Thiamin is also one of the heat sensitive vitamin but the losses of other water soluble vitamin is relatively low and rarely exceeds 5-10 per cent excluding blanching losses.

Oil soluble nutrients like vitamin A, D, E, K and essential fatty acids are mostly contained within the dry matter of the food and they are not concentrated during drying. However, water is a solvent for heavy metal catalysts that promotes oxidation of unsaturated nutrients. As water is removed the catalyst becomes more reactive and the rate of oxidation accelerates. Fat soluble vitamins are lost by interaction with peroxides produced by fat oxidation. Losses during storage are minimized by lowering the oxygen concentration and the storage temperature and by exclusion of light. Drum drying of milk results denaturation of milk proteins leading to reduction in solubility of the milk powder and losses in clotting ability. Spray drying does not affect the biological value of milk proteins. The biological value and digestibility of proteins in most food does not change substantially because of drying.

Changes due to Baking and Roasting

The purpose of baking and roasting is to alter the sensory properties of food, to improve the palatability and to extend the range of tastes, aromas and texture in food produced from similar raw materials. Baking also destroys enzymes, microorganisms and lowers the water activity of the food up to some extent.

A baking changes the texture of the food being produced mostly determined by the nature of food, moisture content and composition of fats, protein and structural carbohydrates and by the temperatures and duration of heating. A characteristic of many baked food is formation of dry crust which contains moist bulk of many foods. When meat is heated, fats melt and become dispersed as oil through the food or drain out as a component of "drip losses" Collagen is solubilised below the surface, to form gelatin. Oils are dispersed through the channels produced in the meat. Proteins become denatured and lose their water holding capacity and contact. These forces out additional fats and water and toughen and shrink the food. Further increase in temperature causes destruction of micro enzymes and inactivation of enzymes. The surface dries and texture become crisper and harder as a porous crust is formed by coagulation and degradation and pyrolysis of proteins. In cereals foods, changes to the granular structure of starch, gelatinization and dehydration produce the characteristic texture of the crust. Rapid heating in baking produces impermeable crust which seals in moisture and fat and protects nutrient and flavor components from degradation. A steep moisture gradient is formed between the moist interior (high a_w) and hygroscopic exterior (low a_w) of the food.

Maillard browning interaction between sugars and amino acids occurs due to severe heating conditions on the surface layers of the food. The high temperature and low moisture content on the surface layers also causes caramelization of sugars and oxidation of fatty acids to aldehydes, lactones, ketones, alcohols and esters. The maillard reaction and strecker degradation produce different aromas according to the free amino acids and sugar present in the particular food (Fennema, 1996).

Each amino acid produces a characteristic aroma when heated with a given sugar, owing to the production of a specific aldehyde. Different aromas are produced, depending on the type of sugar and the heating conditions used (for example the amino acid proline can produce aromas of potato, mushroom or burnt egg, when heated with different sugars and at different temperatures). Further heating degrades some of the volatiles produced by the above mechanisms to produce burnt or smoky aromas. There are therefore a very large number of component aromas produced during baking. The type of aroma depends on the particular combination of fats, amino acids and sugars present in the surface layers of food, the temperature, and moisture content of the food throughout the heating period and the time of heating.

Some baked foods (for example bread and meat) are important components of the diet in many countries and are therefore an important source of proteins, vitamins and minerals. For example, lysine is the limiting amino acid in wheat flour and its destruction by baking is therefore nutritionally important. Other baked foods (for example nuts, biscuits, cocoa, coffee and snack foods) are less important in the diet, and nutritional losses are therefore less significant. The main nutritional changes during baking occur at the surface of foods, and the ratio of surface area to volume is therefore an important factor in determining the effect on overall nutritional loss. With the exception of vitamin C, which is added to bread dough as an improver and is destroyed during baking, other vitamin losses are relatively small. In chemically leavened doughs the alkaline conditions cause the release of niacin which is bound to polysaccharides and polypeptides and therefore increase its concentration. Thiamine is the most important heat-labile vitamin in both cereal foods and meats, and losses are higher. In cereal foods the extent of thiamine loss is determined by the temperature of baking and the pH of the food. Loss of thiamine in pan bread is approximately 15 per cent (Bender, 1978) but in cakes or biscuits that are chemically leavened by sodium bicarbonate, the losses increase to 50–95 per cent. During baking, the physical state of proteins and fats is altered, and starch is gelatinized and hydrolyzed to dextrin and then reducing sugars. The loss of amino acids and reducing sugars in Maillard browning reactions causes a small reduction of lysine. The extent of loss is increased by higher temperatures, longer baking times and larger amounts of reducing sugars. The amylase activity of flour, the addition of sugar to dough, the use of fungal amylases and steam injection into ovens to gelatinize the surface starch and to improve crust colour all therefore affect the nutritive value of the proteins to some extent.

Changes due to Frying

Frying is an unusual unit operation in that the product of one food process (cooking oil) is used as the heat transfer medium in another. The effect of frying on foods therefore involves both the effect on the oil, which in turn influences the quality of the food, and the direct effect of heat on the fried product.

Prolonged heating of oils at the high temperatures used in frying, in the presence of moisture and oxygen released from foods, cause oxidation of the oil to form a range of volatile carbonyls, hydroxy acids, keto acids and epoxy acids. These cause unpleasant flavors and darkening of the oil. The various breakdown products are classified as volatile decomposition products (VDP) and non-volatile decomposition products (NVDP). VDPs have a lower molecular weight than the oil and are lost in vapour from the fryer.

NVDPs are formed by oxidation and polymerization of the oil and form sediments on the sides and at the base of the fryer. Polymerization in the absence of oxygen produces cyclic compounds and high-molecular-weight polymers, which increase the viscosity of the oil. This lowers the surface heat transfer coefficient during frying and increases the amount of oil entrained by the food. Many of these compounds are polar and they slow the evaporation of water and generate foam. However, these polar compounds also have beneficial effects in frying: they add flavour to the fried food, contribute towards the characteristic golden brown colour and optimum fat retention. Oil that has been used for a short period gives improved frying compared to fresh oil because these polar compounds promote better contact between the oil and both water on the product surface and vapour leaving the product. This results in improved and more uniform heat transfer and flavour absorption (Blumenthal, 1991). However, quality deteriorates when oil is used for a longer period. In continuous commercial production, oil that is entrained in the product is replaced continuously and NVDPs are filtered out, thus keeping the oil quality at an optimum level (Nielsen, 1993). Oxidation of fat-soluble vitamins in the oil results in a loss of nutritional value. Retinol, carotenoids and tocopherols are each destroyed and contribute to the changes in flavour and colour of the oil. However, the preferential oxidation of tocopherols has a protective (antioxidant) effect on the oil. This is particularly important as most frying oils are of vegetable origin and contain a large proportion of unsaturated fats which are readily oxidized. The essential fatty acid, linoleic acid, is readily lost and therefore changes the balance of saturated and unsaturated fatty acids in the oil (Kilgore and Bailey, 1970).

The main purpose of frying is the development of characteristic colors, flavors and aromas in the crust of fried foods. These eating qualities are developed by a combination of Maillard and compounds absorbed from the oil.

In many fried foods, oil can account for up to 45 per cent of the product (Saguy and Pinthus, 1995). Where fried foods form a large part of the diet, excess fat consumption can be an important source of ill-health, and are a key contributor to obesity, coronary heart disease and perhaps some types of cancer. These risks and consumer trends towards lower fat products is creating pressure on processors to alter processing conditions to reduce the amount of oil absorbed or entrained in their products.

Changes to protein quality occur as a result of Maillard reactions with amino acids in the crust. The effect of frying on the nutritional value of foods depends on the type of process used. High oil temperatures produce rapid crust formation and seal the food surface. This reduces the extent of changes to the bulk of the food, and therefore retains a high proportion of the nutrients. In addition, these foods are usually consumed shortly after frying and there are few losses during storage. The vitamin accumulates as dehydroascorbic acid (DAA) owing to the lower moisture content whereas, in boiling, DAA is hydrolyzed to 2, 3-diketogluconic acid and therefore becomes unavailable. Frying operations that are intended to dry the food and to extend the shelf life cause substantially higher losses of nutrients, particularly fat-soluble vitamins. Heat- or oxygen-sensitive water soluble vitamins are also destroyed by frying under these conditions.

Changes due to Chilling

The process of chilling foods to their correct storage temperature causes little or no reduction in the eating quality or nutritional properties of food. The most significant effect of chilling on the sensory characteristics of processed foods is hardening due to solidification of fats and oils. Chemical, biochemical and physical changes during refrigerated storage may lead to loss of quality, and in many instances it is these changes rather than micro-biological growth that limit the shelf life of chilled foods. These changes include enzymic browning, lipolysis, colour and flavour deterioration in some products and retrogradation of starch to cause staling of baked products (which occurs more rapidly at refrigeration temperatures than at room temperature). Lipid oxidation is one of the main causes of quality loss in cook–chilled products, and cooked meats in particular rapidly develop an oxidized flavour termed 'warmed-over flavour' (WOF) (Brown, 2000). Physico-chemical changes including migration of oils from mayonnaise to cabbage in chilled coleslaw, syneresis in sauces and gravies due to changes in starch thickeners, evaporation of moisture from unpackaged chilled meats and cheeses, more rapid staling of sandwich bread at reduced temperatures and moisture migration from sandwich fillings may each result in quality deterioration. Vitamin C losses in cook–pasteurize–chill procedures are lower than cooked–chilled foods.

Changes due to Freezing

The main effect of freezing on food quality is damage caused to cells by ice crystal growth. Freezing causes negligible changes to pigments, flavors or nutritionally important components, although these may be lost in preparation procedures or deteriorate later during frozen storage. Food emulsions can be destabilized by freezing, and proteins are sometimes precipitated from solution, which prevents the widespread use of frozen milk. In baked goods a high proportion of amylopectin is needed in the starch to prevent retrogradation and staling during slow freezing and frozen storage. There are important differences in resistance to freezing damage between animal and plant tissues. Meats have a more flexible fibrous structure which separates during freezing instead of breaking, and the texture is not seriously damaged. In fruits and vegetables, the more rigid cell structure may be damaged by ice crystals. The extent of damage depends on the size of the crystals and hence on the rate of heat transfer. However, differences in the variety and quality of raw materials and the degree of control over pre-freezing treatments both have a substantially greater effect on food quality than changes caused by correctly operated freezing, frozen storage and thawing procedures. During slow freezing, ice crystals grow in intercellular spaces and deform and rupture adjacent cell walls. Ice crystals have a lower water vapour pressure than regions within the cells, and water therefore moves from the cells to the growing crystals. Cells become dehydrated and permanently damaged by the increased solute concentration and a collapsed and deformed cell structure. On thawing, cells do not regain their original shape and turgidity. The food is softened and cellular material leaks out from ruptured cells (termed 'drip loss'). In fast freezing, smaller ice crystals form within both cells and intercellular spaces. There is little physical damage to cells, and water vapour pressure gradients are not formed; hence there is minimal dehydration of the cells. The texture of the food is thus retained to a greater extent. However, very high freezing rates may cause stresses within some foods that result in splitting or cracking of the tissues (Spiess, 1980).

Conclusion

Post harvest handling of the crop is one of the most important aspects to be taken care to ensure the maximum utilization of the crop and minimum losses. Too early harvesting and too late harvesting will cause the losses in quality and quantity of the crop therefore timely harvesting is necessary. After harvesting, to make the food consumable and available throughout the year with the purpose to increase its shelf life without adversely affecting its quality different food operations and processing is done. These processing includes blanching, sterilization, size reduction, extrusion, chilling, freezing, baking *etc.* changes the nutritional, textural, flavor and physical characteristics of the food. Processes like mixing, cleaning, sorting, freeze-drying *etc.* do not involve heat; have little or no effect on the nutritional quality of foods. Changes in texture caused by loss of moisture, or fat, formation or breakdown of emulsion and gels, hydrolysis of polymeric carbohydrates, and coagulation or hydrolysis of proteins. The basic motto of the food is to satisfy the human nutritional, physical, and psychological needs could only be fulfilled if the food has been handled and processed carefully and scientifically without causing much loss in terms of its quality and quantity.

REFERENCES

Bender, A.E. (1978) Food Processing and Nutrition. London: Academic Press. p. 13

Blumenthal, M.M. (1991). A new look at the chemistry and physics of deep-fat frying. Food technology (USA).

Brown, H.M. (2000). Non-microbiological factors affecting quality and safety. In: M. Stringer and C. Dennis (Eds) Chilled Foods, 2nd edn. Ellis Horwood Ltd, Chichester, Ch. 8.

Cardello, A.V. (1995). Food quality: relativity, context and consumer expectations. Food quality and preference, 6(3), 163-170.

Fellows, P.J. (2009). Food processing technology: principles and practice. Elsevier.

Fenemma, O.R. (1996). Food Chemistry, 3rd edn. Marcel Dekker, New York, pp. 171–174.

Harper, J.M. and Clark, J.P. (1979). Food extrusion. Critical Reviews in Food Science & Nutrition, 11(2), 155-215.

Josephson, E.S., Thomas, M.H. and Calhoun, W.K. (1975). Effects of treatment of foods with ionizing radiation. In: R. S. Harris and E. Karmas (eds) Nutritional Evaluation of Food Processing. AVI, Westport, Connecticut, pp. 393–411.

Kilgore, L. and Bailey, M. (1970). Degradation of linoleic acid during potato frying. J. Am. Diet. Assoc. 56, 130–132.

Lea, C.H. (1958). Chemical changes in the preparation and storage of dehydrated foods. In: Proceedings of Fundamental Aspects of Thermal Dehydration of Foodstuffs, Aberdeen, 25–27 March, 1958. Society of Chemical Industry, London, pp. 178–194.

Nielsen, K. (1993). Frying oils technology. In: A. Turner (ed.) Food Technology International Europe. Sterling Publications International, London, pp. 127–132.

Rolls, B.A. and Porter, J.W.G. (1973). Some effects of processing and storage on the nutritive value of milk and milk products. Proc. Nutr. Soc. 32, 9–15.

Saguy, I.S. and Pinthus, E.J. (1995). Oil uptake during deep-fat frying: factors and mechanism. Fd. Technology 4, 142–145 and 152.

Seiler, K. (1984) .Extrusion cooking and food processing. Food Trade Rev. March 124–125, 127.

Selmean, J.D. (1987). The blanching process, In: S. Thorne (ed.) Developments in Food Preservation, Vol. 4. Elsevier Applied Science, Barking, Essex, pp. 205–249.

Spiess, W.E.L. (1980). Impact of freezing rates on product quality of deep-frozen foods. In: P. Linko, Y. Mallki, J. Olkku and J. Larinkari (eds) Food Process Engineering. Applied Science, London, pp. 689–694.

Spurgeon, D. (1976). Hidden harvest: a systems approach to postharvest technology. IDRC, Ottawa, ON, CA.

Chapter 15

Principles and Methods of Food Processing for Improvement of Shelf Life

Aasima Rafiq

Introduction

Food processing refers to the application of techniques to foods in a systematic manner that transforms animal, plant, and marine materials into intermediate or finished value-added food products that are safer to eat. Such transformation requires the application of labor, energy, machinery, and scientific knowledge to a step (unit operation) or a series of steps (process) in achieving the desired transformation.

Food is preserved either by manipulating the levels of food ingredients or components to inhibit the growth of microorganisms or destroy them, for example, by keeping the food low in moisture content (low water activity), high in sugar or salt content, or at a low pH (less than 5). The aim of could be considered four-fold (1) extending the period during which food remains wholesome (microbial and biochemical), (2) providing (supplementing) nutrients required for health, (3) providing variety and convenience in diet, and (4) adding value.

Processing is important to prevent the post harvest losses occurring in grains, field crops, fruits, vegetables and spice crops by introducing improved, cost effective technologies so that the incomes of the farming sector will be increased. Processing/preservation by drying, concentration, freezing, cryogenic freezing, fermentation, irradiation, canning, sterilization and pasteurization are introduced in this chapter.

Drying

Drying can be described by three processes operating simultaneously *viz.* energy transfer from an external source to the water or organic solvent, phase transformation of water/solvent from a liquid- like state to a vapour state and transfer vapour generated away from the product and out of the drying equipment

Drying application is an important operation for the production of consistent, stable, free-flowing materials for formulation, packaging, storage and transport. Particle attrition or agglomeration can result in major differences in particle size distribution (PSD), compressibility and flow characteristics.

The purpose of drying is to reduce the moisture level of wet granules to keep the residual moisture low enough (preferably as a range) to prevent product deterioration and ensure free flowing properties. Various equipments used for drying are based on either direct heating static solids bed dryers or direct heating moving solids bed dryers or fluid bed dryer or indirect conduction dryers.

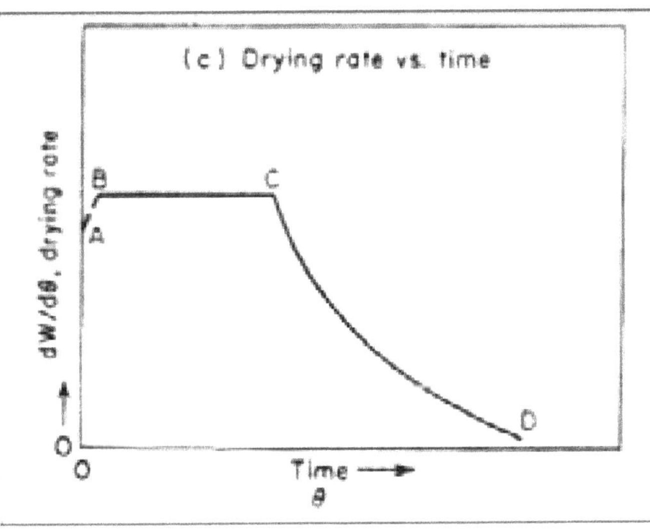

Figure 1: Periods of Drying.

Periods of Drying (Figure 1)

1. Warm up period (A-B)
2. Constant Rate Period (B-C)
3. Falling Rate Period (C-D)

Classification of Dryers (Table 1)

Dryers Can be Classified According to

1. Heat transferring methods

- ✰ Direct: Fluidized, Tray, Spray, Rotary Dryers, *etc.*
- ✰ Indirect: Cone, Tumble, Pan Dryers, *etc.*

2. Continuous/Batch processing:
   - ✰ Batch: small quantities/long residence time
   - ✰ Continuous: large quantities/small residence time

Table 1: Classification of Dryers

Criterion	*Types*
Mode of operation	Batch Continuous
Heat input-type	Convection, conduction, radiation Electromagnetic fields, combination of heat transfer modes Intermittent or continuous Adiabatic or non-adiabatic
State of material in dryer	Stationary Moving, agitated, dispersed
Operating pressure	Vacuum Atmospheric
Drying medium (convection)	Air Superheated steam Flue gases
Drying temperature	Below boiling temperature Above Boiling temperature Below freezing point
Relative motion between drying medium and drying solids	Co-current Counter-current Mixed flow
Number of stages	Single Multi-stage

Concentration

Foods are concentrated for many of the same reasons that they are dehydrated; concentration can be a form of preservation but this is true only for some foods. Food concentration is the heating food until it boils and removing the water or partially freezing food and removing water in the form of ice crystal.

Concentration of liquid foods is a vital operation in many food processes. Concentration is deferent from dehydration; generally, foods that are concentrated remain in the liquid state, whereas drying produces solid or semisolid foods with significantly lower water content. Nearly all liquid foods which are dehydrated are concentrated before they are dried. In the early stages of water removal, moisture can be more economically removed in highly efficient evaporators than in dehydration equipment. Increased viscosity from concentration often is needed to prevent liquids from running off drying surfaces or to facilitate foaming or puffing.

Some concentrated foods are desirable components of diet in their own right. For example, concentration of fruit juices plus sugar yields jelly. Many concentrated foods, such as frozen orange juice concentrate and canned soups are easily recognized because of need to add water before they are consumed.

More Common Concentrated Foods include Evaporated and Sweetened Condensed

- Milks
- Fruits and vegetable juices
- Nectars
- Sugar syrups and flavored syrups
- Jams and jellies
- Tomato paste and many type of fruit purees made by bakers, candy makers *etc.*

Benefits

- Concentration reduces weight and volume and results in immediate economic advantages.
- It is prior to concentrate the liquid food before dehydration because in the early stages of water removal, moisture can be more economically removed in highly efficient evaporators than in dehydration equipment.
- Increased viscosity from concentration often is needed to prevent liquids from running off drying surfaces or to facilitate foaming or puffing.
- Concentrated forms have become desirable components of diet in their own right.

Methods of Concentration

Solar Concentration

- Uses solar energy
- Used to derive salt from seawater in earlier times
- Being practiced today in united states in manmade lagoons
- Slow process and suitable only for concentrating salt solutions

Open Cattles

- Heated by steam
- Being used for some jellies and jams for certain types of soups
- High temperatures and long concentration times causes damage to food

- Thickening and burn on of product to cattle wall gradually lower the efficiency of heat transfer and slow concentration process
- Widely used in manufacture of maple syrup

Flash Evaporators

- Subdivides food material and brings it into direct contact with the heating medium to speed up concentration process.
- Superheated steam at 150°C is used

Some Other Methods

- Thin film evaporators
- Vacuum evaporator
- Freeze concentration
- Ultra filtration and reverse osmosis

Changes During Concentration

Concentration processes that expose food to 100°C or higher temperatures for prolonged period can cause major changes in organoleptic and nutritional properties. Also these changes are different for different food products. Two more common changes are:

- Cooked flavors and
- Darkening of color

Some Other Changes

- Heat induced reactions
- In case of sugar, crystallization of sugar that can result in gritty, sugar jellies or jams.
- Crystallization of lactose due to over concentration "sandiness". in case of certain milks
- Proteins can be easily denatured and precipitated from solution due to high concentration of salts and minerals in solution with protein.
- Proteins can be easily denatured and precipitated fro solution due to high concentration of salts and minerals in solution with protein.
- The gelation of concentrated milk and other proteinaceous foods.
- Microbial destruction, largely dependent on temperature

Freezing

Freezing is a unit operation in which temperature of the food is reduced below its freezing point generally to -18°C or below. This allows the change in state and leads to preservation of taste, texture, and nutritional value in foods. It retards the growth of microorganisms and provides a significant extended shelf life. The material to be frozen first cools down to the temperature at which nucleation starts. Before ice can form, a nucleus, is required upon which the crystal can grow; the process of producing this seed is defined as nucleation. Once the first crystal appears in the solution, a phase change occurs from liquid to solid with further crystal growth (Figure 2).

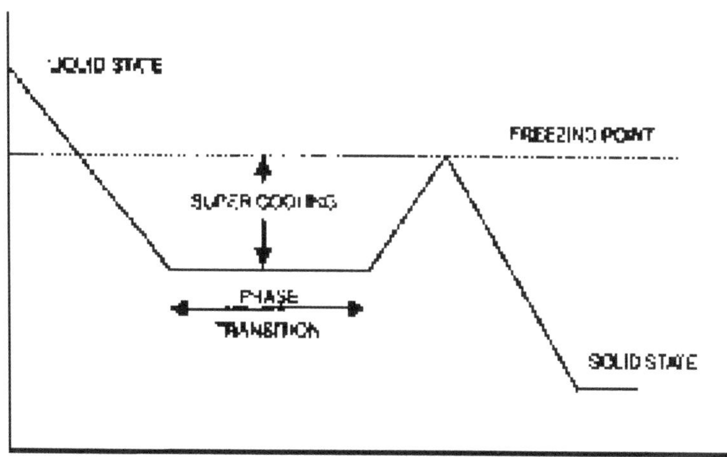

Figure 2: Freezing Process for Foods

Freezing rate is defined as the ratio of difference between initial and final temperature of product to freezing time. Freezing time is defined as time required to lower product temperature from its initial temperature to a given temperature at its thermal center. Freezing point is defined as the temperature at which the first ice crystal appears and the liquid at that temperature is in equilibrium with the solid.

The quality of a frozen-food is influenced by storage conditions.t The changes in quality decrease as temperature is decreased, maintaining low storage temperatures increases the cost of frozen-food storage. Higher temperatures in frozen-food storage must be avoided due to the sensitivity of the frozen-food to temperature. Experience has established that a frozen-food storage temperature of -18° accepted as a safe storage temperature for extended shelf life of a frozen food. An increase in the product temperature results in conversion of ice to liquid state, with the possibility of re-crystallization when the temperature decreases. Small ice crystals will tend to melt as the temperature rises and change back to ice when the temperature is lowered. The re-crystallization results in an increase in ice crystal size and the impacts on quality.

Types of Freezing

Fast Freezing

Quick or fast freezing occurs at –25°C or less. Ice crystals are small and do not damage food cells.

Slow Freezing

Slow freezing occurs at -24 °C or above. Ice crystals are big and damage the food cells causing loss of texture, nutrients, colour and flavour on thawing.

Types of Freezers

Air Blast Freezer

Either still air or forced air is used. Air is recirculated over food at between 30°C and 40°C at a velocity of 1.5–6.0 m/s. In batch equipment, food is stacked on trays in rooms or cabinets. Continuous equipment consists of trolleys stacked with trays of food or on conveyor belts which carry the food through an insulated tunnel.

Belt Freezer

Have a continuous flexible mesh belt which is formed into spiral tiers and carries food up through a refrigerated chamber.

Cold air or sprays of liquid nitrogen are directed down through the belt stack in a counter current flow, which reduces weight losses due to evaporation of moisture.

Plate Freezer

It consists of a vertical or horizontal stack of hollow plates, through which refrigerant is pumped at 40°C.

Slight pressure is applied to improve the contact between surfaces of the food and the plates and thereby increases the rate of heat transfer.

Cryogenic Freezer

In cryogenic freezing food is exposed to an atmosphere below - 60°C through direct contact with liquefied gases such as nitrogen or CO_2.

Liquid Nitrogen Freezer Liquid Carbon Dioxide Freezer

When the CO_2 gas is released to the atmosphere at -70 °C, half of the gas becomes dry-ice snow and the other half stays in the form of vapor. This unusual property of liquid carbon dioxide is used in a variety of freezing systems, one of which is a prefreezing treatment before the product is exposed to nitrogen spray.

Immersion Freezer

The immersion freezer consists of a tank with a cooled freezing media, such as glycol, glycerol, sodium chloride, calcium chloride, and mixtures of salt and sugar. The product is immersed in this solution or sprayed while being conveyed through the freezer, resulting in fast temperature reduction through direct heat exchange.

Effects of Freezing

- Damage caused to cells by ice crystal growth.
- Freezing causes negligible changes to pigments, flavors or nutritionally important components.
- When water in the cells freezes, an expansion occurs and ice crystals cause the cell walls to rupture. Consequently, the texture of the produc

is generally much softer after thawing when compared to non- frozen produce.

- Chemical changes that can cause spoilage and deterioration of fresh fruits and vegetables will continue after harvesting.
- Development of rancid oxidative flavors through contact of the frozen product with air.

Factors Affecting the Quality of Frozen Foods During Storage

- Freezer burn
- Freezing Burn and Dehydration
- Change in the Color of Food
- Water absorption and redistribution
- Re-crystallization
- Drip loss
- Re-crystallization of Ice Inside the Product
- Biological and Chemical Changes
- Protein denaturation
- Lipid oxidation

Cryogenic Freezing

Food products have to be frozen as gently as possible so that when defrosted they can be served fresh and appetizing. A prerequisite for this is "fast cold". The water contained in the cells must freeze very quickly (faster than 5 cm/h) in order to ensure the formation of very small ice crystals that do not damage the cell material. A rapid cooling method called cryogenic freezing is operated which makes use of either liquid nitrogen or liquid carbon dioxide for freezing of foods. According to Smith and Hui (2004) cryogenics freezing is the freezing at very low temperatures (below –150°C, –238°F or 123 K).

Cryogenic freezers (Figure 3) are characterized by a change of state in the refrigerant (or cryogen) as heat is absorbed from the freezing food. The heat from the food therefore provides the latent heat of vaporization or sublimation of the cryogen. The cryogen is in intimate contact with the food and rapidly removes heat from all surfaces of the food to produce high heat transfer coefficients and rapid freezing. The two most common refrigerants are liquid nitrogen and solid or liquid carbon dioxide. In cryogenic freezing of sea foods like prawn, they are first dipped in liquid nitrogen to freeze the outside layer. This prevents the prawns sticking together and from sticking to the freezer belts and then passed through a tunnel where nitrogen gas is sprayed downwards. This produces small crystals, and little moisture loss (Rahman,2007).

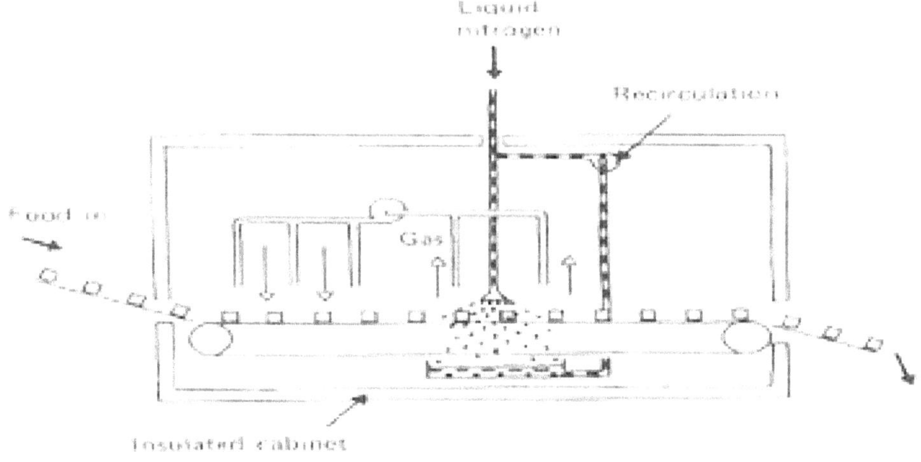

Figure 3: Cryogenic Freezer.
(*Source*: www.fao.org).

Fermentation

In biochemical sense the term fermentation refers to the metabolic process in which organic compounds (particularly carbohydrates) are broken down to release energy without the involvement of terminal electron acceptor such as oxygen. Partial oxidation of the substrate occurs so that only relatively small amount of ATP energy is released compared with the energy generated if a terminal electron acceptor is involved. Partial oxidation of a carbohydrate can give rise to a variety of organic compounds. The compounds produced by microorganisms vary from organism to organism and are produced via different metabolic pathways. The term fermentation can also be applied to any industrial process that produces a material that is useful to humans and if the process depends on the activity of one or more microorganisms. These processes, known as industrial fermentations, are usually carried out on large scale and in vessels in which organism are normally grown in liquid media.

Different Type of Fermentation

1. Homo-fermentation
2. Heterofermentation

Nutritional Value of Fermented Foods

There is a significant increase in the soluble fraction of food during fermentation. The quantity and quality of food proteins is expressed by biological value, and often the content of water soluble vitamins is increased, while the anti-nutritional factors show a decline during fermentation (Potter and Hotchkiss, 1995). Fermentation also results in a lower proportion of dry matter in food and concentrations of vitamins, minerals and protein appear to increase when measured on dry weight basis.

Single as well as mixed culture fermentation of pearl millet flour with yeast and lactobacilli significantly increased the total amount of soluble sugars, reducing and non reducing sugar content of anti nutritional factors to a safe level in comparison with other methods of processing.

The protein efficiency ratio (PER) of wheat increases on fermentation, partly due to increase in availability of lysine. The fermentation process raises the PER value of wheat and soyabeans mixture to a level which is comparable to that of casein. Fermentation may not increase the content of protein and amino acids unless ammonia or urea is added as nitrogen source to the fermentation media. The relative nutritional value (RNV) of maize increased from 65 per centto 81 per cent when it is germinated, and fermentation of flour made of the germinated maize gives a further increase in RNV to 87 per cent (Potter and Hotchkiss, 1995).

During fermentation certain microorganisms produce vitamins at higher rate than others do. Fermented milk products in general show an increase in folic acid content and slight decrease in vitamin B12 while other B vitamins are affected only slightly in comparison to raw milk. The levels of vitamin B12, riboflavin and folacin are increased by lactic acid fermentation of maize flour, while the level of pyridoxine is decreased.

The mineral content is not affected by fermentation unless some salts are added to the product during fermentation or by leaching when liquid portion is separated from the fermented food. Sometimes, when fermentation is carried out in metal container some minerals are solublised by the fermented product, which may cause an increase in mineral content. Phytate content in bread is lowered when the amount of yeast or the fermentation time is raised. Some important fermentation products are shown in Table 2.

Table 2: Some Important Fermentation Products

Product	*Organism*	*Use*
Ethanol	*Saccharomyces cerevisiae*	Industrial solvents, beverages
Glycerol	*Saccharomyces cerevisiae*	Production of explosives
Lactic acid	*Lactobacillus bulgaricus*	Food and pharmaceutical
Acetone and butanol	*Clostridium acetobutylicum*	Solvents
-amylase	*Bacillus subtilis*	Starch hydrolysis

Irradiation

Food irradiation is a process of exposing foods, either prepackaged or in bulk to very high-energy, invisible light waves (radiation) such as gamma rays, X-rays or electron beams. Electromagnetic rays, when made to bombard against materials, they can knock off an electron from an atom or molecule causing ionization. For this reason, these are often called ionizing irradiation. The X-rays and gamma-rays are very short wavelength radiations that have very high associated energy levels (Rahman, 2007).

Gamma rays come from the spontaneous disintegration of radionuclides. Cobalt-60 the choice for gamma radiation source produced by neutron bombardment in a nuclear reactor of the metal cobalt- 59, then doubly encapsulated in stainless steel pencils to prevent any leakage during its use in an irradiator. Cobalt-60 has a half-life of 5.3 years, highly penetrating and can be used to treat full boxes of fresh or frozen food. Over 80 per cent of the cobalt-60 available in the world market is produced in Canada. Other producers are the Russian, Republic of China, India and South Africa. Cesium 137 is the only other gamma-emitting radionuclide suitable for industrial processing of materials. It can be obtained by reprocessing spent, or used, nuclear fuel elements and has a half-life of 30 years. There is no supply of commercial quantities of cesium-137 (Heldman and Hartel, 1998).

Electron Beams have low associated energy levels, which are too low to be practical value in preservation, they need to be accelerated (in cyclotrons, linear accelerators *etc.*) to make them acquire the required energy. Since electrons cannot penetrate very far into food, compared with gamma radiation or X-rays, they can be used only for treatment of thin packages of food and free flowing or falling grains (Heldman and Hartel, 1998).

The international unit of measurement of irradiation dose is the Gray (Gy). One Gray represents one joule of energy absorbed per kilogram of irradiated product. One Gy is equivalent to 100 rad (radiation absorbed dose). The desired dose is achieved by the time of exposure and by the location of the product relative to the source. It also depends upon the mass, bulk density and thickness of the food. The maximum dose of 10 kGy recommended by the Codex General Standard for Irradiated Foods is equivalent to the heat energy required to increase the temperature of water by 2.4°C.

- Irradiation is often referred to as a "cold pasteurization" process as it can accomplish the same objective as thermal pasteurization of liquid foods. Foods that are commonly irradiated include wheat flour, potatoes, pork, fruits and vegetables, herbs and spices, meat and onions. Normally, irradiation dose used for different foods are meat - 4.5 kGy, grains - 0.2 – 0.5 kGy, Fruits & Vegetables 1 kGy (Potter and Hotchkiss, 1995).

Food Irradiation Plant is authorized by Atomic Energy Regulatory Board. Facility could be private, public or joint sector company. Cost of Irradiation Facility varies between Rs. 6-8 Crores excluding land cost. Cost varies based on type of food. In India, Bhabha Atomic Research Centre which has a Radiation Processing Plant at Vashi, Navi Mumbai, KRUSHAK Lasalgaon near Nashik.

Advantages

- It can kill many insects and pests that infest foods.
- It can delay or stop normal ripening and decay processes so that foods can be stored for longer.
- It can kill dangerous micro organisms in foods.

Disadvantages

- ☆ It can only be used on a very limited range of foods.
- ☆ It is still a relatively expensive technology.
- ☆ Vitamin E levels can be reduced by 25 per cent after irradiation and vitamin C by 5-10 per cent.
- ☆ It is ineffective against viruses.

Canning

Canning means, the preservation of food in permanent, hermetically sealed containers (of metal, glass, thermostable plastic, or a multilayered flexible pouch) through agency of heat. Heating is the principle factor to destroy the microorganisms and the permanent sealing is to prevent re-infection. In canning practice "commercial sterility" is achieved by giving a degree of heat treatment sufficient to kill non-sporing bacteria and all spores that might germinate and grow during storage without refrigeration. In practice, complete sterility is seldom achieved, in the fact that certain microorganisms form spores, which may be heat-resistant. To destroy the spores of certain thermophiles would require a degree of heating which would greatly lower the organoleptic characteristics of canned meat products (Park *et al.*, 2014).

The process of canning of fruits and vegetables can be seen in the Figures 4 and 5. A typical commercial canning operation may employ the following general processes: washing, sorting/grading, preparation, container filling, exhausting, container sealing, heat sterilization, cooling, labeling/casing, and storage for shipment. One of the major differences in the sequence of operations between fruit and vegetable canning is the blanching operation. Most of the fruits are not blanched prior to can filling whereas many of the vegetables undergo this step. Canned vegetables generally require more severe processing than do fruits because the vegetables have much lower acidity and contain more heat-resistant soil organisms. Many vegetables also require more cooking than fruits to develop their most desirable flavor and texture. The methods used in the cooking step vary widely among facilities. With many fruits, preliminary treatment steps (*e.g.*, peeling, coring, halving, pitting) occur prior to any heating or cooking step but with vegetables, these treatment steps often occur after the vegetable has been blanched. For both fruits and vegetables, peeling is done either by a mechanical peeler, steam peeling, or lye peeling. The choice depends upon the type of fruit or vegetable or the choice of the company. One of the major differences in the sequence of operations between fruit and vegetable canning is the blanching operation. Most of the fruits are not blanched prior to can filling whereas many of the vegetables undergo this step. Canned vegetables generally require more severe processing than do fruits because the vegetables have much lower acidity and contain more heat-resistant soil organisms (Potter and Hotchkiss, 1995). Many vegetables also require more cooking than fruits to develop their most desirable flavor and texture. The methods used in the cooking step vary widely among facilities. With many fruits, preliminary treatment steps (e. g., peeling, coring, halving, pitting) occur prior to any heating or cooking step but with vegetables, these treatment steps often occur after the vegetable has been blanched.

For both fruits and vegetables, peeling is done either by a mechanical peeler, steam peeling, or lye peeling. The choice depends upon the type of fruit or vegetable or the choice of the company (Teixeira, 2014).

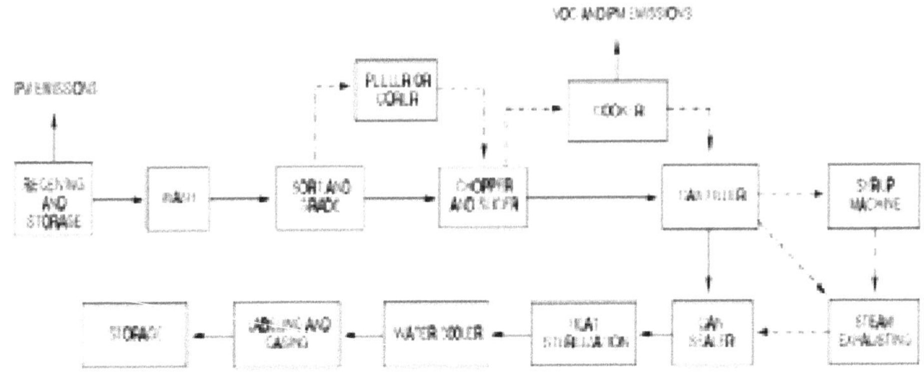

Figure 4: Process Diagram for Canning of Fruits.
(*Source*: http://ecoursesonline.iasri.res.in)

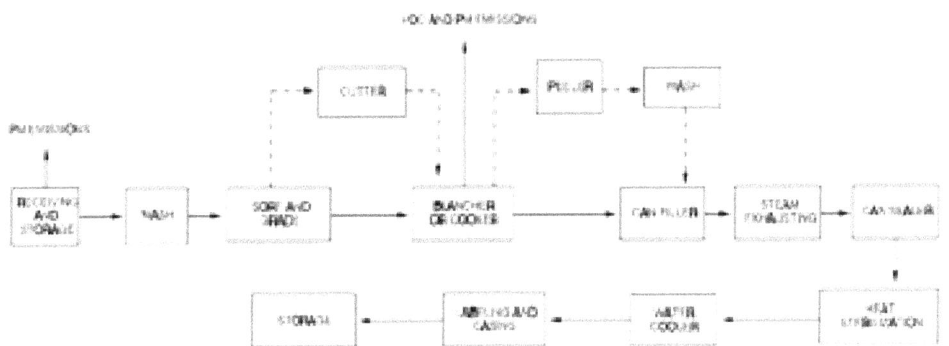

Figure 5: Process Diagram for Canning of Vegetables.
(*Source*: http://ecoursesonline.iasri.res.in)

The container plays a vital role in food canning, it must be: 1) Capable of being hermetically sealed to prevent entry of microorganisms. 2) Impermeable to liquids and gases, including water vapour. 3) Maintain the state of biological stability (*i.e*, commercial sterility) that was induced by the thermal process alone or in combination with other chemical and physical processes. 4) Physically protect the contents against damage during transportation, storage and distribution.

For meat and poultry preservation, wide varieties of materials are used for manufacture of cans. Yet metal containers remain the most frequent used package for canning foods. Steel: Tinplate, tin-free steel, and nickel-plated steel coated with a very thin film of tin are the materials used to manufacture metal food cans. The amount of tin used being only about 1.5 per cent of the can's weight and should not contain more than 1 per cent lead. It is used to prevent rusting.

To prevent interaction between meat product and the metal, cans are coated on the inside with an organic material. Two general kinds of organic coatings are used in the food industry: (i) acid resistant and (ii) sulfur-resistant. Acid-resistant coated cans are used primarily for fruit. Meat products are generally packed in cans that have been lined with sulfur-resistant materials. Some important food groups according to their pH content has been shown in Table 3.

Table 3: Important Food Groups

	Group	*pH Range*	*Examples*
(a)	Low acid foods	5.0 - 6.8.	Meat, fish, poultry, dairy
(b)	Acid foods	4.5-3.7	Fruits such as pear, oranges, apricots and tomatoes.
(c)	High acid foods	<3.7	Jams, Jellies, pickled products and fermented foods

Advantages of Metal Containers

The Metal Ones are Preferable as:

1. It has a high conductivity of heat.
2. It cannot easily be broken.
3. Being opaque, so any possible bad effects of light on food stuffs are avoided.
4. Be able to withstand the stresses imposed during thermal processing and cooling.
5. Be able to withstand the subsequent handling, which includes transportation, storage and distribution

Pasteurization

Thermal pasteurization is a relatively mild heat treatment given to foods with the purpose of destroying selected vegetative species of microorganisms, such as the many pathogens that cause food-borne illness, as well as to accomplish inactivation of enzymes. Pasteurization does not eliminate all vegetative microorganisms, nor does it eliminate more heat-resistant spore-forming bacteria. Therefore, pasteurized foods are not shelf-stable and must be stored under refrigeration and/or with modified-atmosphere packaging, which slow the growth of microorganisms that still remain viable in the product and will ultimately cause spoilage (Rahman, 2007).

Depending on the type of product, the shelf-life of pasteurized foods could range from several days (milk) to several weeks or more (fruit juices). Because only relatively mild heat treatment is involved, the sensory quality and nutritive value of the food are minimally affected. The severity of heat treatment used (time and temperature) and the length of shelf-life achieved depends on the nature of the product, pH, heat resistance of the target microorganisms, sensitivity of the product quality to heat damage, and the method of heating.

Pasteurization Equipment Systems

Liquid products that are capable of being pumped through tubular pipe lines, like milk and fruit juices, are thermally pasteurized by heating/holding/cooling while the products are flowing through a system of heat exchangers and hold tubes. These systems deliver a high-temperature short-time (HTST) process sufficient to inactivate target microorganisms while causing minimal heat damage to product quality. Unpasteurized liquid products are held cold in 'raw product' refrigerated tanks until ready to be pasteurized. As raw product is pumped through the heat exchanger system, the product temperature is quickly raised to the required pasteurization temperature as it flows through the 'heating' heat exchanger, sent through an insulated holding tube to receive the required length of time at the pasteurizing temperature, and quickly cooled and chilled through the 'cooling' heat exchanger to the refrigerated temperature for storage in a 'pasteurized product' refrigerated tank until ready for filling. Pasteurizing temperatures fall in the range of 60–80°C, well below the boiling point of water at atmospheric pressure. Since water is typically used as the heat exchange medium in these systems, there is no need to operate under pressure. Either plate heat exchangers or tubular heat exchangers can be used for relatively thin (low-viscosity) liquids. For viscous liquids, a scraped surface heat exchanger can be used to promote faster heat transfer and minimize surface fouling problems (Fellows, 2000).

Sterilization

Sterilization is making a substance free from all micro organisms both in vegetative and sporing states. Sterilisation of foods by the application of heat can either be in sealed containers or by continuous flow techniques. Whatever the process, the main concerns are with food safety and quality. Sterilization process is complete when no viable microorganisms are present in the product–a viable organisms being one that is able to reproduce when exposed to conditions optimum for its growth. Temperatures slightly above the maximum for bacterial growth result in the death of vegetative bacterial cells, whereas bacterial spores can survive much higher temperatures. Since bacterial spores are far more heat resistant than are vegetative cells, they are of primary concern in most sterilization processes.

Sterilization is not a good term to apply to thermal processing of foods since the criterion of success is the inability of microorganisms and their spores to grow under conditions normally encountered in storage. This means that there could be some dormant non pathogenic microorganisms in the food (any pathogenic microorganisms are dead) but that environmental conditions are such that the organisms do not reproduce. Foods which have been thermally processed with this as the criterion are referred to as commercially sterile, bacterially inactive, or partially sterile (Fellows, 2000)..

The thermal conditions needed to produce commercial sterility depend on many factors, including (1) nature of the food (*e.g.*, pH); (2) storage conditions of the food following the thermal process; (3) heat resistance of the microorganisms or spore; (4) heat transfer characteristics of the food, its container, and the heating medium; and (5) initial load of microorganisms. Various agents used in sterilization process has been shown in Table 4.

Table 4: Various Agents Used in Sterilization

Physical Agents	*Chemical Agents*
Sunlight	**Alcohols**: Ethyle, Isopropyl, Trichlorobutanol.
Drying	**Aldehydes**: Formaldehyde, Glutaraldehyde.
Dry heat: Flaming, Incineration, Hot air.	**Dyes**
Moist heat: Pasteurisation, boiling, steam under normal pressure, steam under high pressure.	**Halogens**
	Phenols
Filteration: Candles, asbestos pads, membranes	**Surface active agents**
Radiation	**Metallic salts**
Ultrasonic and sonic vibrations	**Gases**: Ethylene, oxide, formaldehyde, beta propiolactone

Conclusion

Food processors utilize several processing techniques (drying, concentration, freezing, cryogenic freezing, fermentation, irradiation, canning, sterilization and pasteurization) to transform raw food materials to produce microbiologically safe, consumer-desired, extended shelf life, convenient and value-added foods. Incorporation of knowledge from several disciplines including physics, chemistry, biology, nutrition, engineering and sensory sciences are important for successful food processing.

Food processing operation chosen can influence the extent of changes in product quality (color, texture, and flavor) attributes. Preservation of food shows some merits like it minimizes the risk of toxins and it doesn't get spoiled easily like fresh foods. On the other hand it's harmful on some nutritional basis too. The extent of nutrient and quality retention in processed food depends upon intensity of treatment applied, type of nutrient or food quality attribute, food composition, and storage conditions. Efforts must be made to introduce lean manufacturing concepts developed in other industries to food processing to reduce energy and water use and allow the production of healthy processed foods at affordable cost.

REFERENCES

Fellows, P.J. (2000). Food Processing Technology: Principles and Practices, 2nd edn. CRC Press, Boca Raton, Florida.

Heldman, D.R. and Hartel, R.W. (1998). Principles of Food Processing. A Chapman & Hall Food Science Book An Aspen Publication Aspen Publishers, Inc. Gaithersburg, Maryland.

http: //ecoursesonline.iasri.res.in/mod/page/view.php?id=795

Park, S.H., Lamsal, B.P., and Balasubramaniam, V.M. (2014). Principles of Food Processing in Food Processing: Principles and Applications, Second Edition. Edited by Stephanie Clark, Stephanie Jung, and Buddhi Lamsal. © John Wiley & Sons, Ltd. Published 2014 by John Wiley & Sons, Ltd.

Potter, N.N. and Hotchkiss, J.H. (1995). Food Science, 5th edn.

Rahman, M.S. (2007). Handbook of Food Preservation. CRC Press, Florida, USA. pp. 3-18, 447- 476, 635-665.

Smith, J.S. and Hui, Y.H. (2004). Food Processing: Principles and Applications, I-29, Blackwell Publishing.

Teixeira, Arthur, A. (2014). "Thermal Food Preservation Techniques (Pasteurization, Sterilization, Canning and Blanching)", Conventional and Advanced Food Processing Technologies.

www.fao.org.

Chapter 16

Recent Methods in Handling, Packaging and Storage of different Fruits and Vegetable Crops

S.K. Vishwakarma, Sanjeev Kumar and M.P. Yadav

Introduction

India's wide diversity in climate and soil provides much for growing a wide variety of horticultural produce than in most of the countries in the world. Currently, India is the second largest producer of fruits and vegetables in the world. In India fruits and vegetables are grown only in 6-7 per cent of gross cropped area but contribute more than 18.8 per cent of the gross value of output and 52 per cent export earnings from total agricultural produce. Fruits and vegetables have been shown to earn 25-30 times more foreign exchange per unit area than cereals due to higher yields and higher price available in the international market. But India loses about 30-40 per cent of the harvested fruits and vegetables during handling, storage, transportation. Post harvest management of fresh produce using emerging techniques and technology is one of the biggest opportunities in India.

The quality of fresh produce depends primarily upon the selection and careful handling of good product. Such measures as harvesting at optimal maturity, minimizing injury due to handling, reducing microbial infection through proper sanitation and maintaining optimum temperature and humidity are important in maintaining post harvest quality of fruits and vegetables the consumer. Post harvest interfaces with the fresh produce supply, Cleaning, washing, drying, grading. Package fabrication, product development, production operations, quality control and assurance, ware housing. Distribution, marketing, disposal and the recycling function from obtaining fresh harvested produce through final disposal or recycling of materials, post harvest technology is involved with all aspects of the distribution system. Products today are much more technically demanding, requiring greater protection in processing, transport and storage. There have also been substantial changes in the demographic character and lifestyle of consumers. The controlled atmosphere (CA) storage is the most advanced, economical and widely accepted method for storage of fruits throughout the world till today. While controlled atmosphere (CA) storage is regulated by controlling of temperature, humidity, CO_2, O_2 and gaseous composition (Chattopadhyay *et al.*, 1989).

Today greater numbers of people have at a higher level of education and more disposable income than ever before. Such income is spent for higher quality, more luxurious and processed packaged products that more convenient to use. Therefore, post harvest technology's role in society is becoming increasingly important.

Factors of Shelf Life of Products

1. The environment the product is exposed to during cleaning, grading, processing, packaging, distribution and marketing.
2. The properties of the package used for the product packaging.

The shelf life is critical to marketing success of fresh fruits and vegetables. Shelf life is defined as the length of time that a container or material in a container will remain in a saleable or acceptable condition under specified conditions of storage and distribution system. The use life is defined as the length of time that a subjected to a great many environmental influences. These include humidity, oxygen, light; dust, and temperature, biological elements include human beings, micro organisms, rodents, insects and other biological pests spoilage can be caused by mold, bacteria and yeast. During distribution, physical forces cause damage to the biological products. The major physical forces include impact, shock, vibration, compression and friction. Both Shock or impact damage and vibration can result in extensive product breakage and loss of quality. Compression causes crushing or collapsing of packaged products in storage in ware house and in transpiration. Commercially produced fruits experience a series of mechanical handling procedures before being selected by the consumer. It is generally agreed that every handling operation of fruits offers an opportunity for mechanical injury to occur. Because bruises, cuts and punctures are cumulative and irreversible in nature, the effects are intensified throughout each step of the handling process from tree to consumer (Singh, 2008).

Several investigators have identified the major causes of damage to packaged fruits as being due to impact; static loading and vibration encountered during handling and transportation operations. There is general agreement that, regardless of the method of transport, the most severe shocks likely to be encountered in transit result from handling operations, especially dropping of the package onto a non-resilient horizontal surface.

The handling of fruits has always been potential source of damage. The availability of equipment that can be sequentially positioned to mechanically perform the operations required to convey. Clean, wax, sort, size and pack fruits have resulted in modern high capacity packing lines that often have 3 to 8 fruits moving past a fixed point of a single lane. The mechanical operations can cause bruise, cut and puncture damage to fruits, thus reducing the fruit quality and grade. High quality and grade of fruits on the tree is the goal of the fruit grower. Consumers are willing to pay a price corresponding to quality and grade. If the fruits are damaged before the consumer buys them, potential income is lost for everyone in the chain, and the shelf-life of the fruits before the consumer buys them, potential income is lost for everyone in the chain, and the shelf-life of the fruits is reduced.

Therefore, there is a need to minimize the number of transfer points and the potential and kinetic energy of the fruits at each transfer points to minimize the mechanical damage in the system.

Harvest Maturity Indices

The principles which dictate at which stage of maturity a fruit or vegetable should be harvested are crucial to its subsequent storage, marketable life and quality. Correct stage of maturity of fruit is very important at harvesting time to reduce post harvest losses. This maturity is indicated by several parameters based on visual, physical and chemical indices. These not only help in deciding a harvest date but also to sort out the harvested fruits into lots of varying maturity for their optimum use. Maturity standards for mango, guava, sweet oranges, grape, litchi and ber have been worked out. The Bureau of Indian Standards has suggested guidelines including maturity standards in bananas. In mango, TSS should be 8± 1 Brix and acidity 3.5 ± 0.2 per cent besides several physical characteristics. For banana, judgment of degree of maturity is based on number of days from flower emergence, pulp/peel ratio, and colour of pulp and odor. Central fruit of normal shape from the outside row of second or third hand is sampled. In case of grapes, the sample should be drawn from the central clusters from different vines in an orchard. Banana fruit bunches mature over a period of 90 to 150 days of maturity after flowering in Sapota, fruits take nearly 10 months (41 Weeks) from fruit-set to reach harvest maturity. Papaya fruits of Coorg Honey Dew, Pink Flesh Sweet, Sunrise, Thailand and Washington varieties take 145 to 165 days to attain ripe stage from the date of flowering. Pollinated spikes of jack-fruit take about 100 to 120 days to develop into fully mature syncarpous fruits. Mango fruits take nearly 12 to 15 weeks from fruit set to reach harvest maturity. Guava fruits take about 120 to 150 days to reach harvest maturity (Singh, 2008).

Harvesting of Fruits and Vegetables

In general, manual harvesting of fruits is followed in India. However, different fruits require different methods of harvesting. Green and mature mangoes are harvested manually with the help of bamboo poles with a net attached and are lowered to the ground in a basket with the help of a rope. Mango harvesters have been developed at Konkan Krishi Vidyapeeth (KKV), Dapoli, Indian Institute of Horticultural Research, Bangalore, and Central Institute of Sub-Tropical Horticulture (CISH), Lucknow. The harvester is harvests the fruits with pedicel of 1 to 2 cm in length. This stops sap bleeding and reduces instant infection and increases shelf life of fruits by 2 to 4 days. Mature banana bunches are cut at the stalk end, retaining 30 cm length with the help of a sharp sickle, Citrus fruits are plucked or clipped manually and are collected in baskets. Grape bunches are clipped with scissors. Papaya is harvested by twisting the fruit till it snaps off. Pineapple is picked by giving a downward bend until the peduncle breaks off. Bunches of litchi and date are harvested by their stalks. Ber is harvested by shaking the trees. Hydraulically assisted fruit pickling platform could be used for picking fruits form the height of 6m. Mango and Bael fruits are being harvested very comfortably and economically by this machine at CISH, Lucknow.

Storage of Fruits and Vegetables

Storage losses of fresh fruits in India are high due to high temperature and humidity. Fruits and Vegetables are storage at low temperatures, immediately after harvest reduces the rate of respiration resulting in reduction of buildup of the respiration heat, thermal decomposition, and microbial spoilage and also helps in retention of quality and freshness for a longer period. The storage life is governed by several factors *e.g.* variety, stage of maturity, rate of cooling, storage temperature, relative humidity, rate of accumulation of CO_2, pre-packing, air-packing, air-distribution systems *etc.* The optimum refrigerated storage requirements for different fruits have been worked out *e.g.* 1.7-3.3 °C for apple, 12.8 °C for banana, 0-1.7 °C for grapes, 8.3-10 °C for guava, 10-15 °C for mango, 5.5-7.2 °C for orange and 8.3-10 °C for pineapple. The mango variety Dashehari can be stored 2 to 3 weeks at 12-14 °C and 85-90 per cent relative humidity, in cold storage conditions and Chousa, Langra and Amarpali are stored at 10 °C, 15 °C and 13 °C respectively (Singh, 2008).

Based on the principle of evaporative cooling, low cost, low energy cool chambers have also been developed. The chamber wall of the cool chamber is kept continuously wet by sprinkling of water for evaporative cooling to reduce the temperature by 17-18 °C even, during the peak summer in North India to enable storage of mango, ber and oranges. Storage of fruits in controlled or modified atmosphere is an important innovation. Combined with low temperature, it markedly retards respiratory activity and delays softening, senescence and changes in quality and colour of the stored fruits. The storage life of mango was only 6 to 10 days at ambient temperature but it increased to 3-6 weeks with gas storage at 7.5 per cent CO_2 level. Alphanso variety of mango could be kept for 35 days at 8.3 days at 8.3-10°C and raspuri mango for 40 days at 5.5-7.2°C. Fruits could also be packed in polythene bags which create a modified atmosphere with more CO_2 and less O_2 than in air for prolonged storage as shown in oranges and banana.

Cool stores play an important role in storage of fruits. Rising energy costs, mounting construction equipment and maintenance expenditure is impeding the growth of this industry. It is better to partially substitute the sophisticated high cost, energy intensive, refrigerated system with simple, low cost, energy efficient, evaporative cooling system for short term storage of fruits at farm level, pre cooling before transit and storage and also improving its use during transit. Although the development of cool storage facilities in India is very inadequate however, the significant numbers of pre-cooling units as well as cool stores have been established in grape and mango growing areas and the facilities are being extended to marketing yards. It is a matter of great satisfaction that refrigerated containers are used in transport and shipment of mango, grape *etc.*, particularly for export.

Grading

Systematic grading is a pre-requisite for efficient marketing systems as a well designed programme on grading and standardization brings about an overall improvement not only in the marketing system but also results in quality consciousness. In case of apple, mechanical grading has already been introduced in H.P., J&K and U.P. through packing centers which have been equipped with mechanical graders. Size grader for Nagpur mandarin is in use at the National Research Centre for Citrus, Nagpur. The mechanical sizing is done either on weight or dimension basis. The size grader can be adjusted according to size as per demand for a particular size. The mechanical grading of other fruits in India is still in the infancy and the grading equipments are either not available or have not been put into large scale commercial use in the field.

Packaging

The present packaging systems for fresh fruits in our country is unsuitable and unscientific. The use of traditional forms of packages like bamboo baskets is still prevalent. The other types of packages generally used are wooden boxes and gunny sacks. The use of corrugated fibre board boxes is limited. The use of baskets besides being unhygienic also does not allow adequate aeration and prevents convenient handling and stocking. Wooden packages are also not conducive for packaging of fresh fruits. Considering the long-term needs of eco-systems and to achieve an overall economy, other alternative available like corrugated fibre board boxes, corrugated polypro-pylene board boxes, plastic trays/crates/woven sacks, molded pulp trays/thermoformed plastic trays and stretch film and shrink wrapping would have to be considered. CISH has developed telescopic type C.F.B. boxes for internal and export markets of mango and guava having 0.5 per cent ventilation of surface area of the box.

Recent Advances in Post Harvest Technology

1. An expansion in the international fruit trade and growing consumer demand for high quality subtropical and tropical fruits have created considerable interest and investment in the development of new or improved post harvest technologies. This is evident in most developing countries, where ongoing research and development on post harvest technologies is given much emphasis.
2. In formulating such research and development programmes, priorities must first be set in terms of fruit types and research directed accordingly towards suitable technologies. Scientific understanding of biological and physiological process is important in studying the post harvest handling of any tropical fruit. Most tropical fruits tend to be climacteric (the marked and sudden rise in the respiratory rate of fruit just prior to full ripening),

with a limited post harvest life. Among the climacteric fruits the peak of respiration can occur before or after or coincide with the peak of ethylene production. This knowledge is important for commercial applications such as for the calculation of heat loads and cooling capacity, and the management of mixed loads.

3. Generally, the emphasis in post harvest technology is on reduction in the rate of ripening, reduction of mechanical injury to prevent pathogen entry, requirements of storage and suitable packaging for the export market.
4. There has been considerable work done on the reduction of ethylene biosynthesis which hastens ripening in fruit. Chemicals in the form of 1-aminvinyl glycine and 1-methylcycloprophene (MCP) have been used to reduce ethylene production, which slows down ripening and enhances shelf life for a number on tropical fruits, including banana. MCP was also found to work synergistically with enzymatically hydrolyzed chitosan solution to prevent pathogenic fungi growth and effectively prolonged the life of fruits by 80 to 100 per cent, even at room temperature. The use of gamma radiation has also been developed to slow down the ripening process and to disinfect and pasteurize the fruit surface. Techniques like waxing have also been used to reduce the ripening rate.
5. The other area which is being actively researched now is slow ripening by gene manipulation. This field of biotechnology is now the focus of much research to reduce the problem of losses due to poor shelf life of tropical fruits.
6. In storage under cold conditions, the use of modified atmosphere packaging (MAP) (Yumbye, *et al.*, 2014) has been used for temperate fruits. However, for tropical fruits, pineapples seem to respond well, but for mango results have been quite inconsistent. Perhaps more studies need to be done on the use of MAP using perforated plastic films for tropical fruits to ensure less chilling injury during export. It has been reported that chilling injury was reduced in Tommy Atkins and Keith mangoes stored in perforated bags at 12°C.

Conclusion

Post harvest handling is the final stage in the process of producing high quality fresh produce. Being able to maintain a level of freshness from the field to the dinner table presents many challenges. A grower, who can meet these challenges, will be able to expand his or her marketing opportunities and be better able to compete in the marketplace. Cool stores play an important role in storage of fruits. Rising energy costs, mounting construction equipment and maintenance expenditure is impeding the growth of this industry. It is better to partially substitute the sophisticated high cost, energy intensive, refrigerated system with simple, low cost, energy efficient, evaporative cooling system for short term storage of fruits at farm level, pre cooling before transit and storage and also improving its use during transit. The use of modified atmosphere packaging has been used for temperate fruits. However, for tropical fruits, pineapples seem to respond well, but for mango results have been quite inconsistent. Perhaps more studies need to be done on the use of MAP using perforated plastic films for tropical fruits to ensure less chilling injury during export.

REFERENCES

Akamine, E.K. and Goo, T. (1971). Respiration of gamma irradiated fresh fruit. *J. Food Sci.* 36: 1074.

Chattopadhayay, T.K. (2012). Storage of fruits. *A Textbook of Pomology* (1): 201-211.

Chattopadhayay, T.K., Paria, N.C. and Jana, S.C. (1989). *Scientific Fruit and Vegetable Preservation*. Modem book Agency Pvt. Ltd. Calcatta, India : 1-68

Kitinoja, L., Saran, S., Roy, S.K. and Kadere, A.A. (2011). Postharvest technology for developing countries: Challenges and opportunities in research, outreach and advocacy. *J. Food Sc. and Agril.* 91: 597-603.

Mohamed, S., Taufik, B. and Karim, M.N.A. (1996). Effect of Modified Atmosphere Packaging on the physiological characteristics of ciku at various storage temperatures. *Food Sc. and Agril.* 70: 231-240.

Singh, B.P. (2008). Advance in handling packaging and storage of sub tropical fruits. Winter school on hi-tech production of subtropical fruits: 247-251.

Singh, M.D. (2008). Advance in handling packaging and storage of sub tropical fruits. Winter school on hi-tech production of subtropical fruits: 228-230.

Singh, Vishal, Hedayetullah, Md., Zaman, Praveen and Meher, Jagamohan (2014). Post harvest technology of Fruits and vegetables: An overview. *J. Postharvest Tech.* 2(2): 124-135.

Wijewardane, N.A. (2014). Effect of precooling combined with exogenous oxalic acid application on storage quality of mango (*Mangifera indica*). *J. Postharvest Tech.* 2(1): 45-53.

Yumbye, P., Ambuko, J., Shibairo, S. and Owino, W. (2014). Effect of modified atmosphere packaging (MAP) on the shelf life and postharvest quality of purple passion fruit (*Passiflora edulis* Sims). *J. Postharvest Tech.* 2(1): 25-36.

Chapter 17

Processing and Packaging Technique for Nuts and Oilseeds

Manashi Das Purkayastha and Rimki Boruah

Introduction

Oilseeds have been the backbone of agricultural economy of India since long. Indian vegetable oil economy is the fourth largest in the world next to USA, China and Brazil. It is the second largest producer of groundnut after China and the third largest producer of rapeseed after China and Canada. Oilseed crops play the second important role in the Indian agricultural economy next to food grains in terms of area and production. The major oilseeds cultivated in our country are Groundnut, Rapeseed and Mustard, Castor seed, Sesame, Nigerseed, Linseed, Safflower, Sunflower and Soybean. However, Groundnut, Rapeseed/Mustard, Soybean and Sunflower account for a major chunk of the edible oil output, which contributes more than 70 per cent of domestic availability of vegetable oils in the country. Oil palm, coconut, cotton seed, rice bran, tree borne oilseeds and solvent extracted oil are the secondary source of vegetable oils that contributes about 30 per cent of total domestic availability of vegetable oils (Khalid *et al.*, 2017).

The oils and fats are composed of mixtures of triglycerides of various fatty acids. The fats and oils can be broadly classified in to edible and non edible classes. The edible oil is main source of fat taken in daily meals and is used for cooking purposes and salad dressings. The oils are sold and consumed as pure oil from a single oilseed plant or fruit plant, for example, olive, sunflower, groundnut, soybean, mustard or corn oils, or are marketed and used as blended oils, which are generally designated as edible, cooking, frying, table or salad oil (Swetman, 2008). Non-edible oils are used in soap industry, cosmetics, paint, varnishes and plasticizers (example, castor oil).

This chapter outlines the various unit operations involved in the post harvest technology of oilseeds and edible nuts. Some oilseeds have acquired great significance in the large-scale industrial production of edible oils. So, the processing of these individual oils and their products has been described briefly.

Edible Oil Processing

The commercial edible oil processing system is usually different from that performed by small-scale oil producers using traditional methods. In traditional or water assisted extraction of oil, the pre-ground, pre-heated oilseed or fruit flesh (for example, fresh coconut, olive, palm fruit, shea nut) is either mixed or boiled in water. The softened pulp is squeezed manually to release the oil-water emulsion. Salt can be added to break the emulsion and the oil that floats to the surface is skimmed off. However, the extraction efficiencies are low and labour requirements are high (Axtell and Noble, 2006; Khalid *et al.*, 2017).

Now-a-days, in a typical edible oil processing plant, oil is extracted from the seed first using mechanical extraction (expeller or press) followed by chemical extraction (hexane solvent). By using both methods less than 1 per cent of the oil is left in the meal. Figure 1 shows a simplified diagram of commercial oilseed processing.

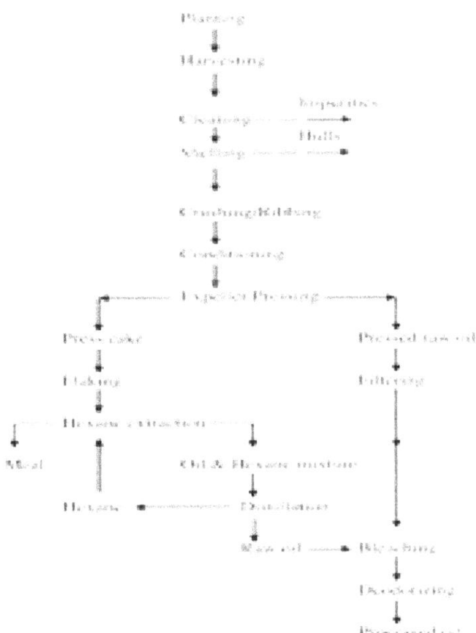

Figure 1: Schematic Showing the Stages in Commercial Edible Oilseed Processing.

Equipment Required for Oil Processing

The specific equipment required will depend on the particular crop being processed, the final oil quality required and the scale of operation. Basically, oilseeds pre-treatment production line includes two parts, one is oilseed pre-treatment (prepress) line, and the other is oilseed press machine (or expeller). Oilseeds pre-treatment part is used for cleaning and adjusting texture of the materials, improving oil yield rate, *etc.* It includes magnetic drum, cleaning sieves, sheller, crusher, flaking roll, cooker, softening pot, puffing machine or expander, *etc.* (Dunford, 2008a; Dunford, 2008b). Table 1 gives a list of different equipments used for pre-treatment of oilseeds, along with their specific roles.

Table 1: Devices Used in Oilseeds Pre-treatment Production Line

Sl.No.	*Equipment*	*Function(s)*
1	Destoner and Magnetic drum separator	It uses the magnetic power of Plate magnets, drum magnets or electromagnets to remove the metal impurities, such as the iron nail and small steel from the harvest. Seeds need to be separated from stones by gravity. Special "destoners" are available to remove stones and mud balls.
2	Cleaning sieves	Foreign materials in seeds, namely mud, dust, leaves and other contaminants, are typically separated out by a combination of rotating or vibrating coarse screens, reels, aspiration and different suction system. This process is commonly referred to as scalping. Any mouldy nuts should be discarded as these can cause aflatoxin poisoning.
3	Teeth roll sheller unit	Some raw materials (for example groundnuts, sunflower seeds) need dehusking (or decortication). The oilseeds are passed between a pair of toothed roller to break open their shells or hull. Knife, disk and impact type dehullers are widely used. This step mainly reduces wear in the expeller as the husks are abrasive. Dehulling reduces fiber and increases protein content of the meal.
4	Crusher	Most oilseeds (eg copra, palm kernels and groundnuts) need grinding in mills before oil extraction. Crusher or grinder reduces the big size kernel to several pieces by pounding action. This is a preparatory step required prior to flaking. A cracking mill consists of two sets of cylindrical corrugated rolls in series. The rolls rotate at differential speeds to break apart seed cells containing oil.
5	Flaking roll	This assists in rupturing the oil bearing cells by making the kernel into thin flakes (0.01-0.015 inch or 0.25-0.37 mm) for better absorption of moisture and heat during cooking and efficient solvent extraction process. A flaking mill has two large diameter rolls turning in opposite direction and forced together by hydraulic cylinders. As the seeds are pulled through the flaking mill, they are stretched and flattened.
6	Cooker or Softening pot	Its main function is wet-heating of the ground oilseeds or their flakes, and is mostly required for canola, soybean and groundnut processing. Oilseeds are conditioned by live steam to 80-90°C using a seed scorcher, which in-turn inactivates most of the enzymes and decreases the viscosity of the oil, allowing it to flow more freely during pressing.
7	Bulking machine	It is mainly used for processing rice bran by applying high temperature and pressure. During this process, the materials are puffed; the internal moisture is evaporated quickly and formed into loose grainy structure. This aids in solvent extraction of oil from the rice bran.
8	Expander (optional)	The dry flake can also be transferred to the Expanders (extruders) where water and steam is introduced into the product and mixed in the chambers of the extruder. The wet cake is then extruded from the expander at approximately 104ºC, sized and cut via a die head. The expander collets are discharged through die openings, which undergo flash evaporation of water, and therefore, the resulting collets have highly porous texture. The expanding process is used when increased bed drainage or percolation is desired in the solvent extraction process.

Sl.No.	Equipment	Function(s)
9	Plate type dryer	All oil-bearing materials need to have the correct moisture content to maximise the oil yield. After puffing, the moisture content is about 11 per cent-13 per cent, which is not suitable for further solvent extraction. So moisture is reduced to 7 per cent-9 per cent using dryer (when the mixture stops sticking together and forms free flowing flour).

Mechanical Expression

Expression is the process of mechanically pressing liquid out of liquid containing solids. After oilseed pre-treatment line, the materials are sent to various types of compression devices/equipments, such as ghanis, oil-press and oil-expeller, to obtain crude oil. The former two machines have batch mode of operation; while expelling is a continuous method and is employed in large-scale industries.

The use of ghanis is prevalent in India and other Asian countries. A ghani consists of a large wooden mortar fixed to the ground and a pestle rotated either by animal traction or (more commonly) a motor (Figure 2). Seed is fed slowly into the mortar and the pressure exerted by the pestle against the side of the mortar breaks the seeds and releases the oil. Oil runs through a hole at the bottom of the mortar where it is collected (Axtell and Noble, 2006).

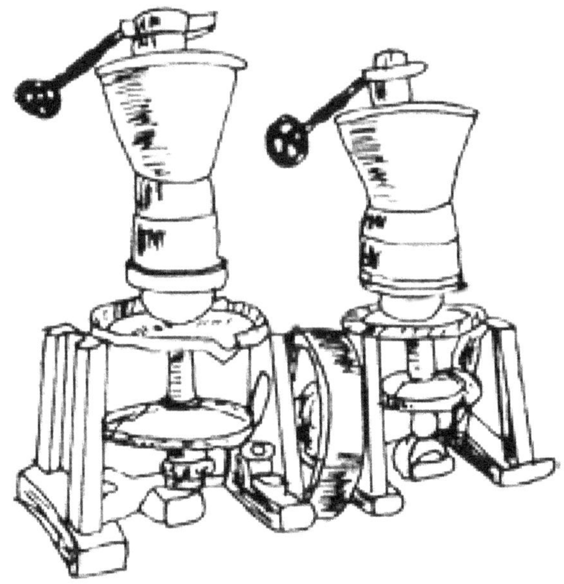

Figure 2: Motor-Operated Ghanis.
Adapted from Shukla *et al.* (1992).

There are many types of oil press (such as Spindle press; Bridge press, also known as a screw press; Ram press; Hydraulic press), but all work on a similar principle. Raw materials are placed in a heavy perforated or slotted cage and a metal plunger is used to press out the oil. Great pressure (5-28 MPa for 15-20 min) is exerted on the oilseed fed through the plunger to squeeze out oil. Though oilseeds and nuts are usually processed "cold" (*i.e.*, without additional heating during expression) in India (temperature below 49°C), the seeds are heated by friction (Axtell and Noble, 2006). Oil flows through the perforations in the casing and is collected underneath. The main difference in the design is the method used to move the plunger, which can either be a screw thread (in bridge press) or a hydraulic jack/pump (in hydraulic press). Bridge press consists of a motor driven screw which rotates horizontally inside a perforated barrel-shaped cage. Hydraulic types are more expensive, need more maintenance, and risk contaminating oil with poisonous hydraulic fluid. Ram presses use a long pivoted lever mechanism which moves a piston backward and forward inside a cylindrical metal cage. An adjustable pressure cone controls the width of the gap in such a way that it produces high pressure on the piston, which in-turn compresses the seed in the cage to liberate oil. The ram press has a low seed throughput (Dunford, 2008b; Moore, 2002; Pighinelli and Gambetta, 2012).

An expeller consists of a helical thread or screw (worm assembly) which rotates concentrically within a perforated cylinder (the cage or barrel). The barrel is usually formed by a series of axially-placed lining bars contained within a robust frame (Figure 3). Pre-treated oilseeds enter one end of the barrel through the feed inlet and are conveyed by the rotating worm assembly to the adjustable discharge end called choke. The screw grinds, crushes and presses out the oil as the seeds pass through the barrel, with gradually increasing pressure, while the cake is contained within the barrel. There are two distinct expeller designs – a single cylinder press that expels the oilcake out in pellet form and a cage-style screw press that expels the meal out in large flakes (Dunford, 2008b; Moore, 2002).

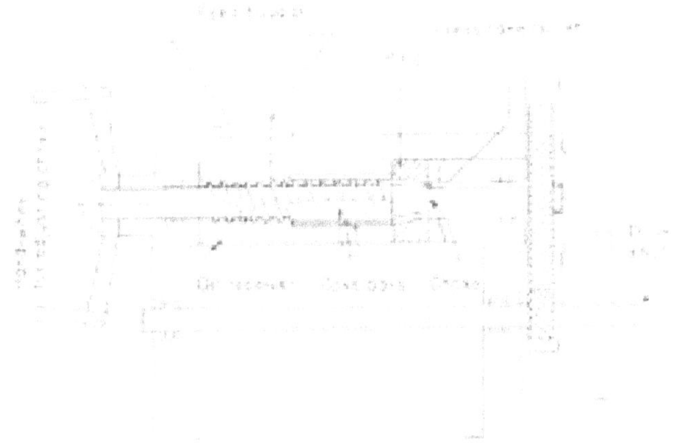

Figure 3: Pictorial Representation of a Screw Press. Adapted from Kurki *et al.* (2008).

Solvent Extraction

The residual oil in the resultant oilcake is 4–7 per cent. The flaking step provides the enlarged surface area needed for efficient solvent extraction. The extraction is performed using petroleum ether or n-hexane (boiling point 60–70 °C) as solvent. It is a diffusion operation in which the solvent selectively dissolves miscible components (oil) from other substances. Immersion and Percolation are the most common types of extraction though Percolation extraction is more common in modern process plants. The Percolation method utilizes a bed composed of oilseed flakes and counter-current flow of the flakes and miscella (the oil-solvent mixture). Three types of extractors are commonly found in the industry: 1) Rotary or Deep Bed, 2) Horizontal Belt, and 3) Continuous Loop. A large percentage of the modern process plants utilize the Horizontal Continuous Loop though some of the older facilities still rely on the Rotary or Deep Bed technology. The Horizontal Continuous Loop units utilize a conveyor system which moves the miscella through different stages of solvent introduction and screening, and finally a product dump cell. The Rotary Deep Bed units consist of a series of concentrically arranged cells, which are filled with oil-bearing material. Each cell is filled consecutively and brought in contact with decreasingly concentrated miscella. The freshly charged solids are brought into contact first with the most concentrated miscella and the fully extracted solids contact fresh solvent before discharging (Dunford, 2008b; Hamm *et al.*, 2013; Moore, 2002).

Solvent removal from the miscella is achieved by distillation under vacuum, which removes 85 per cent solvent. This is followed by stripping with steam for removing the remaining solvent in miscella, and the solvent is discharged to condenser for cooling and future recycling. The maximum amount of solvent remaining in the oil is <0.1 per cent. The oil-free meals are then steamed (desolventizing) and then dry heated (toasting) to remove the residual solvent. Majority of the cake and meal are incorporated in animal feed rations. This process is only carried out in large scale production plants and is not suitable for small or medium scale entrepreneurs, as it requires substantial capital cost and there is also a health and safety risk from using inflammable solvents. Where large amounts of oilseed cake are available, solvent extraction becomes a commercially-viable option to extract the residual oil left in the cake (Dunford, 2008b; Moore, 2002).

Clarification and Filtration

The crude expelled oil now contains a range of contaminants including fine pulp and fibre from the plant material, which is termed foots. Foots are collected on a sieve or filtered though a fine filter cloth under pressure, which are continuously scraped and re-processed in the expeller (Dunford, 2008a). Crude oil also contains small quantities of water, resins, colors and bacteria which make it dark in colour and opaque. These materials are removed by clarification – either by letting the oil stands undisturbed for a few days and then siphoning off the upper layer, or by using a clarifier. A clarifier consists of a holding tank/drum, wherein the oil is heated briefly to 100°C to drive off traces of water and destroy naturally occurring enzymes and contaminating bacteria (Kurki *et al.*, 2008).

Refining (RBD Process)

The oil then goes for additional refining stages of degumming, neutralizing, discoloring and deodorizing to create a final product that is consistent in taste, colour and stability. As a result, edible oils purchased in stores are generally tasteless, odorless, and colorless regardless of the original oilseed type or quality, and are known as "RBD" oils (Refined, Bleached and Deodorized). The RBD process is carried out in large-scale operations, which confers commercial grade specifications to the oil (marketable for cooking and frying purposes). RBD technique removes the components (water, transition metals (iron, copper), polar lipids, chlorophyll) that increase the rate of oil oxidation (rancidity) especially during prolonged storage and frying. However, RBD is not always beneficial to nutrition; crude oil contains more vitamin E, trace elements, phytosterols, which are removed unintentionally, and the net oil loss is ~3 per cent-5 per cent during the refining process. As such, after processing an anti-oxidant is added to the processed oil to replace compounds that were removed during processing. Thus, the trade-off is that crude oil is healthier but less stable. These refining stages are as follows:

Degumming

Degumming removes phosphatides coextracted with the oil, which tend to separate from the oil as a sludge during storage. The phosphatide content of crude oil varies, but is usually in the order of 1.25 per cent, or measured as phosphorus, 500 ppm. Two degumming methods are in use: (a) using water to precipitate phosphatides and; (b) using an organic acid such as 2-3 per cent aqueous citric, malic, or phosphoric acid solution (super-degumming). After mixing the heated oil (80°C) with these reagents, the oil is centrifuged, filtered or settled to separate the precipitated gummy residue. Once removed, these by-products are added to the meal fraction to make it more nutritious product for animal feed (Hamm *et al.*, 2013).

The most recent development in degumming uses both acid and aqueous sodium hydroxide, rather than acid and water, especially with lower quality oils. Phosphatides as well as some of the other impurities are removed and, if sufficient alkali is applied to saponify the free fatty acids, fully refined oil is obtainable (Moore, 2002).

Neutralization (Alkali Refining)

Degummed oil is further purified by alkali refining (especially for water degummed oil). Alkali refining is the most common process used, wherein the oil is first contacted with about 0.05-0.1 per cent of concentrated phosphoric acid in a high intensity mixer to help precipitate phosphatides. It is then contacted with an approximately 12 per cent aqueous solution of sodium hydroxide (caustic soda) to neutralize the free fatty acids of crude oil and any excess phosphoric acid (if present), and to precipitate phosphatides. Sodium hydroxide solution not only neutralizes free fatty acid, but also can act with other substances like protein, mucus, phospholipids and pigment. Temperatures and contact times may vary from about 90°C for few seconds (short-mix process) to about 40°C for 15 minutes (long-mix process). The oil-soap mixture is then heated to about 90°C and is filtered or centrifuged to separate the aqueous soap phase. The resulting soap-stock also contains the precipitated phosphatides and some triglyceride oil. The centrifuged oil must be further contacted with 5-10 per cent hot water to reduce soap level from about 400 ppm to <50 ppm. This is also called stripping, which is a similar process to de-odorizing. Free fatty acids are reduced to <0.05 per cent and phospho-

<2 ppm (Hamm *et al.*, 2013). The soap phase can be acidulated and used as a feed ingredient (Moore, 2002).

Bleaching (Decolonization)

Some oils have a very dark colour (*e.g.* rapeseed oil, rice bran oil, *etc.*) which can be unpopular with consumers. The appearance of the oil can be lightened by bleaching. Degummed oil having phosphorus content below 50 ppm is mixed with bleaching clay (1-3 per cent) and is heated to a high temperature (90-110°C) in absence of oxygen. About 5-30 min of contact time is given, while the oil/clay slurry is progressively dried to about 0.1 per cent moisture content (Hamm *et al.*, 2013). Bleaching clay may be either natural clay or activated with an acid wash. Filtering or centrifuging removes the clay particles, to which the colorant compounds (chlorophylloids and carotenoids) and other contaminants (such as soap, trace metals and sulphur compounds) are adsorbed. It delivers bleached oil ready for deodorizing and chlorophyll is reduced to the concentration required for bleached oil (<25 ppb). This level is innocuous in respect to oxidation and colour of the oil (Moore, 2002).

Deodorizing

This is a distillation process during which volatile compounds that produce unacceptable flavor and odors (including free fatty acids) are eliminated through the process of sparging, *i.e.* bubbling steam (2-4 per cent) through the oil, under a vacuum. The oil, which must be thoroughly purified in the previous process steps, is heated to 260-265°C under very high vacuum (2-4mm Hg pressure) to exclude air (Moore, 2002). This increases the volatility of the compounds to be removed. High temperature and low pressure of the deodorization tower or pot also dehydrates the residual moisture (~0.5 per cent) in the refined oil (Hamm *et al.*, 2013). Deodorizing can also be achieved by treatment with activated charcoal. Because deodorization is the last process normally carried out on edible oils, this step may be delayed until other processes, such as hydrogenation of the oil, have been done (Kurki *et al.*, 2008).

Winterization (Dewaxing)

Winterization is needed to remove the wax in the processing of rice bran, canola, sunflower and corn germ oils. Over time glycerides (wax) can degrade, releasing fatty acids into the oil increasing the acidity levels and reducing the quality. In some cases, it is desirable to dewax the oil to avoid a hazy appearance. The process is carried out by chilling oil in a continuous heat exchanger to 5°C and metering about 0.1 per cent of a filter aid into the chilled oil stream on the way to a filter. This reduces the wax content to <50 ppm, which no longer produces a visible haze (Hamm *et al.*, 2013). The oil may also be allowed to stand for a definite duration at a series of low temperatures, so that naturally occurring glycerides (with higher melting points) solidify and can be removed from the oil by filtering thereafter. This step is also carried out with palm oil to separate the oil into two separate products: a solid fat (stearin) and a liquid oil (olein) (Moore, 2002).

Hydrogenation

This process changes the melting behaviour of oils and improves the oxidative stability. Hydrogen is added to the double bonds of unsaturated fatty acids at temperatures of 160-200°C and pressures of 100-300 kPa in the presence of a nickel catalyst to facilitate the reaction. Chlorophyll, phosphatides, soaps, and especially sulphur compounds "poison" the catalyst, so these compounds must be removed prior to hydrogenation. It is possible to hydrogenate fatty acids with multiple double bonds preferentially. These are the most easily oxidized and their concentration must, therefore, be reduced or eliminated for best effect on stability. Melting behaviour is most usually evaluated by determining the proportion of solid fat in a sample over the temperature range of interest, usually 10-40°C (Hamm *et al.*, 2013). Another consequence of hydrogenation is the formation of isomeric fatty acids. Positional isomers as well as geometric, or *trans* isomers are formed. The role of *trans* isomer fatty acids in the diet will continue to be controversial for some time. It is applied to several oils such as cottonseed, marine oil, soybean and canola. Very highly hydrogenated oils are rarely used in margarine and baking shortenings, because of the tendency to form large β-crystals over time. This impairs eating properties and baking performance (Moore, 2002).

Inter Esterification

Interesterification is another process for changing the melting properties of fats and oils. The position of fatty acids in the triglyceride is changed randomly. When the process is applied to blends of different oils, triglycerides with fatty acid compositions are formed that can be sufficiently different from the original oils to change the melting characteristics of the resulting fat in a way not achievable by mere blending. The process represents an alternative to the use of partially hydrogenated fats to manufacture products of a variety of melting properties (Moore, 2002). In modern plants, the process is carried out in a heated, agitated tank, equipped with a vacuum system and a water spray. The oil must be very dry and low in free fatty acids. Sodium methoxide (a strong base, 0.05 per cent) is added to the heated oil (150°C) as a powder under agitation and contacted for about 30 min to catalyze the reaction. The oil is then cooled to about 90°C, phosphoric acid is added to neutralize the base. The oil is then washed, or alternatively, bleached (Hamm *et al.*, 2013).

Supercritical Fluid Extraction Technique

An innovative and modern extraction technique is Supercritical Fluid Extraction (SFE) which is considered to be an environment friendly. A fluid is considered supercritical when it presents diffusivity similar to gas and density comparable to liquids (Dunford, 2008b).

In supercritical carbon dioxide ($SCCO_2$) technique, carbon dioxide is utilized above its critical pressure (7.3 MPa) and temperature (31°C) as solvent. The unique advantage of $SC\text{-}CO_2$ is the easy removal of solvent from the extract. When pressure is released from the system, carbon dioxide returns to the gas phase and oil precipitates out from CO_2-oil mixture. Carbon dioxide is recycled; hence, CO2 released from the system is not an environmental issue and is also non-toxic to human health (Pighinelli and Gambetta, 2012).

The pictorial representation of this system is presented in Figure 4. A diaphragm compressor is used to compress the CO_2 to the desired pressure. Grains are placed in the preheated vessel and then the extraction with CO_2 begins at a rate of 40 g/min for 3 h. The extraction is held at 50°C and pressure of 30MPa. Extracts are collected in another vial attached to the depressurization valve and cooled in a bath maintained at 0°C. $SCCO_2$ under pressure has a higher density resulting in higher dissolvability of oils (Pighinelli and Gambetta, 2012).

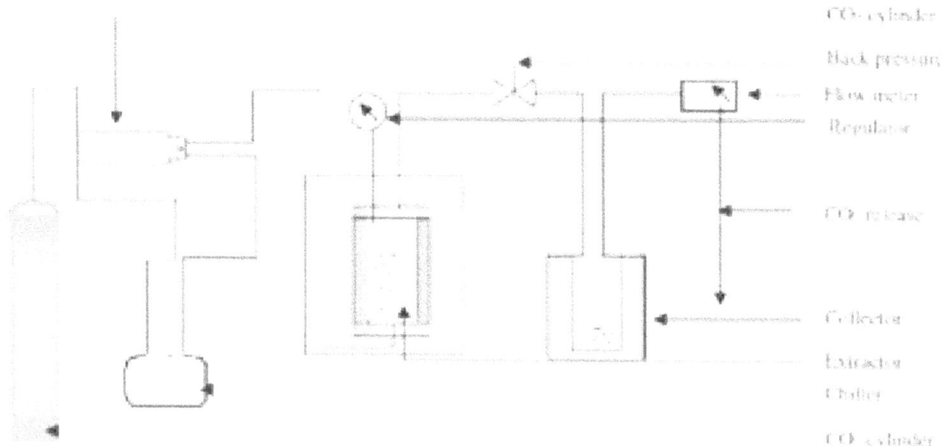

Figure 4: Pictorial Representation of a Supercritical Carbon Dioxide Technique.

The use of supercritical CO_2 for the extraction of oilseeds such as soybean, cottonseed, corn germ, rapeseed, safflower and sunflower has been the choice for the majority of edible applications. The oil yields were 38.8 per cent to solvent extraction, 35.3 per cent for the supercritical extraction and 25.5 per cent for expelling. In terms of extracting the components like omega-6-fatty acid and omega-3-fatty acid, supercritical extraction was more efficient than solvent and expelling extractions. The oils obtained with $SCCO_2$ were shown to have similar quality compared with hexane extracts; they also had lighter colour and lower iron and phospholipid content, resulting in a lower refining loss and reduction of subsequent refining steps (Dunford, 2008b). Moreover, lower temperature and smaller particle size favors the extraction efficiency. This has an added advantage of producing lesser free fatty acids as compared to traditional oil extraction methods. Another advantage is related to protein quality of the extracted meal that can be comparable with that of hexane-extracted meals.

Despite the many advantages of $SCCO_2$ extraction and also the high volume of research carried out, commercial-scale $SCCO_2$ extraction of oilseeds is not readily accepted. The process based on $SCCO_2$ extraction is simpler than conventional hexane extraction in terms of eliminating the need for hexane evaporators and meal desolventizer, but $SCCO_2$ process has some disadvantages such as high equipment costs and the inability to achieve continuous processing of high volumes of oilseeds under SCF conditions. Those two reasons are considered major impediments to commercialization of the $SCCO_2$ process (Pighinelli and Gambetta, 2012).

Enzymatic Oil Extraction Method

Aqueous enzymatic extraction is also a safe and environmental friendly technique. The basic principle in this technique is to break the cell wall of plant materials with the help of enzymes. In this technique low temperature and low pH conditions are used to extract the oil. Various types of enzymes are used like pectinase, hemicellulase, -amylase and many others. It is often accomplished using ultrasonication or high speed homogenization. The ultrasonic extraction has several advantages over tradition methods such as lower extraction temperatures, shorter processing time, decreased solvent quantities and improved oil yield (Dunford, 2008b).

Oils of Plant Origin (Seed Oils and Fruit Pulp Oils)

With regard to the processes used to recover plant oils, it is practical to divide them into fruit pulp and seed oils. While only two fruits are of economic importance in oil production (olive oil and palm oil), the number of oilseed sources is enormous.

Soybean

Soybean is today the most widely cultivated oilseed in the world. Soybean contains about 40 per cent protein and 20 per cent oil. Soy protein is the most economical protein produced from the soy meal.

Soya Oil

Conventionally soybeans are dried in the field, as the pods are self-opening at maturity. Once the 'pod filling' stage of the standing crop is reached, reduction in moisture content is very fast. For minimum loss, the crop needs to be harvested at 15-18 per cent moisture level (wb). Heap drying of soybean is not recommended as the grains get infested with fungus (1-1.5°/s) arid germination also gets reduced (92–71 per cent). For mechanical drying of soybean and its products, continuous flow heated sand medium drier, tray type natural convection drier, modified natural convection drier and multi-purpose driers have been developed. Traditional approach of threshing *i.e.*, crushing of soybean crops with bullock or a tractor is not suitable. Fines, pods, sticks, stones and other impurities must be removed. This is particularly true when producing high-protein meal. A number of threshers (multicrop thresher, spike tooth thresher) and cleaner-cum-graders (Pedal Operated Air Screen Grain Cleaner, Pedal cum-power operated cleaner, Hand Operated Double Screen Grain Cleaner) have been developed (Shukla *et al.*, 1992).

Cleaning is carried out on a multi-deck screener. The cleaner is composed of screens which has opening sizes that are usually designed to segregate three fractions: (a) the oversized impurities or pods; foreign materials, such as leaves, sticks, most stones and so on (b) The clean fraction, which passes through the first screen, and is also subject to aspiration through a current of air, for removing light particles, mainly loose hulls, and dust (c) the small fraction – fines and sand – passes through the bottom screen. It is prudent to employ destoners, which might be a gravity table that uses the difference in density to separate heavy stones from lighter seeds. A simpler way of separating stones is to use a current of air, which carries seeds but not stones (Shukla *et al.*, 1992).

Cracking followed by dehulling is the next operation where grain splits in to two halves and the husk and germ are separated to produce clean whole soybean dal. Clean soybeans are cracked into pieces in cracking mills. These are generally equipped with two pairs of corrugated rolls: the top pair breaks the beans into halves or quarters, the bottom pair into quarters or eighths. Each mill is equipped with a feeder, which distributes the beans uniformly over the lengths of the rolls, at a constant and adjustable rate. It is essential that soybean hull which is about 7–8 per cent of the bean weight, is removed. For this, soybean is dried (at 65–70°C), ideally to 9.5-10.5 per cent moisture. It is then stored for some time (12-48 h). This is called tempering, during which duration the dry hull loosens itself from the kernel. Simple mechanism of rubbing the soybean seed/grain between two surfaces can detach the hull, followed by aspirating the lighter hulls with a current of air (Dunford, 2008a). In India, several types of dehullers have been developed for soybean: rotor concave type, hand grinder, manually operated, power operated and cylinder-concave type dehullers/decorticator.

The cracked soybeans or grits proceed to the cooker–conditioner. The purpose of cooking–conditioning is to heat the grits to 60–70 °C for 10-15 minutes or by application of steam to wet beans and soften them. Raw soybean contains some anti-nutritional factors which could be inactivated/eliminated by this wet-heat treatment. A slight moisture adjustment, to 10–11 per cent, can also be made for good flaking. Two types of cooker are used by the industry. The vertical stack cooker consists of a succession of horizontal steam-heated cooking pans stacked in a vertical shell. Each successive compartment is equipped with agitating arms, connected to a central rotating shaft. The grits enter at the top and are swept successively over each cooking pan, from top to bottom. Each pan has an opening which allows the cracks to flow from one stage to the next. Large-capacity soybean plants tend to use horizontal rotary cookers instead of a vertical stack. These are made of a cylindrical shell housing parallel steam-heated tubes over their entire length. The grits are continuously introduced into the cylinder at one end of the machine and gradually progress to the discharge at the opposite end after a definite residence time (Hamm *et al.*, 2013). In some models, the tubes rotate inside the shell; in others the entire shell rotates along with the tubes. The rotation of the tubes continually lifts the material, with the discharge end being lower than the feed end, so that the material advances every time it drops (Figure 5).

The hot, soft cracks leaving the conditioner are finally sent to a heavy roller flaking machine. It is equipped with one pair of large-diameter smooth rolls, which laminate the grits into flakes. Oil cells are weakened in the process, and the oil becomes accessible to the solvent in the extractor. Optimum flake thickness varies between 0.2-0.38 mm. High roll pressure is necessary to form flakes of uniform thickness. This pressure is applied by means of a hydraulic system. The feeder

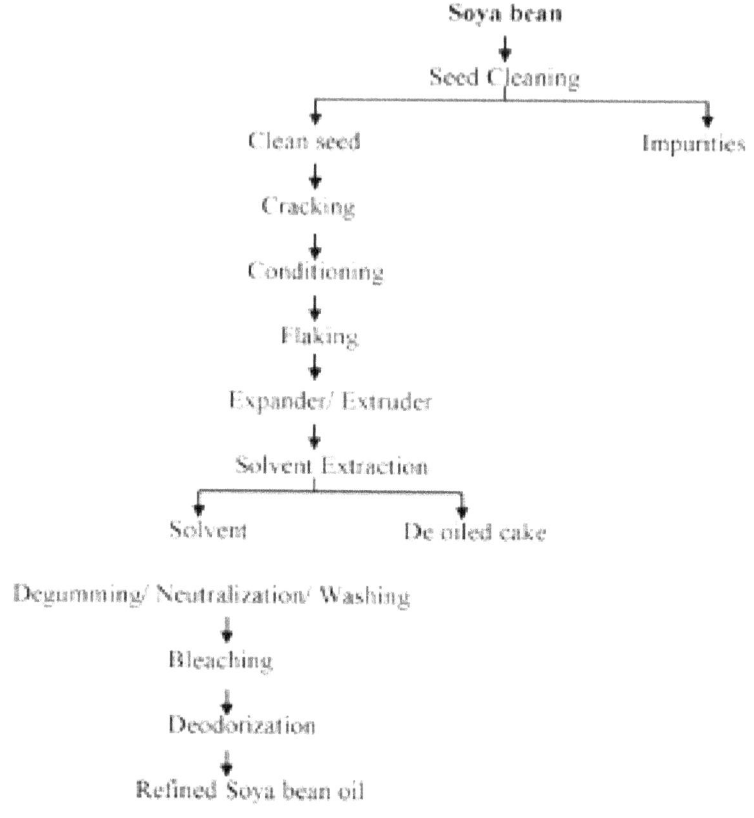

Figure 5: Process Flowchart for the Production of Soya Oil. Adapted from Shukla *et al.* (1992).

distributes the material uniformly over the length of the rolls, at a constant and adjustable rate. The surface moisture of the flakes is reduced to 7-8 per cent. In case of higher moisture levels (above 9 per cent), the oil recovery goes down mainly because of the plasticizing effect of the soy-meal in the screw barrel assembly which causes poor compression (Hamm *et al.*, 2013). These soy-flakes are either sent for solvent extraction or can be directly consumed in daily diet to increase protein content.

The expander is an optional additional step in soybean preparation. It is an extrusion cooking technique which consists in heating the flakes in a few seconds by mixing with live steam. The equipment consists of a horizontal barrel through which the soybean is pushed forward by means of a rotating worm/screw assembly (Hamm *et al.*, 2013). Live steam is introduced inside the barrel and mixed with the product, raising its temperature and moisture, as well as the pressure inside the barrel. Then the material is pushed through either a die plate with several orifices or a hydraulic cone. Due to the high pressure reached inside the equipment, an expansion phenomenon of the product and a flash evaporation of the water in the product take place at the outlet. As a result, the product has a 'sponge'-like texture. The expander has following advantages:

- ☆ Increases the bulk density of the material in the extractor.
- ☆ Produces a higher percolation in the extractor.
- ☆ Reduces the solvent retention of the extracted material entering the meal desolventizer.
- ☆ Enables the same residual oil content to be achieved after extraction, with higher full miscella concentration.

Soybean oil is conventionally recovered by solvent extraction. Dehulled, rolled, steam conditioned flakes are conventionally subjected to direct solvent extraction which recovers about 99 per cent of the oil. Mechanical deoiling of soybean using hydraulic press and/or screw-press has been applied as a preparatory step prior to solvent extraction due to inability of mechanical process to remove major oil fraction from seeds. Extruding and pouring the hot extrudate (expanded soyabean) in the expeller or screw press results in 70-80 per cent oil recovery and a bland, light golden edible cake (for the production of soy-flour).

Soy Flour

The standard process for producing soy flour for human consumption is shown in Figure 6. Milling (size-reduction) is done by stone burr mill and hammer mill, and then ground to pass through a 200 mesh screen. The flour produced by stone burr mills are used in conventional dishes whereas fine grade flour produced from hammer mills may be used for bakery purposes. In addition, soybean is sometime wet-milled into paste to prepare various products such as soybadi, snacks, soy paneer, *etc.*, for which multipurpose grain mill, screw type wet grinder, plate type wet grinder and deoiled cake grinder (bar-type hammer mill) are used (Shukla *et al.*, 1992).

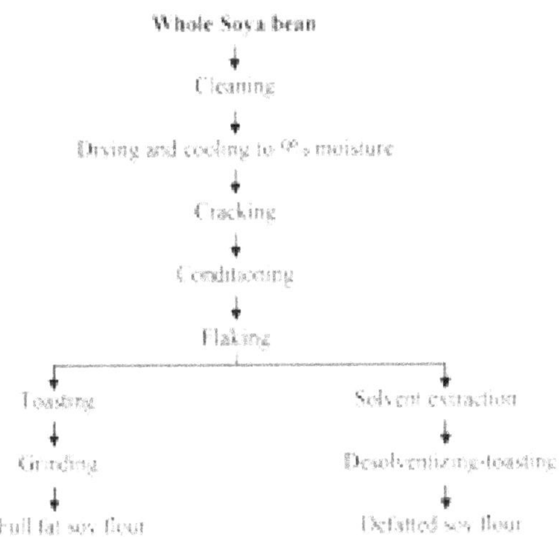

Figure 6: Process Flowchart for the Production of Full-Fat and Defatted Soy Flour. Adapted From Shukla *et al.* (1992).

Soy-Protein Concentrates and Isolates

Soy-protein concentrates and isolates are produced from defatted flour, by extracting the proteins with an aqueous medium or buffer (pH ≥ 7). The aqueous extract is separated from the fibrous residue by centrifugation. The pH of the clarified extract is then adjusted to about 4.5 with food grade acid to precipitate the proteins at their isoelectric points. The protein curd is concentrated and washed with water. The concentrated protein can be dried as such or neutralized with food grade alkali and dried. This process usually yields soy protein concentrate, containing not less than 70 per cent protein (Shukla *et al.*, 1992). If the protein curd is purified with ultrafiltration or nanofiltration, removing the low molecular weight carbohydrates, minerals and other major constituents, the resulting product is called soy protein isolate, which contains more than 90 per cent protein.

Soy Milk and Soy Paneer

There are several processes for producing soy milk, all of which aim at heat treating the soybeans to inactivate trypsin inhibitors and lipoxygenase, tenderization of beans, maceration (seed to water ratio=1:10) and straining out a white colloidal solution, which mimics the appearance of milk. This white watery colloidal dispersion of soybean juice, is called soy-milk (6 per cent of total solids), has a smooth mouthfeel and characteristic 'beany' flavour. The solid residue left after soy-milk extraction is called okra, having high total solids (55 per cent) and proteins (62.7 per cent), is used in bakery products and composite flours (Shukla *et al.*, 1992).

Coagulation of soy milk yields a white, soft gelatinous mass which has bland taste and unique body and texture resembling paneer obtained from milk in appearence as well as physico-chemical characteristics. This is known as soy-paneer or tofu. Citric acid as coagulant gives the maximum yield at 75°C with 74 per cent moisture, 15.5 per cent protein and 3.9 per cent fat. The product possesses fragile texture (Shukla *et al.*, 1992).

Extruded Soy Products

Extrusion processing of soybean, with or without blending with other cereals and legumes, produces several products such as soy-cereal based weaning food, texturized soy products *etc.* These products have good flavour, oxidative stability arid high nutritive value. These can be consumed as soy-fortified snacks/flakes or curry additives (Shukla *et al.*, 1992).

Peanut (Groundnut)

Groundnut is a major oilseed of India. It yields 70 per cent kernel and 28 per cent oil. The oil can be used for cooking, they can be used as a shortening or as a base for confectioneries and they can be used to make peanut butter. Oil contains high amounts of energy and fat-soluble vitamins (A, D, E, and K) and essential fatty acids. The oil content of the kernels is between 45-55 per cent (FAO, 2007b).

Groundnut Oil

The process of removing groundnut pod from the plants or haulms is known as stripping. The most common method of stripping is to pull out the pods from the plants manually. Two types of manually operated strippers namely; comb type and drum type, are available. In dryer, moisture passes quickly from kernels to pods. So, over-drying of pods must be avoided, as over-heated kernels are extremely brittle, can change colour and lose their flavour. Drying to 8 per cent or less is critical to prevent the growth of *Aspergillus flavus* (FAO, 2007a). In case of mechanical driers, portable trailer bins, radial drying bins and vertical flow bins are all suitable for unshelled groundnuts. In India, groundnut is traditionally dried in sun which requires 48-72 h for reducing the moisture content of groundnut pods from 26 per cent to 9-7 per cent (Shukla *et al.*, 1992). Continuous flow dryers are normally suitable for larger producers. The best drying temperature for groundnut is 36-58°C (Hamm *et al.*, 2013). Dried groundnuts should be stored in their shells, if intending to store for a long time. Good storage facilities are needed to reduce losses; a temperature of 2-4°C with a relative humidity of 65 per cent allows storage for atleast 2 years with little loss of viability. Small amount of shelled dried seed, treated with some insecticide could be successfully stored for one year without significant loss of viability in laminated polyethylene bags. Next, decortication of pods is done by either traditional hand shelling or using decorticators (with feeder-cum–separator attachment) for removal and separation of the kernels; among which a motorized rubber tire groundnut sheller has been found to be the most efficient. Layered sieves, having a speed of 250 rpm with 50 mm stroke, are used for grading of kernels (FAO, 2007b). The groundnuts are then sorted and graded to check for damaged, shrivelled and moldy seeds which may carry aflatoxin. Moldy and infected nuts must be removed and destroyed (FAO, 2007a). Only graded nuts are dried, crushed and sent for oil expression. There are two ways to crush the nuts - traditional motor operated ghanis or using a roller mill (Shukla *et al.*, 1992). The roller mill consists of two mild steel rollers mounted in a frame. A seed hopper containing the seed is mounted above the rollers. A chute takes the crushed seed from the rollers to the collecting bag. One of the metal rollers is fixed while the position of the second roller can be adjusted so that the gap between the two rollers can be adjusted according to the size of the seed being milled. The roller gap can easily be adjusted by inserting mild steel plates or spacers between the rollers during adjustment. Groundnut crushing is a two stage operation; first at a 2mm gap and then at a 1mm roller gap to produce a fine flour (Hamm *et al.*, 2013). Different varieties of groundnut may be larger than this and the roller settings may need further adjustment to get the best performance. Water is added to the flour to assist in rupturing the oil-bearing cells and also to assure even heating of the groundnut flour. With groundnuts it is usual to add 10 per cent water by weight. The flour is transferred into the heating pans, to assists in breaking the oil cell walls and to decrease the viscosity of oil. The time of heating depends on the temperature of flour and its water content. The mixture should be heated for about 10-15 minutes to reduce the water content and the temperature is raised to about 90°C (FAO, 2007b). When the flour and water mixture is initially heated, the mixture is sticky and stick together to form a lump or ball, when pressed, indicating the presence of moisture in the flour. As heating continues, it can be noticed that the material gradually becomes hotter and free-flowing (the handfeel test), and is ready for pressing.

The press should be filled quickly so that the flour does not cool down too much. Press plates are used to equalize the pressure throughout the press cage and to help the flow of oil. They also make it easier to empty the oilcake from the cage by dividing it into several pieces. The filled cage is positioned under the pressure plate and pressing is carried out, till no more oil flows out. Some of the husk is added back to the cake (10 per cent by weight) to make it easy for pressing the oil out (FAO, 2007a, b).

The oil released from the cage flows onto the collecting tray. The discs of oil cake will be stuck to the press plates and can be separated by a metal scraper. The oil is filtered, allowed to stand for 48-72 hours to clarify. Then, most of the clear oil can be carefully poured off or siphoned and heated (90-110°C) (Hamm *et al.*, 2013). If groundnut oil is heated without being left to stand for 48-72 hours, the oil colour will be dull and will foam when cooking. The extracted meal is still very rich: 41–50 per cent protein content. Shells and hulls are used as fertilizers or as combustible feed for the boiler.

Peanut Butter

Peanut butter is spreadable paste, made from a preparation of crushed groundnuts, roasted and mixed with 5–7 per cent groundnut oil and salt. The peanuts are first shelled and cleaned. They are then roasted at 218°C for 40-60 minutes either on trays in an oven, with intermittent mixing or in a rotary roaster which allows each nut to become uniformly browned. After roasting the skin of the nuts are loose and removed by gentle brushing. A simple winnower to remove the skin from the nuts can be made by allowing the nuts to fall in a gentle stream in front of an electric fan. The heavy nuts will fall straight down while the lighter skin will be blown away. Next, nuts are ground in a mill that may be powered by hand or with a motor. The most commonly used mill is an adjustable plate mill. The tighter the distance between the plates, the finer the texture of the butter (Shukla *et al.*, 1992). The milling process may have to be repeated to obtain the desired texture. Adjustments on the mill can produce varying textures. Salt (2 per cent by weight) and antioxidant are added at this stage, and is then packed in jars. To avoid separation of the oil and the settling out of the solids within the peanut butter after few days of storing, a stabilizer called glyceryl monstearate (GMS) can be added to it at the rate of 2-3 per cent by weight, as a premix (FAO, 2007b).

Rapeseed (and Mustard)

Rapeseed ranks fifth among the major oil seeds of the world. Rapeseed oil is obtained from *Brassica napus*, commonly called oilseed rape or colza, and from *Brassica rapa*, known as turnip rape. The oil from mustard is known for its pungent flavour which is developed during milling.

The seeds have a spherical shape, are very small (1.5–2.0mm diameter), are coloured from black-brown to red-black and have a weight of 3–5 mg. They contain 35–45 per cent oil, 19–23 per cent protein, 10–15 per cent cellulose fibre, 3–4 per cent ash and carbohydrates. Other than in Asia, most rapeseed crops grown today are 'double low' or 'double zero' ('00') varieties, meaning they contain low levels (<2 per cent) of erucic acid in the oil portion and low levels (<30 μmol/g) of antinutritional compounds called glucosinolates in the meal portion. The low-erucic acid, low-glucosinolate rapeseed varieties is registered/traded under the name of 'Canola' in 1978 by the Western Canadian Council Crushers Association (Mag, 1994).

To achieve maximum yield, most industrial plants use a two-step extraction process for rapeseed and canola: mechanical pressing first (called pre-pressing), followed by solvent extraction. A 'prepress' is used to reduce the oil content to around 20 per cent. The press cake is then treated with hexane to produce a meal with 0.7–1.5 per cent oil. At small capacities, high-pressure pressing – either full or double pressing – is a potential alternative to solvent extraction. Depending on the seed type and the process used, residual oil contents of between 6 and 12 per cent may be expected.

Rapeseed Oil

In India, rapeseeds are usually harvested at a moisture content of about 30–35 per cent. After harvesting, it is left for some days in the field along with the plants. The plants dry to 20-25 per cent moisture level in the field at which the seeds are threshed. However, the optimum moisture content for threshing is 12–20 per cent (Mag, 1994; Shukla *et al.*, 1992).

For separation of dust, dirt, stones, chaff *etc.*, the pedal/power operated air screen cleaners, Aspiration, indent cleaning, sieving, vibrating screen or flat rotary screen can be used. After cleaning the rapeseed impurities should be less than 0.5 per cent. Dehulling of the seed is, at present, not a commercial process (Mag, 1994).

In many extraction plants, the cleaned seed is first heated to about 30-40°C to prevent shattering. This is especially important with very cold seed. Otherwise, the cleaned seeds are sent for flaking. Usually, two sets of roller mills are used, with the second set adjusted to a tighter clearance than the first. Because rapeseeds are very small, special care must be taken to ensure that all seeds pass between the two rolls, and not through the side. Because of their small size, rapeseeds are usually not cracked before flaking. The thickness of the flake is important, with an optimum of 0.3-0.38 mm (Mag, 1994). Flakes thinner than 0.2 mm, are very fragile while flakes thicker than 0.4 mm result in lower oil yield (Hamm *et al.*, 2013).

The flaked seeds are conveyed to the cooker–conditioner. The purpose of cooking–conditioning is to heat the flakes in order to soften the oil cells, sterilize the seed by destroying enzyme activity (myrosinase, which hydrolyzes glucosinolates; and lipases, which hydrolyze triglycerides and phosphatides), and facilitate the release of the oil. This enzyme can hydrolyze the small amounts of glucosinolates in canola and produce undesirable breakdown products which affect both oil and meal quality. The conditioner is a large horizontal rotary or multistage vertical stack vessel. In either case, the heating medium is steam at 6–10 bar gauge pressure. The cooking cycle usually lasts 15-20 minutes and the temperatures normally range between 80° and 105°C, with an optimum of about 88°C, and dried down to 3–5 per cent moisture content (Hamm *et al.*, 2013). Drying in the cooker helps transmit mechanical pressure to the product during subsequent mechanical extraction in the screw press.

Cold pressing of rapeseed should be done if the moisture level of seed ranges between 9–10 per cent. In tribal areas of India, a local expeller, known as petula is used for extraction of oil from mustard seeds. This equipment consists of two wooden planks and four wooden logs. Oilseed is loaded in jute cloth, steamed (5–10 minutes) and then pressed in between two planks. The capacity of petula ranges between 1.5 to 2.5 kg seed/batch (Mag, 1994). The traditional bullock operated ghani, over head type power driven gheni and portable power ghanies are also used in rural areas. Addition of 12–14 per cent water in oilseed during such expression yields maximum amount of oil.

In some parts of the country, power driven rotary mills are used for extraction of oil from rapeseed/mustard, which is very common. In this mill, both the pestle and mortar are made of wrought iron bucket, rolled of 1–3 mm steel sheet is fitted on the mortar to serve as seed container. A ring of 25 cm inner diameter is seated on a saucer (which serves as oil bowl) to form the mortar. The top periphery of the mortar is fitted with vertical wooden pegs (15 cm long) which forms replaceable scrapers. The pestle is obliquely placed on the taper ring such that it leaves a clearance of 0.125-0.5 mm between the round ring depending upon the seed. The pestle rotates due to friction with the rotation of the taper ring. Mortar is made to rotate from a shaft by means of a pinion working in a bevelled wheel, fixed to its lowest position. A power driven rotary mill rotates at a speed of 14–16 rpm (Shukla *et al.*, 1992). The bucket has about 45 cm diameter at the top, tappering to about 30 cm at the bottom. Pestle is about 75 cm long and the height of the rotary, mill from the bottom plate to the top is around 1 m. The oil flows from the tapper ring dripping on to a plate placed below it. It yields oilcake with 10–12 per cent residual of oil after two crushings. Here, initial moisture level should be in the range of 10-12 per cent for maximum recovery of oil (Mag, 1994).

Canola oil can be made into products like margarine and shortening through processes called interesterification or hydrogenation, which solidifies the oil. Industry makes a wide variety of canola based hydrogenated oil stocks for tailor-made shortenings (Hamm *et al.*, 2013). It may also be mixed with other oils that are more solid by nature, including palm kernel oil. This process results in a semi-solid product that does not need to be further processed.

Cottonseed

Cotton is one of the most important commercial crops of India and is the single largest natural source of fibre. Cottonseed consists mainly of inner kernel, usually called 'meats'. This meaty portion contains all the oil as well as proteins. The meat is enclosed in a tough fibered shell called Hull. The seed is a byproduct of the cotton ginning industry and the cotton fibre is used for textiles. The seed from ginning remains covered with short fibres called 'lints', adhering to the shell. The oil content of cotton seed is 15 per cent-25 per cent, and after decortications, it is 30 per cent-40 per cent (Agarwal *et al.*, 2003).

Cottonseed Oil

Cottonseed oil has a mild taste. It is generally clear with a light golden colour, after refining. If uncooked meats are used for extraction, the resulting crude cotton oil has a red colour, coming from the gossypol pigment, and oils must be refined to remove gossypol.

Cottonseed oil is described by scientists as being "naturally hydrogenated" because of the levels of oleic, palmitic, and stearic acids in it. The oil has a ratio of 2:1 of polyunsaturated to saturated fatty acids and generally consists of 65-70 per cent unsaturated fatty acids including 18-24 per cent monounsaturated (oleic) and 42-52 per cent polyunsaturated (linoleic) and 26-35 per cent saturated (palmitic and stearic) (Agarwal *et al.*, 2003). This renders it stable frying oil without the need for additional processing that could lead to the formation of *trans* fatty acids. When it is partially hydrogenated, its monounsaturated fatty acids actually increase. Thus, its light, non oily consistency and high smoke point make it most desirable for stir-fry cooking, as well as for frying. An additional benefit that accrues from Cottonseed Oil is its high level of antioxidants- tocopherols, which contribute to its long shelf-life.

Sunflower

Sunflower (*Helianthus annus* L.) is one of the oldest native crops of North-America, grown and cultivated as a food crop. It is a robust oilseed crop, the oil of which has a mild taste, pleasant flavour, good keeping quality with acceptable amounts of vitamins A, D and E. Sunflower seeds contain 70–75 per cent pure kernels (containing 55–65 per cent oil and 52–57 per cent protein) and 25–30 per cent pure hulls (containing 1.5–3.0 per cent oil). As with rapeseed, the high oil content of sunflower seed requires a two step process – mechanical pressing followed by solvent extraction – to fully recover the oil.

Sunflower Oil

The sunflower seeds are crushed under conventional conditions. Expellers are suitable for oil extraction from sunflower. Two-three crushings are usually adopted. Cooked seeds are mechanically pressed to 18–22 per cent oil and then sent to the solvent extractor. Maximum extraction efficiency is obtained when seed moisture ranges between 8.5 to 8.67 per cent. Ghani oil was darker in colour and has relatively higher free fatty acids content, probably due to practice of sprinkling water in the feed while crushing. Hydraulic and expeller pressed oils are of light amber colour and have low free fatty acids.

Sunflower seed press cake is hard and must be broken into pieces. The average size of cake pieces going to the extractor should be around 6 mm, and the proportion of fines limited to 10 per cent (Hamm *et al.*, 2013). The refined oil is pale yellow and has good keeping quality with little tendency for flavour reversion. The residual meal, left after oil extraction, is used as a high grade protein supplement for livestock. This meal is also used as a nitrogenous fertilizer. The seed heads and stalks could be used as a dry season fodder while hull is a good source of fuel for use in furnaces.

Safflower

Safflower (*Carthamus tinctorius* L.) is a valuable oilseed crop that is still to be exploited for its greater potential for oil and protein cake. Safflower seed resembles sunflower seed, except for its ivory-like colour. There are two categories of safflower varieties, the type that produces oil which is high in monounsaturated fatty acids (15 per cent oleic acid), and those with high concentrations of polyunsaturated fatty acids (60–80 per cent linoleic acid). Of all the commercially available edible oils, it has the highest content of polyunsaturated fatty acids. Traditionally known as source of dye in ancient India, the safflower has attained considerable importance as an oilseed crop lately. It is cultivated in many states of India and numerous varieties of this crop are under cultivation, varying in oil (35-45 per cent) and dye contents. Safflower seed is also grown in some areas for bird feeding (Khalid *et al.*, 2017).

Safflower Oil

In India, safflower oil has been extracted using *ghani*. This process leaves about 10-15 per cent oil in the residual cake. Currently ~40 per cent of the safflower seeds are crushed using *ghani* extraction technique around the world. A further extraction using expellers can reduce the residual oil to 6-8 per cent (Shukla *et al.*, 1992).

Processing of Spent Oilcake

The meal could also be used for isolation of protein; however, safflower flour contains lignan glycosides which impart a strong bitter flavour and has cathartic activity. These can be eliminated or reduced to a low level in the preparation of concentrates or isolates: de-oiled cake are ground to get 75μ size flour and treated with aqueous ethanol for debittering (Figure 7). Protein isolate can be prepared by dissolving pre-treated safflower flour in water at pH 9. The slurry is then filtered and acidified to pH 4 to precipitate most of the proteins (85-90 per cent) (Khalid *et al.*, 2017).

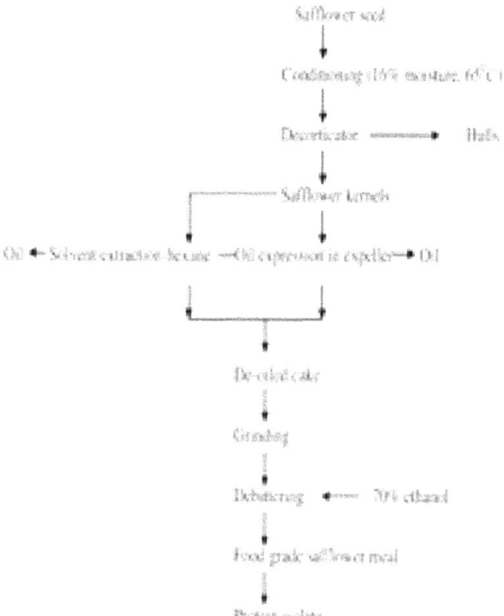

Figure 7: Process Flowchart for the Production of Edible Safflower Protein. Adapted from Shukla *et al.* (1992).

Linseed

Linseed oil is obtained from the seed of the flax plant (*Linum usitatissimum* L.). The linseed (flax) plant exists in two main varieties; one gives a high yield of seed and the other fibre. Flax species used for oil production are different to and smaller (0.3m high) than the ones used for the production of textile (1.5m high). Linseeds contain about 38–45 per cent oil and ~25 per cent proteins. Linseed oil's main characteristic is that it contains 45–58 per cent linolenic acid, which makes it the best known siccative oil. It has the property of absorbing the oxygen from ambient air and forming a solid but elastic material known as linoxyne. It is this characteristic that makes it interesting to the paint and varnish industries (Shukla *et al.*, 1992).

Linseeds contain a small percentage of a cyanogenous glucoside called linamarin and an enzyme called linasis. At 40–50°C and in the presence of moisture, the enzyme acts on the linamarin to release cyanhydric acid (HCN). During cold pressing, the linasis and the linamarin are not eliminated, and feeding such type of press-cake to animals may kill them. To avoid poisoning cattle, the linasis must be inactivated by heat; boiling of seed or cake (10 min) is reported to ensure this. Alternatively, the heat treatment in the cooker and in the press permits the elimination of toxic elements. Cakes or meals are thus no longer dangerous for livestock feed; they actually make excellent feed and fertilizer (Shukla *et al.*, 1992).

Coconut Oil

The coconut palm tree is found on all tropical coasts. Coconut oil is widely used as edible oil or for industrial applications. The coconut is made of a very hard shell containing white, sweet liquid called coco milk. When the fruit has matured, nearly all the milk has been transformed into albumen. It is this albumen that makes the copra, after appropriate drying. The oil content of coconuts is approximately 65 per cent in the dried copra or 35 per cent if the fresh meat is used (Swetman, 2008).

Sesame Seed Oil

Sesame oil is a kind of edible oil derived from sesame (*Sesamum indicum*) seeds. It is mainly used as a flavour enhancer in China, Korea, Japan, Middle East and Southeast Asia for cuisine, and also cooking oil in South India. The fruit consists of a pod with an elliptic form, containing up to four longitudinal cells, each divided into two parallel cells, which may contain 15–20 seeds each. Sesame seeds contain 45–55 per cent oil and 19–25 per cent protein. Sesame seed, however, is mostly consumed as seed, in the bakery and pastry industry (Belitz *et al.*, 2009). Only a small quantity is crushed to produce oil. The seeds are white, yellow, dark red or black. The sesame seed trade divides them into two categories (Warra, 2011):

1. Black sesame, containing 15–25 per cent of black seeds.
2. White sesame, which contains more than 95 per cent white or yellow seeds. The oil from white sesame seeds is of better quality.

The traditional method applied for the extraction of sesame oil is hot water flotation. This was practiced mainly in Uganda and Sudan. Sesame oil is excellent, semi-siccative edible oil. It has a nut-like sweet flavour and contains high linoleic acid and antioxidants (like sesamol, semamolin, *etc.*). So, it is appreciated and known for its high resistance to rancidness and its agreeable taste. Other uses of sesame oil are that it is used as a salad or cooking oil, shortening, making margarine and soap, paints, cosmetics and perfumes (Warra, 2011).

Rice Bran Oil

Bran is a hard outer layer of grain as shown in Figure 8. The grain plus husks and bracts is called 'paddy'; the decorticated rice is named 'cargo rice' and the rice cleaned from its integument is called 'white rice' or 'polished rice'. Rice bran oil content ranges from 12-25 per cent depending upon the quality of the bran. It has the ideal ratio of saturated, mono unsaturated and poly-unsaturated fatty acids and is the closest to World Health Organization recommendation. It has a good balance between oleic and linoleic acid, with a low linolenic acid content.

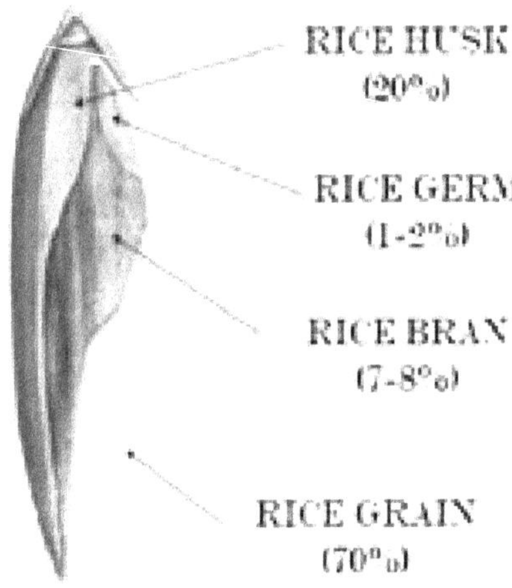

Figure 8: Structure of Rice with its different Layers.

Rice bran oil is a vegetable oil produced from the outer brown layer of rice which contains highest amount of all natural vitamin-E and a unique component called oryzanol that is linked with increase in good cholesterol and lowering down of the bad cholesterol and triglycerides. This oil is extensively used in Japan, Korea, China, Taiwan and Thailand as premium edible oil. It is the conventional and the most favorite cooking medium of the Japanese and is popularly known as "Heart Oil" in Japan. In India rice bran accounts for 7 –8 per cent of the rice produced and the recovery of rice bran oil from rice bran is usually 15-16.5 per cent (Hamm *et al.*, 2013).

The bran is obtained from the polishing of rice during milling. Rice bran oil is mainly obtained by direct solvent extraction of the bran.

As rice bran oil (RBO) contains 5–10 per cent waxes and stearins, the crude oil requires refining and dewaxing (winterizing) for obtaining high quality edible oil. Therefore it is said to be the most difficult oil among all vegetable oils, to refine. Different degumming processes are available for crude rice bran oil. Enzymatic degumming is the best process already available today for reducing the phosphorous content of crude RBO below 5 ppm (Hamm *et al.*, 2013). It converts non-hydratable lecithin (gums) to water soluble lysolecithin, which is separated by centrifugation. Deodorization of rice bran oil can be carried out in the normal manner by heating the oil to temperatures 200-250°C under high vacuum, stripping out the undesirable volatiles, in a current of dry steam. Any free fatty acids, peroxides and certain proportion of natural tocopherol antioxidants are also removed. The edible grade bran oil is coloured leaf-green. Appearance of rice bran oil ranges from cloudy to clear depending on the degree of dewaxing and winterization processes applied.

Palm Fruit and Kernel Oils

The palm oil is the world's second largest edible vegetable oil, only ranking after soybean oil. Palm oil is obtained from *Elaeis guineensis*, a tree found in Guinea, South East Asia, West Africa, Central and South America. Malaysia and Indonesia are the principal producers of palm oil. The fruit cluster contains about 3000–6000 fruits. The fruits provide two different oils, the first from the mesocarp (palm oil) and the second from the seed/kernel (palm kernel oil). Today palm oil is extensively used in the production of shortenings and margarine, for deep frying and as liquid edible oil.

Palm Oil

Immediately after arrival at the processing plant, the fruit bunches are unloaded from the trucks through special transfer hoppers into cages, which are then introduced into an autoclave. Sterilisation is conducted in large steam autoclaves to stop lipase enzyme activity in the fruit bunch and to ease mechanical stripping of the kernels. Each autoclave generally contains 3–10 cages of 1.5–7.0 tonnes of fruit bunch. Saturated steam is injected until a pressure of 2.5–3.0 kg/cm^2 is reached. Each autoclave normally performs 12–14 operations per day. The bunches are processed for 45–60 min, according to their size and ripeness.

The sterilized bunches are then fed continuously into a rotary drum thresher or beater arm thresher, which strips and separates the fruits from the bunch stalks. Rotary strippers comprise a drum made up of equally spaced longitudinal U-bars, which revolve at low speed. Longitudinal flat bars are located radially inside the drum. The clusters are introduced at one end of the stripper and are raised by centrifugal force and by the flat bars, eventually falling on the U-bars. The impact frees the fruits, which pass through the U-bars and are finally removed by a screw conveyor. The empty bunches are discharged at the other end of the drum.

From the thresher, the fruits are conveyed into a digester. The purpose of digestion is to separate the fruit pulp from the nuts and to cause the oil-bearing cells to burst, which facilitates oil extraction by pressing. Therefore, this operation is of the utmost importance as it affects the oil yield. The best digestion conditions are attained by macerating the fruits at a temperature of 90–100°C for about 20 min. The digester is a vertical cylindrical vessel provided with a double steam jacket and a central paddle stirrer (Hamm *et al.*, 2013).

The mash, consisting of digested pulp and palm kernel nuts, is then transferred to the continuous screw press, which separates the solid portion, including the fibres and nuts, from the liquid phase (water and oil). Continuous screw presses are made up of a perforated cage or barrel of horizontal cylindrical design, in which the pulp is subjected to increasing pressure by a variable pitch worm, which causes the oil to come out. The counter pressure is adjusted by means of a mobile cone, which regulates the press cake thickness (Dunford, 2008b). Two products are obtained at the outlet of the press:

- ☆ A mixture of water, oil and solid impurities (vegetable residue)
- ☆ A press cake, which contains fibres and nuts

Crude oil is sent to the clarification section, wherein water and impurities are removed from the oil to yield a clean and dry product. A clarification plant conventionally comprises a continuous decantation or settling tank, which separates oil, water and impurities by the use of difference in their specific gravities. Continuous decantation tanks are of rectangular horizontal or cylindrical vertical design. The sludge from crude oil can be treated in centrifuges for reducing liquid effluent quantities. Finally, the oil is dried to below 0.1 per cent moisture by using vacuum dryers (Dunford, 2008b). Crude oil has high carotene content and, hence, the colour of the oil is yellow to red. During refining, the palm oil color is destroyed by bleaching and the free fatty acids are removed. Decolouring agent (floridin or acticarbon) are pumped into the oil, continuously stirred and heated for 25 min. After that, the oil-floridin mixture is filtered to separate them (Hamm *et al.*, 2013).

Fractionation of palm oil is important for getting different grade oil. Oil is stored in a storage tank (10-20°C higher than the melting point of the fat) with constant stirring so as to avoid solid fat sedimentation and to ensure even crude oil to be sent to the plant. Before the RBD palm oil been sent to the crystallizer, it has to be processed though a plate heat exchanger, which will heat the oil from 45°C to 70°C approximately, fully melting the solid oil. The oil is cooled to ensure crystallization (Hamm *et al.*, 2013). Oil slurry gradually enters though the filter plate and liquid fat flows out. When filter is stuffed with solid fat, it stops the ingress of the crystallized palm oil. The crystal obtained is filtered. The filter cake afterwards will be heated to melt and will be then pumped into the solid fat storage tank.

Palm Kernel Oil

The nuts are separated from the fibrous material, which in-turn are obtained from the continuous screw presses or depericarper (a vertical column through which a hot air current is circulated). The centrifugal sieves are also used to separate the nut, fibre and dregs. The fibrous material is sucked into a duct, then separated from the air in a cyclone, dried to 15 per cent moisture and finally conveyed to the boiler house to be used as fuel. After separation from the fibre, the fresh nuts are dried to detach the kernels from the stone shells. The moisture content of the nuts is reduced from about 16 per cent to 10–12 per cent. The dried nuts pass through a nut grading screen which can classify the nuts according to size, which are then transferred to two or more crackers.

The cracked mixture consists of free kernels, shells, unbroken nuts, partly cracked nuts and dust (fibre, thin shells, bits of broken kernels). These are segregated by a dry separation system, followed by a clay bath or hydro cyclone. The dry separation system consists of two double-stage winnowing plants, in which separation is carried out by sucking air through a vertical column (Dunford, 2008b). This way, a maximum number of fibres, shell and dirt particles are removed without carrying the kernels away. The clay bath separation system makes use of the difference in specific gravity between the kernels (1.06–1.16) and the shells (1.30–1.35). In the hydro-cyclones, separation is carried out in a water stream, rotating at high speed inside a cyclone (Hamm *et al.*, 2013).

After separation, the kernels contain about 20 per cent moisture, which is reduced to 7 per cent by drying. The kernel is crushed and the oil is squeezed out under high pressure and temperature ranging from 70-100 °C. Palm kernel leaves a meal, which is normally used as animal feed in the form of pellets. The shells, another byproduct of the oil palm, are utilised as fuel. The ash of the empty bunches can be used as fertiliser due to its potash content.

Olive Oil

Olive oil is obtained from *Olea europaea sativa*. The main producing countries are Spain, Italy and Greece, followed by Syria, Tunisia and Turkey. Olive oil is not extracted from seeds, but from the pulp of fruits. The olive fruit is oval-shaped and has an average weight of 4-12 g. The fruit contains pulp (containing 50–60 per cent water and 20–25 per cent oil) and a kernel or pit (containing 30 per cent water and 8–10 per cent oil). The pulp is surrounded by a skin.

Based on the methods of extraction/processing, the oil from olive can be categorized into the following (Belitz *et al.*, 2009):

- Extra virgin olive oil (*extra vierge*): obtained from cold-pressing of the fruit by physical means; pleasant aromatic taste; up to 0.8 per cent free fatty acids (calculated as oleic acid)
- Virgin olive oil (*fine vierge*): Slightly less aromatic in taste; up to 2 per cent free fatty acids
- Lampante oil: obtained from a further hot pressing of disintegrated fruit pulp; much less taste; more than 2 per cent free fatty acids
- Pomace oil: obtained by solvent extraction of the olive pomace (the solid residue left after virgin oil extraction)

Virgin Olive Oil

The harvest of olives is a delicate process. The fruits used for the production of first quality (extra virgin) oil are collected by hand or in some manner that does not damage them. For optimum quality, the fruits should be processed immediately after harvesting.

The olives first go through a leaf-removal and water-washing process, followed by kneading or crushing. Crushing is generally done in a stone mill or hammer mill, in order to tear the flesh and release the oil from the cells that contain it. This produces a finely ground paste. After crushing, the paste is either slowly mixed at ambient temperature for 10–20 min, occasionally by adding common salt to break the oil–water mixture that is formed in the crusher; or the paste is mixed and heated, generally for an hour or more, to allow the oil droplets to merge and increase in size. Cylindrical mixers with a vertical or horizontal shaft equipped with blades are used for this purpose (Belitz *et al.*, 2009).

From the mixer, the paste is spread in layers in a hydraulic press. A single pressing step, lasting 1.0–1.5 hours, is generally applied. The press separates a water–oil liquid mixture from a solid phase, called pomace (containing stones and pulp residue). The separation of oil and water is performed in horizontal separators, called decanters or in centrifuges or by gravity decantation. The initial cold pressing provides virgin olive oil (provence oil). This is then followed by warm pressing at about 40°C (Belitz *et al.*, 2009). Virgin olive oil is extracted from the pulp, and has the unique feature of being edible without further processing, if obtained from good-quality raw material. Oil recovery yield is 86–90 per cent.

Olive Pomace Oil

The olive pomace should be processed as quickly as possible, due to high rate of hydrolysis and enzymatic action. The moisture content of pomace after pressing is first reduced from 35 per cent to 8–10 per cent. Rotary dryers are usually employed. The dried olive pomace contains 45–50 per cent kernel and 50–55 per cent pulp (of which 5–7 per cent is skin). The pulp has 10–15 per cent oil content and the stone contains 1–2 per cent oil (Belitz *et al.*, 2009).

The oil can be solvent extracted with hexane without separating the pulp and stone (batch process). After such extraction, the pulp meal contains 4–5 per cent oil and the stones have 0.3–0.5 per cent residual oil. In another process, the pomace pulp and stones are separated and the pulp is pelletised before extraction. Residual oil content in pulp of less than 1 per cent is achieved when using a continuous percolation extractor. The stones are not extracted and some oil is lost with the pulp sticking to the stone when they are separated. The stones are an excellent fuel for the pomace dryer. The spent pulp meal can be used as animal feed, fertiliser or fuel. For the isolation of refined olive oil, it is then passed through conventional refining stages. The resulting refined oil ("sansa" oil) contains at most 0.3 per cent of free fatty acids (Belitz *et al.*, 2009).

Corn Germ Oil

Corn oil is extracted from the corn germ, which in-turn is a byproduct of corn milling. Corn germ contains about 85 per cent of the total oil of the kernel. However, its oil content varies widely with the degermination method applied. Degermination is mostly carried out in one of two ways:

In dry process, the corn is first cleaned by combination of screens and aspirations, then it is mixed with warm water to bring its moisture content of 21-24 per cent. The grains are held is tempering bins for 1-2 h before going to straight roller mill or mills using degerminator. The germ being tougher than the endosperm tends to be flattened rather than ground by the rolls and then scalded off by the shifters or sieves. The germs are then dried to 14 per cent moisture. Perfect separation of germs from endosperm is not possible by this method, because part of the endosperm stays with them. The germ produced in this way contains only 20-20 per cent fat. In most these mills, fractions rich in germ are not used to recover the oil but are utilized in feeds. The recovery of the oil is less in dry milling process, and therefore, only one-third of the total oil is recovered by this process (Hamm *et al.*, 2013).

The wet process is used in the starch and derivatives industry. First grains are cleaned by series of screens and air separators and then left for steeping in large baths for a period of 30–40 hours at 54°C temperature, during which acidulated water (0.2 per cent sulphuric acid) causes the grain to swell and soften. The steeping operation removes most of the water soluble materials including sugars, proteins and minerals, which are recovered by concentrating the steep water and are used for the preparation of the feeds and other products. The steeped grains then passes to the degerminator, which are mills designed to break up the grain and loosen the germ without disintegrating it. The germs are washed in a large amount of water to remove any adhering starch, finally squeezed and passed into a steam heated rotary drier where moisture content is reduced to 5 per cent (Hamm *et al.*, 2013). This process produces a clean separation of germ from the other soluble, and the dried germs may contain up to 45-50 per cent oil.

Oil is usually extracted from the germ by a combination of expelling in continuous screw presses and solvent extraction of the press-cake. The germs must be cooked, flaked and then pressed. The flakes are sometimes pelletised or expanded before extraction. The initial expeller can recover a little more than half of the oil and subsequent solvent extraction (with hexane) will bring the total yield to about 95 per cent. The solvent is removed from oil by distillation, recovered and re-used. The de-oiled germ is desolventised and combined with fibre and concentrated steepwater to produce corn gluten feed, as it contains about 20 per cent proteins.

Refining of crude corn oil includes degumming to remove corn lecithin (phospholipids), alkali treatment to neutralize free fatty acids, bleaching for colour and trace elements removal, winterization (removal of high melting waxes) and deodorization (steam stripping under vacuum) (Hamm *et al.*, 2013). The crude oil has a colour from golden to dark yellow and a specific taste. The refined oil is used as salad or cooking oil and also in the preparation of margarine.

Almond Oil

Almonds (*Prunus dulcis*) are typically characterized into three phenotypes, which include non-bitter (sweet), semi-bitter, and bitter. It is not technically a nut or fruit, but is a drupe due to its multiple layers. The outer hull of an almond tree's fruit is a thick, green coating. Inside the hull is the endocarp, which is hard and woody. Inside the endocarp lies the seed, or almond. Almond oil is extracted from this ripe seed. It is extracted using cold-pressed method and is available in two different forms *i.e.* sweet almond oil and bitter almond oil. It is a rich source of vitamin E, vitamin B, vitamin A, minerals, and omega fatty acids.

Mechanical Expression of Sweet Almond Oil

The almonds are harvested once the fruit gets ripe, then washed and de-pulped by removing the mesocarp. The obtained seed is dried to a moisture content of 4-6 per cent which is sunsequently sent for dehusking. De-shelling is achieved traditionally using a hammer and a dehusking stone. Almond nuts are generally cracked along the margin to release the brown spindle-shaped kernel from the endocarp. The cracked kernels and other debris are removed by winnowing. After this stage, the whole intact kernels are sorted manually from the broken ones, and are directly marketed as edible nut. For oil extraction, the broken kernels are ground in attrition mill for further size reduction. The crushed kernels are roasted, placed within a cheese-cloth and passed through an expeller to exude out oil. Almond oil produced by a cold press process is darker in colour than a refined process, but contains higher amount of monounsaturated fats and antimicrobial activity (Akubude *et al.*, 2017).

Distillation Extraction of Bitter Almond Oil

In bitter almonds, the kernels accumulate cyanogenic diglucoside amygdalin, which is related to bitterness. Upon disruption of tissue, amygdalin will be degraded by β-glucosidases (emulsin) with release of toxic compounds like benzaldehyde and hydrogen cyanide (prussic acid).

Distillation extraction is used to extract amygdalin from bitter almonds. Emulsin enzymes, a component of the bitter almond, are activated when water is present. During distillation extraction, the kernels are exposed to water or steam to denature β-glucosidases, and to wash away the toxic hydrolysis products (benzaldehyde and hydrogen cyanide), if formed. Distillation also liberates the essential oil simultaneously from the tissues. The oil is cooled and condensed for collection. Amygdalin is not present in the raw oil. This prussic acid, or cyanide, can be removed with an additional distillation, leaving oil of bitter almond free from prussic acid (Akubude *et al.*, 2017).

Shea Nut Butter

The shea tree (*Butyrospermum parkii* or *Vitellaria paradoxa*), grows wild in the equatorial belt of central Africa between Gambia and Sudan and also in Uganda. The kernels contain 42 to 48 per cent oil (butter). The kernel oil extracted has a relatively high melting point and is used in rural areas in the making of foods, soap manufacture and cosmetics. Shea butter is used as an extender in chocolate as its properties are similar to cocoa butter, as baking fat and also as a base for skincare products and moisturising cream (Anonymous, 2012).

The exterior green pulp of the fruit is removed either by burying the fruit in the ground so that the pulp ferments and falls off, or by parboiling or sun drying them, followed by smoking over an open fire for 3 to 4 days. The former methodology takes 12 days or more, while the latter is comparatively faster, and the dried nuts have characteristic smoky flavour and can be stored for long periods (Anonymous, 2012). Shea nuts are mainly exported as smoked kernels and further processing requires decortication by crushing the outer shell to remove kernels.

Nuts are shelled manually by pounding them individually using the mortar and pestle, and the resulting kernel particles are roasted on a metal sheet over a fire, followed by grinding them between two stones to produce a smooth paste. A small amount of water is added to the paste and the mixture agitated by hand using a "paddling" motion (Anonymous, 2012). The mixture is continuously stirred for few hours till it becomes lighter in colour and a white mass of butter starts to float on the top of the mixture.

The butter is then boiled over an open fire until clear. The resulting oil is decanted off the dark brown residue and is washed repeatedly with warm water until clean. The remaining water is removed by heating. Impurities settle out and the butter can be left to cool and solidify. Though this Traditional wet processing of shea is a slow and laborious process, the fat content in the resulting shea butter varies between 25 and 40 per cent of the dry kernel weight (Anonymous, 2012).

Cocoa Butter

Cocoa butter is a byproduct of cocoa bean processing industry. Cocoa beans come from the seeds (containing 50–58 per cent fat) of the cacao pod. Cacao trees (*Theobroma cacao*) produce 20 to 30 pods a year, and each of these pods contain 25 to 40 seeds that are surrounded by a mucilaginous pulp. Each pod produces about 0.043kg of fermented and dried cocoa beans (Asselstine *et al.*, 2016).

There are actually three intermediate cocoa products that may be derived from cocoa processing: cocoa liquor, cocoa butter, cocoa cake (cocoa powder). Cocoa butter contains a high proportion of saturated fat, derived from starch and palmitic acid and contains trace amounts of caffeine and theobromine (Asselstine *et al.*, 2016). It also contains fat soluble antioxidants such as vitamin E in the form of β-tocopherol, -tocopherol and γ-tocopherol.

Processing of Cocoa Pod

Pod development of cocoa from flowering to full maturity stage takes around 5-6 months. Only fully ripen, healthy and undamaged pods are selected and harvested with special knives (Guda and Gadhe, 2017).

The beans are procured by opening the pod longitudinally into two halves, using small cutlass or wooden billet, and the rind is separated. Germinated, black or diseased beans are discarded. The time interval between the harvesting and breaking of the pods plays a critical role in the development of flavour. The minimum time interval between harvesting and pod breaking is around 5-6 days. If the time interval between the harvest and pod breaking is more than 6 days than it affects the flavour of cocoa, which ultimately results in the quality deterioration (Asselstine *et al.*, 2016).

Pulp that surround the bean is liquefied by the application of natural liquefying enzymes present in pulp. This is called fermentation or sweating. Liquefaction of pulp helps in easy removal of pulp from the bean and to develop the flavour and aroma precursors of the chocolate. Fermentation is conducted for 4-7 days in baskets, boxes, trays or a heap covered with banana leaves (or cocoyam leaves). In all cases, the bottom and sides of the box or basket should be covered with leaves to prevent the cocoa from drying out; however the leaves placed at the bottom should be perforated for ensuring drainage of the pulpy liquid produced during fermentation. Insufficient drainage will result in a bad fermentation (Guda and Gadhe, 2017). The beans then change colour (from white/mauve to brown). The inside is light brown or reddish. Well-fermented beans have a shiny appearance, without mould and their cotyledons break easily. They release a chocolate aroma. The cacao fruit pulp is very sweet, hence liquefied fermented pulp is used in the distillery industries to produce distilled alcoholic spirit (Asselstine *et al.*, 2016).

After fermentation, the remnants of the pulp are removed by washing the beans or mixing them with sawdust. The moisture content of the beans at the end of the fermentation is approximately 60 per cent, which is then reduced to 7–7.5 per cent for preventing mould growth during storage and transportation (Asselstine *et al.*, 2016). Drying can be accomplished using sun drying or conventional dryers. Fast drying rate and high drying temperature are avoided because it results in the hardening of the shells, formation of acid in the beans and as the taste of the beans spoils above 55 °C (Guda and Gadhe, 2017). As soon as the shell of the cocoa hardens, the beans are polished using a custom-made polishing machine, for improving the external appearance.

After drying and polishing, the beans are cleaned of any extraneous matter, graded via a mechanical grader and finally packed in food safe jute bags, whose fibers have been created with vegetable oil. Mechanical grader consists of different sized mesh sheets around a rotating cylindrical with helical screw inside to convey the beans. There are 3 grades of cocoa beans established by the Cocoa and coffee Industry Board: Grade I, Grade II and defective. Grade I has a bean count of 80/100g, less than 1 per cent commercial defects (that is. in order of importance, moudly, over-fermented smoky, under –fermented or insects infested beans). Grade II has a bean count of 85/100g with less than 4 per cent commercial defects (Guda and Gadhe, 2017). During gradation, first broken pieces of beans and shell fragments are removed, next flagged beans are removed, then small beans and lastly large grade I beans. Defective beans are not exported.

Processing of Cocoa Bean to Extract Cocoa Butter

Cocoa bean consists of about 85 per cent nib and 15 per cent shell. Dried beans are roasted at controlled condition and then cracked to remove shell. Using controlled roasting (at 98-120°C for 90-95 min) proper chocolate flavour is developed. Cracked beans are de-shelled and cleaned de-shelled nibs are recovered. Cocoa nibs are ground to a cocoa mass or liquor (Naik and Kumar, 2014). Cocoa liquor can be directly used or can mix with other required substances for chocolate making.

Pure cocoa liquors are used to extract cocoa butter. The obtained cocoa mass is pressed using hydraulic press or mechanical press, to extract a pale yellow liquid (cocoa butter) and cocoa cake. The butter obtained is filtered, centrifuged and deodorised by steam distillation (Naik and Kumar, 2014). It is used to produce chocolate or cosmetics due to its characteristic odour and flavour of chocolate. The cocoa cake obtained on completion of pressing is crushed and ground to make cocoa powder. This powder is tempered and stabilised between 18 and 20°C. This powder is also used to produce chocolate, pastries, milk-based drinks or cosmetics (Asselstine *et al.*, 2016).

Processing of Edible Nuts

Nuts are hard-shelled fruits, or the edible kernels of fleshy drupes or berries, or seeds that are traditionally consumed (Wickens, 1995). Most edible "nuts" contain concentrated food reserves for future generations of plants and provide valuable sources of energy, protein, oils, minerals and vitamins suitable for human consumption. Many dessert nuts loose their palatability or otherwise deteriorate if not properly dried or cured or are badly stored, especially those with a high oil content. Among the major nuts walnuts and chestnuts are among the more perishable while pistachios and almonds are among the better keepers.

Dehusking and Washing

Most tree nuts are enveloped in a fleshy or fibrous outer covering or husk which may or may not remain attached to the nut as it ripens and falls to the ground, such as in case of walnut, pecan, almond and macadamia. Few nuts, termed 'sticktights', remain attached to their husks, the proportion of which depends on the cultivar (Wickens, 1995).

The husk has to be removed either mechanically or by hand before nuts can be dried, shelled or otherwise processed. Walnuts especially must be dehusked as soon as they are collected, otherwise the colour and quality of the kernel will be progressively adversely affected the longer the husks remain. Next, hand washing in tubs or machine washing in cylindrical drums may be carried out.

Drying and Dehydrating

Artificially produced heat in dryers is now increasingly replacing sun-drying for curing nuts, especially by large-scale commercial producers. The former gives a more thorough and uniform drying of the nuts, especially in walnuts. Sun-drying is still largely practised by small producers, particularly with almonds. Drying is completed when the kernels can be heard to rattle in the shells or can be broken rather than bent with the fingers. Similarly, chestnuts are traditionally sun-dried or dried over a wood fire in a specially constructed kiln. The curing is complete when the shells may be easily separated from the kernels.

Bleaching

Some nuts intended for marketing in-shell may have their appearance enhanced by bleaching to remove stains due to sun scorch and disease; walnuts are frequently thus treated by immersion for 1-2 minutes in a bleaching solution (Wickens, 1995). Sulphur dioxide or burning sulphur may be used, particularly for almonds that have been blemished during harvesting or drying. Over bleaching can give the shells a sickly white appearance and is liable to soften them and flavour the kernels; too little causes irregular bleaching. Pecans can be considerably improved by removing the outer rough layer and polishing, a process which, if required, makes the nuts more responsive to bleaching and drying. Outer layer may be removed by steel brushes or by the use of revolving drums containing coarse sand. Bleaching is effected by dipping the nuts for 4 minutes in, for example, sodium hypochlorite containing 2 per cent active chlorine (Wickens, 1995). Various dyes may be used to colour the polished and bleached nuts into an attractive brown or reddish brown.

Grading

Mechanical graders, such as a perforated revolving cylinder, are commonly used to sort nuts into various sizes, the smaller nuts being the first to fall. Pecan nuts, due to their ovoid or oblong shape, require special graders consisting of variously spaced rollers.

Imperfect, faulty or broken nuts may be removed by hand as the nuts pass along a continuous belt. A suction machine may also be used to lift blank or imperfectly filled nuts over a trap while the heavier nuts pass on. The low specific gravity of Grade 1 macadamia kernels allows them to be removed by floatation in ordinary water. Grade 2 may be removed using 3 per cent of salt solution, with Grade 3 sinking to the bottom. The nuts are then thoroughly dried (Wickens, 1995).

Cracking

Mechanical cracking is faster but usually results in a higher percentage of broken kernels. Nut-cracking machines require graded nut to be fed to a pair of cracking jaws or rollers with special shaking devices to separate the kernels from the broken shells. The kernels are then passed along a continuous belt for segregation of broken kernels and other debris. Special cracking machines are used for macadamia and pine nuts. The leathery, fibrous shell of oyster nut is easy to cut and a slit is made around the edge of the disc-shaped nut with a knife. Another method of cracking hard-shelled nuts is to heat them in burning straw and then cool rapidly with cold water.

Graded pecan nuts are individually picked up in cups on an endless chain passing through a hopper. Each nut passes to a slot where a piston-like rod exerts pressure at the end of the nut and cracks it, the shells and kernels then being released to a receptacle below (Wickens, 1995). Prior wetting is required to reduce shattering of very dry pecans, the kernels being re-dried after shelling.

In case of cashew nut, the tough leathery shell is removed by preliminary roasting, thus making the shell brittle and responsive to cracking, as well as lessening the danger of blistering from the caustic oils in the fruit. In India, roasting is usually done by hand, placing small quantities of kernels in an open iron pan over a small circular earthenware furnace. Care is required to avoid overcooking or charring of the kernels, while with undercooking, the shell remains tough. Shelling is performed by hand; a wooden mallet and a flat stone are used to crack the shell, after which the kernel is sometimes removed with a wire prong. The kernels are then spread out on wire gauze trays in a hot-air room under controlled temperature in order to loosen the pink or reddish-brown skin before removal and also to remove any excess moisture.

Cooking and Salting

Some nuts are traditionally eaten cooked and salted. Roasting or cooking in oil is said to produce the best flavour. A common method of salting is by immersion in a strong brine solution before allowing the almonds to drain and dry. Peanuts and chestnuts are normally cooked before eating, while almonds are preferred roasted or toasted and salted, likewise the macadamia nut. Macadamia nuts are preferred cooked and salted, either by roasting in an oven or immersion in a vat of hot oil, the latter method is the preferred commercial method as it gives the nut an attractive gold-brown colour. Refined coconut oil is generally used. Excess oil is removed centrifugally and the nuts laid out to cool on wire mesh trays, being salted while they are still lukewarm, using a 15 per cent solution of gum arabic or a special oil to enable the salt grains to adhere more readily to the nut (Wickens, 1995). As the cooked nuts tend to be hygroscopic, they are vacuum packed as soon as they are cool.

Cashew nuts are always marketed cooked as the roasting procedure renders the shell more easily removable, which automatically cooks the kernel. Moreover, the risk of contact with the caustic sap in the fresh fruit is also removed. Cooking also reduces the terebinthine or turpentine flavour in pine nuts. Similarly, in bitter almonds, roasting denatures the emulsin enzymes and amygdalin thus cannot be hydrolyzed to liberate toxic benzaldehyde and hydrogen cyanide (prussic acid) (Akubude *et al.*, 2017).

Storage

Many nuts are stored in the shell for longer periods but there is now a trend towards the cold storage of shelled nuts or kernels. Cold storage in vacuumized containers or the use of inert gases appears to offer little advantage over cold storage for a similar period. Freshly gathered chestnuts have high moisture content and consequently prone to fungal attack. The onset of mould growth during storage may be prevented by immersing the fresh chestnuts in a solution of a fungicide for 48 hours. Gas treatment in autoclaves prior to cod storage, has been effectively used. Cob nuts are sometimes stored in sawdust, which helps to prevent shrivelling (Wickens, 1995).

Packaging of Nuts and Oilseed Products

Nuts and oilseed products are the foods with low moisture content and high percentage of fat and therefore are vulnerable to deterioration such as oxidation and moisture absorption reactions, which leads to organoleptic and textural changes (Lima *et al.*, 2004; Lima and Borges, 2004). Exposure to light and heat, transmission of gases and odours (permeability of packaging material), relative humidity, mechanical stresses and contamination by microorganism are the prime determining factors (Irtwange and Oshodi, 2009). Therefore, the selection of packaging material must be such that it should avoid the action of environmental factors such as ingress of light, moisture, oxygen and microorganisms into the product, should be chemically inert to the packaged food as-well-as assure the integrity of the product during transport and storage (Gadani *et al.*, 2017; Robertson, 1993; Sarantópoulos *et al.*, 2002).

Depending upon the fatty acids composition, each fats and oil has a different degree of susceptibility to oxidation. The oils containing high degree of unsaturated fatty acids are highly prone to oxidative rancidity; whereas oils with high degree of saturated fatty acids are less susceptible. Since natural antioxidants are present in unrefined fats and oils, therefore they are less prone to rancidity than refined oil (where the antioxidants get removed during the process of refining) (Vijayalakshmi, 2005). Likewise, antioxidative phytochemicals in nuts, such as flavonoids, phenolics compounds, luteolin, tocotrienols, isoflavones, ellagic acid, β-carotene (Rainey and Nyquist, 1997; Sajilata and Singhal, 2006), are lost to a profound level during roasting and cooking. This further accentuates the need of proper packaging and storage for enhancing the shelf-life of such fat-rich food commodities. Table 2 reviews the features of the polymers or materials used in packaging of different nuts and oilseed products.

Table 2: Characteristics of the Packaging Materials Used for Nuts and Oilseed Products

Packaging Materials	*Nuts, Oilseeds, Packaged Fats and Oil*	*Polymer used*	*References*
Kraft Paper bags	Cashew nut, walnut, almond, groundnuts	A strong paper that can be bleached white and printed or unbleached and brown. Usually are used in multiple layers or 'plies' to give the necessary strength. Can also be laminated to polythene or wax treated to provide greater moisture protection	Fellows and Barrie (1993); Shakerardekani and Karim (2012)
Cellulose with nitrocellulose	Cashew nut, groundnut, walnut, almond	Made from wood pulp (mostly Eucalyptus). Strong, puncture-resistant film that can be 'dead folded'. Improves barrier properties, makes the film heat sealable and resistant to oils, moisture, air and odors if the plain cellulose is coated with nitrocellulose	Fellows and Barrie (1993); Shin and Selke (2014)
Low-density polyethylene or LDPE	Cashew nut, groundnut	Softer and more flexible and has a lower tensile strength. Useful material for heat sealing since it has a low melting temperature (105–115°C) good impact and tear strength.	Fellows and Barrie (1993); Shin and Selke (2014)
Food-grade poly vinyl chloride	Cashew nut, groundnut, walnut, almond	Produced from vinyl chloride monomers. Bears high toughness and strength good dimensional stability and clarity excellent oil barrier properties and good heat stability.	Shakerardekani and Karim (2012); Shin and Selke (2014)

Packaging Materials	*Nuts, Oilseeds, Packaged Fats and Oil*	*Polymer used*	*References*
Polyamides (PA or nylon) nylon	Cashew nut, groundnut, walnut, almond	Provide excellent optical clarity, oil and chemical resistance good mechanical strength over a wide range of temperatures. Good flavor and gas barriers, but have poor water vapour barrier properties	Shakerardekani and Karim (2012); Shin and Selke (2014)
Polyethylene terephthalate or (PET)	Cashew nut, groundnut, walnut, almond, mustard oil, sunflower oil, soya bean oil, almond oil	High mechanical strength, good chemical resistance, light weight, excellent clarity, and reasonably high barrier properties. Stable over a wide range of temperatures (60°C to 220°C). mostly oriented biaxially to improve its mechanical strength and gas barrier properties.	Shakerardekani and Karim (2012); Shin and Selke (2014)
Polypropylene or PP bags	Groundnut, safflower, sunflower, soybean	Good chemical and grease resistant. Good water vapor barrier property	Shin and Selke (2014)
Metallized Polyethylene Terephthalate (MPET)	Soya bean seeds	Good water vapor and oxygen barrier property.	Chuansin *et al.* (2006)
Aluminium foil (AF)	Soya bean seeds, cocoa butter	Good water vapor and oxygen barrier property.	Chuansin *et al.* (2006); Sedlacekova (2017)
Tinplato Containers	Mustard oil, sunflower oil, soya bean oil	Ensures a longer shelf-life and are sturdy. Suitable for high filling and packaging operations.	Vijayalakshmi (2005)
High Density Polyethylene Containers	Mustard oil, sunflower oil, soya bean oil, coconut oil, almond oil, shea butter, cocoa butter	Provide a moderately long shelf life. Light in weight and are transport-worthy.	Vijayalakshmi (2005)
Glass bottles	Mustard oil, sunflower oil, soya bean oil, almond oil	Excellent protection and barrier property. Used for high-speed operation. Highly fragile and high tare weight.	Vijayalakshmi (2005)
Flexible plastic pouches	Mustard oil, sunflower oil, soya bean oil	Made from laminates or multi-layered films of different compositions. Pouches may be in the form of pillow or as Stand-up-pouches. Are more economical than any other packaging system.	Vijayalakshmi (2005)

Packaging Materials	*Nuts, Oilseeds, Packaged Fats and Oil*	*Polymer used*	*References*
Multi-layer packet (Tetra pack)	Mustard oil, sunflower oil, soya bean oil	Six layered tetra pack is aseptically processed. Allow liquid food to retain colour, texture, natural taste and nutritional value for up to 12 months, without the need for preservatives or refrigeration.	Vijayalakshmi (2005)
Non-recyclable grease proof and leak proof paper moulds	Shea butter, cocoa butter	Can be heated by microwave and conventional oven. Grease and leak proof.	Premazon (2016)
Waxed paper	Cocoa butter	Can be produced either from vegetable oils or mineral based commercially used wax paper is coated with paraffin 28 which is extracted from petroleum good grease barrier property good water-repellent property	Sedlacekova (2017)

The stability of fats and oil are compromised as soon as they are exposed to several extraneous agents and prooxidants. Off-odour and discoloration in fats and oil are due to oxidative rancidity caused by the presence of oxygen, leading to the formation of hydroperoxides and peroxides and subsequently formation of secondary oxidation products like aldehydes and ketones (Figure 9). The rate and intensity of these reactions increases in presence of enzymes (lipoxygenase), metal, light and heat. Apart from the above mentioned deterioration, hydrolytic rancidity initiated by the presence of moisture (even in traces), causes off- flavour development. It is the result of hydrolysis of triglycerides that leads to the formation of glycerol and free fatty acids, and can be triggered by the presence of enzyme lipases. Atmospheric moisture and oxygen may be present in the oil itself and headspace of the package, or may enter the container through its body, seams or seals (including closures and folded areas). Hence, control of these factors by selecting appropriate polymers is crucial. Multi-layer films, flushing of the headspace with inert gases and vacuum packaging are commonly practiced to retard oxidation. Active and Intelligent packaging by inclusion of nanomaterials are anticipated to deter such unfavourable reactions by many-folds (Das Purkayastha and Manhar, 2016). Nonetheless, nanoparticle-based packaging materials are currently under developmental stage and must comply with food safety regulations before being released into the market.

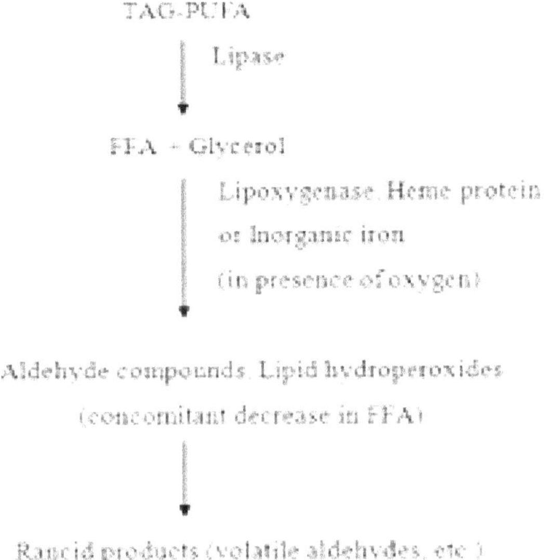

Figure 9: Mechanism of Rancidity in Oil
(TAG: Triacylglycerol; PUFA: Polyunsaturated Fatty Acid; FFA: free Fatty Acid).

Conclusion

Oil-bearing fruits, seeds and nuts occupy an important place in global agriculture by providing vegetable oils and high-protein meals for food, feed, and industrial uses. Vegetable oils are of great importance to human health as well as to the development of oil chemistry. It can be produced from many technological processes, solvent and mechanical. Extruding-expelling process followed by solvent extraction technology is still the most widely used process and the one with high oil yield; however due to the use of chemical (solvent) which affect the environment and human health, new technologies have been researched such as supercritical and enzymatic extractions. The future for such innovative technologies seems to be promising and requires further upscaling to pilot plant or industrial level. Polymer packaging materials under vacuum or inert gas flushing is presently the preferred choice for improving the shelf-life of processed oils and nuts. Knowledgeable efforts by industry, government, and consumers will promote continued improvement, and an understanding of the functional characteristics of packaging will help in maintaining the safety, wholesomeness, and quality of the packaged fats and oils.

REFERENCES

Agarwal, D.K., Singh, P., Chakrabarty, M., Shaikh, A.J., and Gayal, S.G. (2003). Cottonseed oil quality, utilization and processing. In: Technical bulletin for the Central Institute for Cotton Research, Nagpur, 25(4): 1-16. Retrieved from: www.cicr.org.in/pdf/cottonseed_oil.pdf

Akubude, V.C., Maduako, J.N., Egwuonwu, C.C., Olaniyan, A.M., Ajala, E.O., Ozumba, C.I., and Nwosu, C. (2017). Almond processing in Nigeria: Problems and Prospects. In Proceedings of 18th International Conference and 38th Annual General Meeting of the Nigerian Institution of Agricultural Engineers (NIAE): Dynamics of agricultural engineering for food and agro-industrial raw material production for economic recovery in Nigeria, at UMUDIKE, Nigeria, held on October 2017, 38: 177-183.

Anonymous. (2012). Improved Method of Shea Butter Processing. Published by Scientific Animations Without Borders. Retrieved from: https: //answers.practicalaction.org/our-resources/item/improved-method-of-shea-butter-processing

Asselstine, M., Mollo, J.M., Morales, J.M., and Papanikolopoulos, V. (2016). Cocoa Liquor, Butter, and Powder Production. In Senior Design Reports (CBE). Paper 88. University of Pennsylvania. Retrieved from: http: //repository.upenn.edu/cbe_sdr/88

Axtell, B. and Noble, N. (2006). Small-scale Oilseed Processing. Published by Practical Action, UK. pp: 1-10. Retrieved from: https: //answers.practicalaction.org/our-resources/item/small-scale-oilseed-processing

Belitz, H.D., Grosch, W., Schieberle, P. (2009). Edible Fats and Oils. In: Food Chemistry. Springer, Berlin, Heidelberg. pp: 640-653. Retrieved from: https: //doi.org/10.1007/978-3-540-69934-7_15

Chuansin, S., Vearasilp, S., Srichuwong, S., and Pawelzik, E. (2006). Selection of Packaging Materials for Soybean Seed Storage. Conference on International Agricultural Research for Development University of Bonn, Bonn.

Das Purkayastha, M., and Manhar, A.K. (2016). Nanotechnological Applications in Food Packaging, Sensors and Bioactive Delivery Systems. Nanoscience in Food and Agriculture, 2, pp: 59-128.

Dunford, N. (2008a). Oil and oilseed processing I. Food Technology Fact Sheet Robert M. Kerr Food and Agricultural Products Center, 158, 1–4.

Dunford, N. (2008b). Oil and oilseed processing II. Food Technology Fact Sheet Robert M. Kerr Food and Agricultural Products Center, 159, 1–4.

FAO. (2007a). Oilseeds: Oil Seeds Processing Toolkit. Instructional Manual- AGS. Retrieved from: www.fao.org/3/a-au139e.pdf

FAO. (2007b). Groundnut Oil: Oil Seeds Processing Toolkit. Instructional Manual-AGS. ITDG publication (The Manual Screw Press by K Potts and K Machell (ref no 7). Retrieved from: www.fao.org/3/a-au138e.pdf

Fellows, P., and Axtell, B. (1993). Appropriate Food Packaging: Materials and methods for small businesses. Published by International Labour Office Entrepreneurship and Management Development, Geneva, Switzerland. ISBN: 978-1-85339-562-8. https: //doi.org/10.3362/9781780442617

Gadani, B.C., Miléski, K.M.L., Peixoto, L.S., and Agostini, J.D.S. (2017). Physical and chemical characteristics of cashew nut flour stored and packaged with different packages Food Science and Technology, Campinas, 37(4), 657-662.

Guda, P., and Gadhe, S. (2017). Primary Processing of Cocoa. International Journal of Agricultural Science and Research (IJASR), 7(2): 457-462.

Hamm, W., Hamilton, R.J., and Calliauw, G. (2013). Edible oil processing. 2nd Edition. John Wiley and Sons, Ltd, UK. ISBN 978-1-4443-3684-9.

Irtwange, S.V., and Oshodi, A.O. (2009). Shelf-life of roasted cashew nuts as affected by relative humidity, thickness of polythene packaging material and duration of storage. Research Journal of Applied Sciences, Engineering and Technology, 1(3), 149-153.

Khalid, N., Khan, R.S., Hussain, M.I., Farooq, M., Ahmad, A., and Ahmed, I. (2017). A comprehensive characterisation of safflower oil for its potential applications as a bioactive food ingredient - A review. Trends in Food Science and Technology, 66: 176-186

Kurki, A., Bachmann, J., Hill, H., Ruffin, L., Lyons, B., and Rudolf, M. (2008). Oilseed processing for small-scale producers. ATTRA - National Sustainable Agriculture Information Service, pp: 1-16. Retrieved from: www.attra.ncat.org/attra-pub/PDF/oilseed.pdf

Lima, A.C., Garcia, N.H.P., and Lima, J.R. (2004). Obtenção e caracterização dos principais produtos do caju. Boletim CEPPA, 22 (1), 133-144.

Lima, J.R., and Borges, M.F. (2004). Armazenamento de amêndoas de castanha de caju: influência da embalagem e da salga. Revista Ciência Agronômica, 35(1), 104-109.

Mag, T. (1994). Canola: Seed and Oil Processing. Canola Council of Canada, pp: 1-6. Retrieved from:

Moore, J. (2002). Oil seed processing plants: A Guide to Sealing. AESSEAL Environmental Technology, Rotherham, U.K. Retrieved from: https: //arthomson.com/wp-content/uploads/2013/04/Resources-Mechanical-AESSEAL-Guides-OILSEED.pdf

Naik, B., and Kumar, V. (2014) Cocoa Butter and Its Alternatives: A Reveiw. Journal of Bioresource Engineering and Technology, 1: 7-17

Noble, N. (2008). Coconut Processing. Published by Practical Action, UK. pp: 1-4. Retrieved from: https: //answers.practicalaction.org/our-resources/item/coconut-processing

Pighinelli, A.L.M.T., and Gambetta, R. (2012). Oil Presses (Fact sheet). DOI: 10.5772/30699. Retrieved from: https: //www.intechopen.com/books/oilseeds/oil-presses

Premazon, A. (2016). Method for packaging and delivering Shea butter and Cocoa butter. United States Patent Application Publication.

Rainey, C., and Nyquist, L. (1997). Nuts: nutrition and health benefits of daily use. Nutrition Today, 32(4), 157-163. http: //dx.doi. org/10.1097/00017285-199707000-00006.

Robertson, G.L. (1993). Food packaging: principles and practice. Marcel Drekker, New York, pp 680.

Russell, A. (2008). Manual Screw Press. Published by Practical Action, UK. Pg 1-12. Retrieved from: https: //answers.practicalaction.org/our-resources/item/manual-screw-press

Sajilata, M.G., and Singhal, R.S. (2006). Effect of irradiation and storage on the antioxidative activity of cashew nuts. Radiation Physics and Chemistry, 75(2), 297-300. Retrieve from: http: //dx.doi.org/10.1016/j.radphyschem.2005.07.004.

Sedlacekova, Z. (2017). Food packaging materials comparison of materials used for packaging purposes. B. Sc. Thesis submitted to Helsinki Metropolia University of Applied Sciences.

Shakerardekani, A., and Karim, R. (2012). Effect of different types of plastic packaging films on the moisture and aflatoxin contents of pistachio nuts during storage. Journal of Food Science and Technology. DOI 10.1007/s13197-012-0624-0.

Sarantópoulos, C.I., Oliveira, M.L., Coltro, L., Vercelino, A.R.M., and Corrêa, G.E.E. (2002). Embalagens plásticas flexíveis, principais polímeros e avaliação de propriedades. Campinas: CETEA/ITAL.

Shin, J., and Selke, S.E.M. (2014). Food Processing: Principles and Applications, Second Edition. Edited by Stephanie Clark, Stephanie Jung, and Buddhi Lamsal. John Wiley and Sons, Ltd.

Shukla, B.D., Srivastava, P.K., and Gupta, R.K. (1992). Oilseeds Processing Technology. In Technology Mission on Oilseed, Central Institute of Agricultural Engineering, Bhopal, Published by M/s. Maria Industries, Bhopal, India. pp: 1-245. Retrieved from: https: //idl-bnc-idrc.dspacedirect.org/bitstream/handle/10625/11596/IDL-11596.pdf?sequence=1

Spurgeon, T. (2017). Refining linseed oil. In Living Craft- A Painter's Process, Zoetrope, 12th Edition. ISBN: 978-0-615-47394-9

Swetman, T. (2008). Oil Extraction. Published by Practical Action, UK. pp: 1-10. Retrieved from: https: //answers.practicalaction.org/our-resources/item/oil-extraction

Vijayalakshmi, N.S. (2005). Pacakaging aspects of oils, fats and vanaspati. Central Food Technological Research Institute, Mysore.

Warra, A.A. (2011). Sesame (sesamum indicum L.) seed oil methods of extraction and its prospects in cosmetic industry: a review. *Bayero Journal of Pure and Applied Sciences*, 4(2): 164 – 168

Wickens, G.E. (1995). Edible nuts: Non-wood Forest Products 5. FAO - Food and Agriculture Organization of the United Nations, Rome. M-37, ISBN 92-5-103748-5

Wijeratne, W.B., Wang, T., & Johnson, L.A. (2004). Extrusion-Based Oilseed Processing Methods. In: Nutritionally Enhanced Edible Oil (ed. N. T. Dunford and H. B. Dunford). AOCS Press. pp: 21-88.

Chapter 18

Processing and Packaging Techniques for Meat, Fish and Poultry

Ashok Kumar Yadav and Lakshita

Introduction

World's largest livestock population in the world is well owned by India. About 20.5 million people depend upon livestock for their livelihood. Livestock contributes about 16 per cent to the income of small farm households. Livestock and poultry in India plays a condemning aspect in agriculture economy on the terms of milk, meat, eggs *etc.* and provide extensible provisions during period of economic upturns and fender against crop faux pas . Food manufacturers need to produce meat, poultry and fish products which are not only healthy but can also meet the consumer demands. Even in its developing stage Meat industry in India ranks as the top food industry on a global platform. Demand for processed meat and poultry products shall boom on a great exponential ratio due to its increasing consumption rate. India accounts the largest livestock resources in the world.

India accounts 50 per cent of buffaloes, 15 per cent cattle and goat. Livestock contributes 11 per cent to nation economy. Earlier consumption was limited to a small section of consumers. Although the scenario has changed now. The consumption rate is increasing. Still because of ethical reasons slaughtering of cows is banned except in West Bengal and Kerala. Out of all the livestock; buffalo has proved to show its full export potential.

Meat Production

Meat constitutes very high biological value. In Fourth Five Year Plan, eight bacon factories were established. Also development took place in laying advanced slaughterhouses.

Traditionally meat industry is characterized by butcher workers who had poor knowledge, untrained, non- scientific and with poor hygiene practices. Today there are 3600 licesensed slaughterhouses in India. They require modernization. According to an analysis Europe leads in production followed by Asia. Despite of acquiring such vast raw material, India contributes less than 1 per cent to the world meat population. Exports from India comprise of fresh chilled meat, frozen meat and frozen meat products. Figure 1 shows meat supply on a global platform.

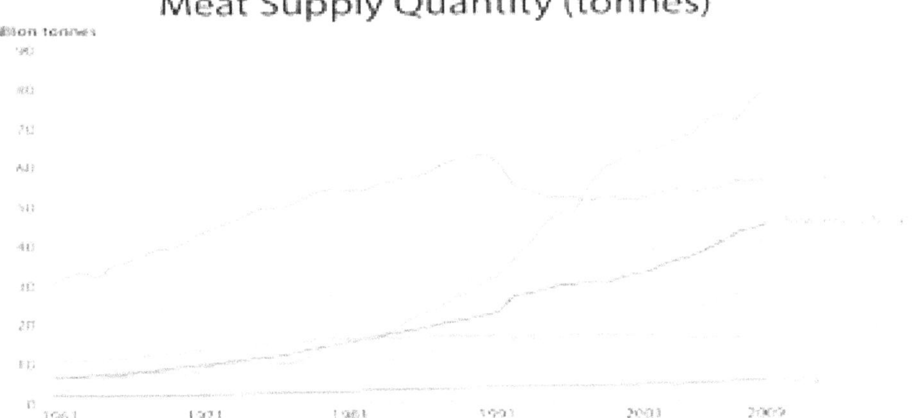

Figure 1: Meat Supply on A Global Platform.

India gains a major portion of profit by exporting buffalo meat, poultry and beef to Gulf countries. One of the biggest reasons is its geographic proximity. Solely for export purpose India needs installation of modern and hygienic slaughterhouses with chilling facilities.

Processing of Meat and Meat Products

Processing refers to treatments involved carried forward by salting that includes chemical and physical change in the natural state of meat. Processing's main purpose is to increase shelf life of the meat.

Basic Processing Procedures

Comminution- refers to subdivision or reduction of raw meat into meat pieces or particles. Comminution is done with the help of meat mincer for coarse ground products whereas bowl chopper is also employed for making fine meat emulsion.

Emulsification- Meat emulsion comprises of a dispersed phase of solid or liquid fat droplets and a continuous phase of water containing salt and proteins.It is very important to maintain low temperature during emulsion formation in order to avoid melting of fat particles, denaturation of soluble proteins and lowering of viscosity. This is done by adding ice flakes instead of chilled water during chopping. The emulsion breakdown can occur due to sudden exposure to high temperature because of coalescence of finely dispersed fat particles into larger ones.

- **Meat extension:** Those non- meat food items that are added in meat products are usually called extenders *i.e.* fillers, binders, emulsifiers or stabilizer. Meat extensions can be sodium alginate, potato starch, flour, soya products *etc.* It should be done on large scale for mass availability of meat products.
- **Preblending:** Mixing of a part or curing ingredients with ground meat in specified proportion is called Prebelnding. This helps to form stable emulsions.
- **Hot Processing:** It refers to the processing of carcass this has various advantages. It improves cooking yield and sensory quality.

Cooking- Cooking temperature and time varies according to the cut. Various methods are applied (Roasting, frying, pressure cooking, hoist heat cooking, stewing, simmering, microwave cooking).

Processing of Poultry Products

India's major portion of poultry food processing comprises of chicken. Consumption style has changed from traditional (tandoori chicken, chicken seekh kabab, chicken kofta, chicken tikka) to modern methods (barbeque, chicken patties, chicken sausages).

Most processing procedures involve cooking, which brings about a number of changes in meat. Cooking coagulates and denatures the meat proteins altering their solubility. It inactivates or destroys the indigenous proteolytic enzymes. Cooking invariably decreases the water content of meat, lowering the water activity level. It deepens the flavour and modifies the texture. In addition, considerable number of micro organisms is killed enhancing the storage life of meat.

Smoking and cooking take place simultaneously in most cured meat products. During smoking, carbonyl groups present in smoke react with amino groups of protein whereas, phenols and polyphenols in smoke could react with sulphydryl group of protein. Both the reactions cause some loss of available amino acids thereby decreasing the nutritive value of protein. Water soluble vitamins may also be affected to some extent. In fact, some destruction of thiamine (Vitamin B1) is inevitable, although effect on riboflavin (Vitamin B2) and niacin may be very little. Smoking process can be nutritionally advantageous because it helps to stabilize the fat soluble vitamins due to anti oxidant properties. Canning process is particularly detrimental to the water soluble vitamins present in meat. In canning, about 20 40 per cent of thiamine, 10 per cent each of riboflavin and niacin, 20 per cent of biotin and 20 30 per cent pantothenic acid are destroyed.

Processing changes the nutritional characteristics of fresh meat to some extent. The percentage of protein is slightly decreased whereas that of fat and minerals is increased. The percentage of minerals is generally increased due to added salt and seasonings. Besides, processed meats have more caloric values as compared to fresh meat due to the addition of fillers, binders and other extenders in the form of cereal flours or skimmed milk powder and frequently some fats. Meat products contain enough of vital minerals such as iron, sodium, potassium and phosphorus. However, these are particularly deficient in calcium. Much of the requirement of iron, which is an absolute necessity for health upkeep, can be made available by the meat products. Anemic patients are usually recommended a liver diet because of its high iron content. A regular intake of iron is must for the proper synthesis of haemoglobin, myoglobin and certain enzymes due to very limited capacity to store iron in the body. All the water soluble vitamins are present in meat products but thiamine, riboflavin and niacin are present in significant quantities. Liver containing meat products are extremely rich in vitamin

Processing of Fish Products

The fish flesh separator works by squeezing the flesh from the skin and bones of fish and passing the flesh through perforations on a stainless steel plate or drum. The skin and bones do not pass through the perforations and are separated by the machine. The comminuted fish flesh can then be used in many food products *e.g.*, fish sticks, sandwich spreads, hors d'oeuvres, *etc.*The flesh separator has, (1) a stainless steel drum (approx 8½ in. in length and 6½ in. in diameter) perforated with closely spaced holes in. in diameter, and (2) a continuous rubber belt (approx 41 in. long and 8¼ in. wide) which runs over a series of moving rollers. Position of the rollers is adjustable to regulate the pressure exerted against the drum by the rubber belt. Headed and gutted fish are fed into the machine and pass between the belt and perforated drum. Pressure applied by the belt on the fish forces the fish flesh through the perforations of the drum while the skin and bones pass to the waste discharge chute. The operator can adjust the pressure exerted by the belt to remove most of the light meat during the first pass through the machine. If it is desired to remove the remaining light meat and dark flesh under the skin, the waste can be passed through the machine again after pressure exerted by the belt has been increased. Alternatively, pressure exerted by the belt can be adjusted to the maximum so that one pass removes all the fish flesh both light and dark.

The loss of drip in fresh and thawed fillets is effectively controlled by the addition of small amounts of sodium tripolyphosphate (TPP) to the fillets prior to either their distribution fresh to retail or in preparation for freezing. If fish fillets are treated with sodium tripolyphosphate, the surface layer of protein is modified so that its ability to hold water is greatly increased. This surface layer of modified protein prevents the escape of fluid from the interior of the fillet, with the result that drip formation is prevented. Sodium tripolyphosphate can be applied to fillets by two methods fillets may be dipped in appropriate concentration of TPP solutions, or the solution can be sprayed directly onto the fillets. With fillets that are to be frozen, the most effective dip solution is 12 per cent TPP containing 4 per cent salt. The drip during thawing of red snapper and sole fillets is reduced about 50 per cent by TPP treatment before freezing.

Raw frozen fresh fish is used, such as hake or other less costly fish which is in plentiful supply. The frozen fish is ground up, and then placed in an open stainless steel mixing kettle. To the ground, fish is added isopropanol solvent and extraction takes place, at well below the ambient temperature. The resulting slurry is separated in a centrifuge, and the isopropanol phase goes to solvent recovery while the solids enter a second stage extractor (a covered, jacketed mixing vessel). Here, extraction takes place at about 170°F, near the boiling point of the solvent. The phases are separated again, and third stage extraction proceeds at 170°F, using fresh (recovered) solvent. After final centrifuging, solid material goes to a vacuum tumbler dryer that removes residual solvent. Once dried, the fish solids enter a mill where they are ground to a light gray powder. There are different processing methods followed for fish products.

Butt Method

In summer all fish must be salted in butts or other water tight containers, but in winter they are often salted in kenches. A butt is a large barrel (formerly a molasses hogshead) and is about 3 ft in diameter and 4 ft high. The salters throw the cod face (flesh side) up into butts and sprinkle salt uniformly over each layer. When coarse salt is used, 6.5 7 bu are required for each butt of fish. If finer salt is used, a slightly larger quantity is often added and in hot weather more salt is required. The fish are piled high above the top of the butt and the last few layers which are exposed are placed with backs up. A pile of salt is placed on top of the fish. The salt and fish settle slowly and within a day or two sink below the top of the butt. After the fish have settled, a bushel or more of salt is placed on top. About 3 weeks' time is required for the completion of the salting process.

Kench Method

During winter or on board schooners, these fish are often salted in kenches. A kench is a regular pile of fish made by laying them on their backs with napes and tails alternating. A considerable quantity of salt is spread over each layer. The top layer of fish is turned with backs up. As the salt extracts the water from the fish, it runs to the floor and is drained off. Since the fish do not stand in brine, it is much more difficult to obtain uniform penetration of salt by the kench method therefore; there is much greater danger of spoilage (souring) by this procedure than by the butt method. About 20 lb of salt are used on each 100 lb of fish.

Drying

When fish are to be dried, they are removed from butts or kenches and washed with sea water or brine to remove any objectionable slime. They are then hauled to a building or room having a good concrete floor. Here they are clenched on frames about 8 in. above the floor. Weights of various kinds are placed on the kenches to press surplus brine out of the fish. The fish drain and slowly dry in the kenches the longer they remain on kenches the less time they must remain on the flakes for final drying.

Salting Mackerel

Salting of mackerel begins at sea aboard the fishing trawler. Each mackerel is split so that it will lie open and flat after the viscera has been removed. The splitting knife is held by the fingers and guided by the thumb and slides along the upper side of the fish. After splitting, each fish goes to a tray where the gibber opens the fish with a jerk causing it to break lengthwise along the lower end of the ribs. Viscera and gills are removed and the fish is thrown into a wash barrel partly filled with clean salt water the fish is thrown into the barrel open and face down. Here, the blood is soaked from the fish. They remain in the salt water until the splitting is finished, which may be 6 8 hours, or even longer. Then the deck is cleaned up and the men proceed to salting. The mackerel are removed from the salt water by emptying the wash barrels onto the clean deck and are rinsed by throwing buckets of clean water over them. They are then dipped into fine salt, such as Liverpool No. 2, and placed in a barrel flesh side down, except that 2 3 bottom layers have the flesh side up. Coarse salts are not used as they give the fish a ragged appearance. The barrels of salted mackerel are then not disturbed until the vessel arrives in port. Here they are removed to a cool storehouse and remain until needed for market. From time to time, additional brine is added to the barrels to replace any loss by leakage or evaporation. This is important as any exposed fish soon rust and cannot be marketed.

Salting Salmon

In dressing salmon for pickling, first remove the head then split the fish along the back ending the cut with a downward curve at the tail. Remove the viscera and of the backbone scrape away the blood, gory, and black stomach membrane. Thoroughly scrub and wash the dressed fish in cold water. Place them in pickling butts with about 15 lb of half ground salt to every 100 lb fish. Lay fish in a tier, flesh side up, sprinkle salt evenly over each tier and repeat until tank is full. Several boards are then laid across the fish with the boards weighted down in order to keep the fish sub merged in the pickle, which will form. Allow the fish to stand in the pickle about 1 week, holding the brine at about 90°F. Remove the fish from the pickle, rub clean with a scrub brush, and repack in market barrels, using 1 sack of salt to every 3 barrels of 200 lb fish. About 40 52 red salmon, 25 35 Coho salmon, 70 80 humpback salmon, 10 14 king salmon, and 25 30 dog salmon will be required to fill each when packing a market barrel of dressed, salted salmon.

Brine Salting

In the preparation of brine salted mullet, the fish should be dressed as soon as possible after removal from nets or seines (within 6 hours at the most). Split fish down the back and along the backbone the heads are cut through so that the fish can be laid out flat. The viscera can also be easily pulled out after cutting through the heads. Roe, in season, is usually separated and dried, sailed or smoked as a profitable by product. Gills are removed and the appearance of the product is improved if the black membrane of the belly cavity is also removed. Heads and foreparts of the backbone are often taken out of larger fish (those weighing more than 1½ lb). After thorough cleaning, the fish are washed in clean sea water or light brine to remove blood and slime. Soaking for ½ hr in brine will make this cleaning.

Repacking can be done any time after the fish are struck. They are graded and sorted for size and condition, and any remaining blood, salt, scales, *etc.*are rinsed off in the brine. Repacking is usually done in smaller kegs or barrels with a layer of salt on the bottom, and on the top with a light sprinkling between fish layers. After the containers are headed" sufficient concentrated brine is added to fill the containers. The product should be refrigerated if it is to be stored for very long.

Bismarck Herring

These are prepared from herring of uniform size. The fish are first washed in a special washing machine, consisting of a large revolving drum equipped with a spray of water. The washed and scaled fish are then cleaned, beheaded, and boned. They are then rinsed with water and brushed inside to remove the black lining of the belly cavity. They are then placed in salt brine for 2 3 hours. Following this, they are put into a vinegar pickle (from 5 6 per cent acetic acid) containing a moderate amount of salt. They remain in the pickle for 2 days after which they are packed tightly in boxes with slices of onion, and some pepper and mustard seed. A vinegar sauce (from 2.2 to 2.4 per cent acetic acid) containing some sugar is added and the box is closed and wrapped for marketing. The herring are usually shipped immediately but, if stored, are kept in cool, dry rooms.

Mustard or Kaiser Friedrich Herring

These are prepared in exactly the same manner as Bismarck herring However, a mustard sauce, instead of sweetened vinegar, is added when the fish are packed. The mustard sauce is usually prepared in special factories and is merely thinned preparatory to use in the marinating factory.

Canned Marinated Herring

These are prepared by washing choice herring of uniform size in a revolving cylindrical screen, which also removes scales. The fish are then dressed by removing heads, tails, and bones by hand they are then rinsed and placed in about 75 per cent brine for 2-3 hours. From the brine tanks they are transferred to a vinegar pickle of 5-6 per cent acidity, containing considerable salt. After about 2 days the fish are ready to pack in cans, in which they are placed in layers, with onions, peppers, and mustard seed on each layer. A small amount of 2¼ per cent vinegar and a little sugar is added to each can. Cans are exhausted, sealed, and processed.

Meat and Poultry Packaging Materials

The packaging materials listed in this section may be safely subjected to irradiation subject to the provisions and to the requirement that no induced radioactivity is detectable in the packaging material itself. Oxygen in the air hastens both the chemical breakdown and microbial spoilage of many foods. To help preserve foods longer, scientists have developed ways to help overcome the effects of oxygen. Vacuum packaging, for example, removes air from packages and produces a vacuum inside. Modified atmosphere packaging (MAP) and controlled atmosphere packaging (CAP) help to preserve foods by replacing some or all of the oxygen in the air inside the package with other gases such as carbon dioxide or nitrogen. Yes, it is safe to freeze meat or poultry directly in its supermarket wrapping, but this type of wrap is permeable to air. Unless you will be using the food in month or two, overwrap packages with airtight heavy-duty foil or freezer wrap. This should protect the product from freezer burn for longer storage possibly. While extremely rare, a toxin produced by Clostridium botulinumis the worst danger in canned goods. Never use food from containers that show possible "botulism" warnings: leaking, bulging, or badly dented cans; cracked jars or jars with loose or bulging lids; canned food with a foul odor; or any container that spurts liquid when opening. Don't taste such food even a minuscule amount of botulinum toxin can be deadly. Can linings might discolor or corrode when metal reacts with high-acid foods such as tomatoes or pineapple. As long as the can is in good shape, the contents should be safe to eat, although the taste, texture and nutritional value of the food can diminish over time.

Cans that freeze accidentally, such as those left in a car or basement in sub-zero temperatures, can present health problems. If the cans are merely swollen – and you are sure the swelling was caused by freezing – the food may still be usable. If seams have rusted or burst, throw the cans out immediately. Discard frozen cans that have been allowed to thaw above 40°F (4.4°C).

Let the intact can thaw in the refrigerator before opening. If the product doesn't look and/or smell normal, throw it out. Do not taste it! If the product does look and/or smell normal, thoroughly cook the contents right away by boiling for 10 to 20 minutes. Products can then be refrigerated or frozen for later use.

Packaging that can be purchased or is available to use in grocery stores (such as produce or meat bags) have been approved by the FDA for food contact. These include: Plastic Wraps and Storage Bags - Consumer plastic wraps and bags are made from three major categories of plastics: polyethylene (PE), polyvinylidene chloride (PVDC) and polyvinyl chloride (PVC). The plastic resins are petroleum derivatives. Plasticizers, colorants or anti-fog compounds may be added. In-store Produce Bags - Typically made from polyethylene or other plastic film, these bags are used for consumer in-store packaging of fruits and vegetables. Do not use for cooking; the thin plastic may melt or burn. Oven Cooking Bags - Both the bags and their closure ties are made from heat-resistant nylon. They can be used in a microwave oven or in a conventional oven set no higher than 400°F (204.4°C).

Aluminum Foil is 98.5 per cent aluminum with the balance primarily from iron and silicon to give strength and puncture resistance. The molten alloy is rolled thin and solidified between large, water-cooled chill rollers. During the final rolling, two layers of foil are passed through the mill at the same time. The side coming in contact with the polished steel rollers becomes shiny; the other side comes out dull. It does not make any difference which side of the foil contacts the food. Freezer Paper - white paper coated on one side with plastic to help keep air out of frozen foods, thus protecting against freezer burn and loss of moisture. Parchment Paper - an odorless and tasteless paper made from cotton fiber and/or pure chemical wood pulps. It may be waxed or coated and is greaseproof or grease resistant. Parchment paper is primarily used in baking as a pan liner or to wrap foods in for cooking. Wax Paper - a triple-waxed tissue paper; made with a food-safe paraffin wax which is forced into the pores of the paper and spread over the outside as a coating. Grocery bags are not intended or formulated for cooking foods. Levels of components such as metal fragments, glue and chemicals may be present at higher-than-acceptable limits and can migrate into the food. These bags may not necessarily be sanitary, particularly since they may be stored under variety of conditions. The use of plastic trash bags for food storage or cooking is also not recommended because they are not food grade plastic and chemicals from them may leach into the food. Pinholes in foil or a blue liquid that may form on the food that has come in contact with the foil are not harmful. These reactions can occur when salt, vinegar, highly acidic or highly spicy foods come in contact with aluminum foil. The product is a harmless aluminum salt and presents no safety problem if consumed, however it can be trimmed off to improve the food's appearance. Some aluminum salts are used in antacid medicines for the treatment of stomach disorders.

Small amounts of chemicals from packaging materials can migrate into foods. It is for this reason that each packaging material must be regulated for a specific use by the FDA. However, sometimes consumers misuse packaging materials in ways not intended or anticipated when the material was regulated for food use. For example, cold food storage containers–such as cottage cheese cartons and margarine tubs–used for refrigerator or freezer storage of foods are intended for those uses only. They have not been tested or approved for another use, including cooking. Do not use these types of containers for heating food. They are not heat stable and chemicals from the plastic may migrate into the food during heating.

Microwave food in packaging materials only if the package directs, and then use only one time. Materials suitable for microwaving include oven bags, wax paper and plastic wrap. Do not let the plastic wrap touch the food, and donor reuse the wrap. Foam insulated trays and plastic wraps on fresh meats in grocery stores are not intended by the manufacturer to be heated and may melt when in contact with hot foods, allowing chemical migration into the food. In addition, chemical migration from packaging material to a food does not necessarily require direct contact. Excessive heat applied to a closed container may drive off chemical gases from the container that can contaminate the enclosed food.

These types of plastic products should not be used in a microwave oven because they are subjected to heat when thawing or reheating. To avoid alchemical migration problem, remove meats from their packaging.

Plastic packaging materials should not be used at all in conventional ovens. They may catch on fire or melt, causing chemical migration into foods. Sometimes these materials are inadvertently cooked with a product. For example, giblets may be accidentally cooked inside the turkey in their packaging or a beef roast may be cooked with the absorbent pad from the fresh meat packaging underneath.

Plastic wrap, foam meat trays, convenience food dishes, and egg cartons have been approved for a specific use and should be considered one-time-use packaging. Bacteria from foods that these packages once contained may remain on the packaging and thus be able to contaminate foods or even hands if reused.

Storage times of meat and poultry products vary depending upon their processing method and packaging. Fresh meat in foam trays and shrink wrap, and opened packages of lunch meats may be refrigerated 3 to 5 days; ground meats, poultry+9 and variety meats, 1 to 2 days. Unopened packages of hot dogs and lunch meats can be stored 2 weeks. If processed meat and poultry products bear "use-by" dates, observe them.

Purpose of Packaging Meat and Meat Products

Packaging should protect the product from contamination and prevent it from spoilage, and at the same time it should:

- Extend shelf life of a product
- Facilitate distribution and display
- Give the product greater consumer appeal
- Facilitate the display of information on the product

Purpose of Packaging of Fish Products

In order to maintain good quality of fresh fish during transportation, fish boxes made of suitable materials should be used. When purchasing fish boxes the six following requirements should be remembered; they should:

- be of a suitable size for the range of fish to be handled or the product to be put into them
- be of a convenient size for manual handling or lifting by mechanical equipment
- be stackable such that the weight of the containers on top rests on the containers underneath and not on the fish
- be constructed of impervious non-staining materials
- be easy to clean
- provide drainage for melted ice

Retail Packaging for Freshwater Fish Products

Basic packaging materials include paper, cartons, sheets of metal, metal foils and many kinds of plastics. Despite the rapid growth in use of plastics, the role of paper and carton as packaging materials does not decrease.

Kraft paper or carton is often laminated with polyethylene or aluminum foil which renders them waterproof. Such material is used for production of trays for packaging of fresh or frozen products. More often, trays are made of plastic materials such as polystyrene or expanded polystyrene. Expanded polystyrene is frequently used but it is partly oxygen-permeable and so those products which are sensitive to rancidity have to be additionally overwrapped or skin-packed with suitable film.

Trays used for packing are generally overwrapped with a protective film, often with PE wrapping which shrinks. The film shrinking is achieved by use of hot air or hot water.

Stretch wrapping is often used for products which are heat-sensitive. The film is stretched over the product manually (very often in the supermarket) or by machine. Foils used as wrapping or bags for packing of trays with product must be puncture-proof, extensible and impervious to gases like oxygen.

Plastics such as polyethylene film or copolymer of ethylene and vinyl acetate are very often used for packing of frozen products. Polyethylene packs can be produced manually using pre-made bags. An impulse or bar sealer is used to seal the bags which are hand-filled.

In order to improve the barrier properties of packages laminates are used, for example polyester/polythene. Products which are particularly sensitive to oxygen are vacuum-packed. During the sealing operation, air is removed from the package. A laminate nylon/polythene is commonly used as packaging material. This type of packaging is used, for instance, for smoked trout which are arranged on a board with, for example, a coated texture. Numerous machines exist for vacuum-packing with single, double or continuous chambers. Vacuum-sealing machines can additionally be equipped with a modified atmosphere packing system (MAP). Immediately on removing the air from the package a mixture of gases is pumped in. Usually this mixture consists of 30 per cent nitrogen, 40 per cent carbon dioxide, and 30 per cent oxygen. In the case of fat fish the oxygen is replaced by nitrogen. This method is increasingly used for packing fresh fish. The MAP products have to be stored at the temperatures lower than 3° C because of *C. botulinum* hazard. MAP packages consist of two kinds of foil. The bottom film is foil-rigid or semi-rigid. This foil is formed by, for example, extrusion and the resultant tray is moved to the packing section. Because of product drip it is placed on an absorbing board. The top web is drawn over the filled trays and sealed round the edges. The pack may be evacuated or gas-flushed before sealing.

Vacuum-skin packaging is becoming more common for packing smoked fish. In this process the wrapper is heated and wrapped over the product, the film molding completely to the product shape and sealing the product completely, forming an extra skin.

Conclusion

When compared with other developing regions, conditions in the majority of traditional Asian slaughterhouses and slaughter slabs are extremely poor. Many of them do not fulfil minimum hygienic requirements to produce safe and wholesome meat. Improvements are needed urgently. Flesh foods are categorized as meat, poultry, or fish. Quality of the various meats is judged on the basis of texture, marbling, and overall palatability. Changes in consumer preferences have led to innovations and developments in new packaging technologies. Active packaging is useful for extending the shelf life of fresh, cooked and other meat products. Forms of active packaging relevant to muscle foods include; oxygen scavengers, carbon dioxide scavengers and emitters, drip absorbent sheets and antimicrobial packaging. Antimicrobial packaging is gaining interest from researchers and industry due to its potential for providing quality and safety benefits. Future research in the area of microbial active packaging should focus on naturally derived antimicrobial agents, bio preservatives and biodegradable packaging technologies.

REFERENCES

Glossary of Packaging Terms, Sixth Edition, Compiled and Published by the Packaging Institute International, 1988, ISBN 0-86512-951-7.

Packaging Foods with Plastics, by Wilmer A. Jenkins and James P. Harrington, Technomic Publishingcompany, Inc., 1991, ISBN 87762-790-8.

Chapter 19

Role of Warehousing Corporation and Food Corporation of India on Post Harvest Conservation

Shikha Bathla and Tanu Jain

Introduction

The warehousing scheme in India is an integrated scheme of scientific storage, rural credit, price stabilization and market intelligence and is intended to supplement the efforts of co-operative institutions. This corporation was established as a statutory body in New Delhi on 2nd March 1957. The Central Warehousing Corporation provides safe and reliable storage facilities for about 120 agricultural and industrial commodities. There are 432 storage warehouses in India and having capacity to store 9.96 million tonnes of food and there are 17 religion offices in India under Central Warehousing Corporation. Warehouses are scientific storage structures especially constructed for the protection of the quantity and quality of stored products. Warehousing may be defined as the assumption of responsibility for the storage of goods. It may be called the protector of national wealth, for the

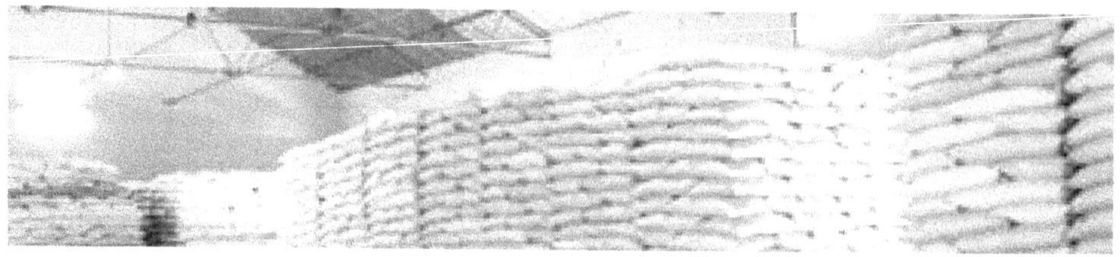

Figure 1: Punjab State Warehousing Corporation.

produce stored in warehouses is preserved and protected against rodents, insects and pests, and against the ill-effect of moisture and dampness.

Functions

- To acquire and build godowns and warehouses at suitable places in India.
- To run warehouses for the storage of agricultural produce, seeds, fertilizers and notified commodities for individuals, co-operatives and other institutions,
- To act as an agent of the govt. for the purchase, sale, storage and distribution of the above commodities.
- To arrange facilities for the transport of above commodities.
- To subscribe to the share capital of state Warehousing corporations and
- To carry out such other functions as may be prescribed under the Act.
- The Central Warehousing Corporation is running air-conditioned godowns at Calcutta, Bombay and Delhi, and provides cold storage facilities at Hyderabad.
- Special storage facilities have been provided by the Central Warehousing Corporation for the preservation of hygroscopic and fragile commodities.
- The corporation has also evolved techniques for the storage of spices, coffee, seeds and other commodities.

Features of Warehouses

- **Scientific Storage**: Here, a large bulk of agricultural commodities may be stored. The product is protected against quantitative and qualitative losses by the use of such methods of preservation as are necessary.
- **Financing**: Warehouses meet the financial needs of the person who stores the product. Nationalized banks advance credit on the security of the warehouse receipt issued for the stored products to the extent of 75 to 80 per cent of their value.
- **Price Stabilization**: Warehouses help in price stabilization of agricultural commodities by checking the tendency to making post harvest sales among the farmers. Farmers or traders can store their products during the post harvest season, when prices are low because of the glut in the market. Warehouse helps in staggering the supplies throughout the year. They thus help in the stabilization of agricultural prices.
- **Market Intelligence**: Warehouses also offer the facility of market information to persons who hold their produce in them. They inform them about the prices prevailing in the period, and advise them on when to market their products. This facility helps in preventing distress sales for immediate money needs or because of lack of proper storage facilities. It gives the producer holding power; he can wait for the emergence of favourable market conditions and get the best value for his product.

Working of Warehouses

Acts

The warehouses (CWC and SWCs) work under the respective Warehousing Acts passed by the Central or State Govt.

Eligibility

Any person may store notified commodities in a warehouse on agreeing to pay the specified charges.

Warehouse Receipt (Warrant)

This is receipt/warrant issued by the warehouse manager/owner to the person storing his produce with them. This receipt mentions the name and location of the warehouse, the date of issue, a description of the commodities, including the grade, weight and approximate value of the produce based on the present prices. In order to educate the farmers on scientific storage of food grains and post harvest loss minimization, CWC introduced its Farmers Extension Service Scheme in 1978-79 wherein the technical staff posted at its warehouses visit the adjoining villages and train the farmers on Post Harvest Technology. The scheme is presently in operation through 305 rural based warehouses. Further, to encourage the farmers and motivate them to avail public warehousing facilities, CWC offers a rebate of 30 per cent on its storage charges for the farmers' stocks. A **Warehouse Receipt**, which is a negotiable instrument, is issued to the farmers, who can obtain institutional credit on pledge of the Warehouse Receipt and thus avoid distress sale and also known as Farmers Extension Service Scheme.

Use of Chemicals

The produce accepted at the warehouse is preserved scientifically and protected against rodents, insects and pests and other infestations. Periodical dusting and fumigation are done at the cost of the warehouse in order to preserve the goods.

Financing

The warehouse receipt serves as a collateral security for the purpose of getting credit.

Delivery of Produce

The warehouse receipt has to be surrendered to the warehouse owner before the withdrawal of the goods. The holder may take delivery of a part of the total produce stored after paying the storage charges.

Types of Warehouse

On the Basis of Ownership

1. Private warehouses: These are owned by individuals, large business houses or wholesalers for the storage of their own stocks. They also store the products of others.

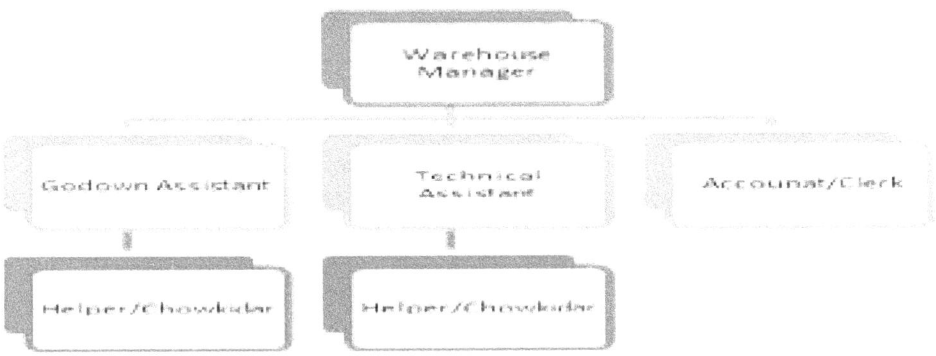

Figure 2: Organization Structure for Warehouse.

2. Public warehouses: These are the warehouses, which are owned by the govt. and are meant for the storage of goods.
3. Bonded warehouses: These warehouses are specially constructed at a seaport or an airport and accept imported goods for storage till the payment of customs by the importer of goods. These warehouses are licensed by the govt. for this purpose. The goods stored in this warehouse are bonded goods. Following services are rendered by bonded warehouses:

On the Basis of Types of Commodities Stored

1. General Warehouses: These are ordinary warehouses used for storage of most of foodgrains, fertilizers, *etc.*
2. Special Commodity Warehouses: These are warehouses, which are specially constructed for the storage of specific commodities like cotton, tobacco, wool and petroleum products.
3. Refrigerated Warehouses: These are warehouses in which temperature is maintained as per requirements and are meant for such perishable commodities as vegetables, fruits, fish, eggs and meat.

On the Basis of Committee

National Cooperative Development Warehousing Board

The National Co-operative Development & Warehousing Board as well as the Central Warehousing Corporation was to be established by statue. The State Warehousing Companies were also to be established by statutes. The composition of the National Co-operative Development and Warehousing Board was also suggested. This was to be a high-powered Board with a Standing Committee. The Board was to act through the All India Warehousing Corporation (Central Warehousing Corporation) for the discharge of such of its functions as they pertained to warehousing. In regard to distribution, the Board's functions were principally to be those of planning, directing and co-ordination of the activities of the All-India Warehousing Corporation and the State Warehousing Companies.

The Development Board was Set Up With the Following Objectives

- ✰ To provide funds to warehousing corporations and the State Governments for financing co-operative societies for the purchase of agricultural produce on behalf of the Central Government.
- ✰ To advance loans and grants to State Governments for financing cooperative societies engaged in the marketing, processing or storage of agricultural produce including contributions to the share capital of these institutions.
- ✰ To subscribe to the share capital of the Central Warehousing Corporation and advance loans to State Warehousing Corporations and the Central Warehousing Corporation.
- ✰ To plan and promote the programmes through co-operative societies for the supply of inputs for the development of agriculture and to administer the National Warehousing Development Fund.

All India Warehousing Corporation-Central Warehousing Corporation

The All-India Warehousing Corporation was to have an authorized share capital of Rs.20.00 crores to be subscribed by the National Cooperative Development & Warehousing Board as well as other institutions like banks, co-operative societies, insurance companies *etc.* A minimum dividend rate guaranteed by the Central Government was also suggested. It was also suggested that the debentures floated by the Corporation should also be guaranteed by the Central Government as to repayment of principal and the interest.

State Warehousing Companies

The share capital contribution of State Warehousing Corporations not less than 50 per cent should be subscribed by the All-India Warehousing Corporation and the rest by the State Government and there should not any other way of contribution to the share capital. The establishment of a State Warehousing Corporations in any particular State was to be effected as soon as the State Government agreed to pay its stipulated part of the share capital. Only two shareholders of these corporations were thus envisaged, namely, the All-India Warehousing Corporation and the State Government. The functions of the State Warehousing Corporations were similar to those recommended for the All-India Warehousing Corporation. The first state Warehousing Corporation was setup by Government of Bihar in the year 1956-57 and the last state Warehousing Corporation was setup in the year 1973 in Meghalaya.

Food Corporation of India

The Food Corporation of India under the Department of Agriculture and Cooperation Government of India was set up to provide price support to producers, to distribute food grains at concessional prices to the poor through the Public Distribution System (PDS) and to ensure national food security by carrying buffer stocks. It is one of the largest corporations in India and probably the largest supply chain management in Asia. It operates through 5 zonal offices and 26 regional offices. Each year, the Food Corporation of India purchases roughly 15-20 per cent of India wheat output and 12-15 per cent of its rice output. The purchases are made from the farmers at the rates declared by the Govt. of India. This rate is called as MSP (Minimum support Price)[1]

The stocks are transported throughout India and issued to the State Government nominees at the rates declared by the Govt of India for further distribution under the Public Distribution System (PDS) for the consumption of the ration card holders. (FCI itself does not directly distribute any stock under PDS and its operations end at the exit of the stock from its depots). The difference between the purchase price and sale price, along with internal costs, are reimbursed by the Union Government in the form of Food Subsidy.

The present procurement policy needs to be drastically revised so as to limit government purchases only for preventing a sharp fall in prices (*i.e.* to support prices) instead of Government buying all that is offered at pre-determined "fair" prices. FCI was set up at a time when the nation was facing severe crisis in food grains and no state was in a position to cope on its own. With inter-state movement of food grains restricted, FCI was the only organization, which was entitled to move food grains freely across state borders, to import food grains and to coordinate food grain availability across the country. However, the position has since changed. States have set up their own agencies for procurement and distribution and free movement of food grains across the state[2]. The role of FCI should, therefore, be limited to minimum buffer stocking, *i.e.* not more than 10 million tonnes. To feed its Public Distribution System, the Government should invite tenders from private grain companies/traders to deliver a specified amount of grain at a place and timing determined by the Government. Procurement by FCI for below the poverty line people and the Public Distribution System should be de-linked from procurement for buffer stocks.

Features of Food Corporation of India

Procurement

The Government policy of procurement of Food grains has broad objectives of ensuring MSP to the farmers and availability of food grains to the weaker sections at affordable prices. It also ensures effective market intervention thereby keeping the prices under check and also adding to the overall food security of the country. FCI, the nodal central agency of Government of India, along with other State Agencies undertakes procurement of wheat and paddy under price support scheme. Coarse grains are procured by State Government Agencies for Central Pool as per the direction issued by Government of India on time to time.

The procurement under Price Support is taken up mainly to ensure remunerative prices to the farmers for their produce which works as an incentive for achieving better production. Before the harvest during each Rabi/Kharif Crop season, the Government of India announces the minimum support prices (MSP) for procurement on the basis of the recommendation of the Commission of Agricultural Costs and Prices (CACP) which along with other factors, takes into consideration the cost of various agricultural inputs and the reasonable margin for the farmers for their produce.

To facilitate procurement of food grains, FCI and various State Agencies in consultation with the State Government establish a large number of purchase centres at various mandi and key points. The number of centres and their locations are decided by the State Governments, based on various parameters, so as to maximize the MSP operations. For instance for Wheat procurement more than 19,000 procurement centers were operated in RMS 2018-19 & for Rice procurement more than 40,000 procurement centres are operating in KMS 2017-18. Such extensive & effective price support operations have resulted in sustaining the income of farmers over a period and in providing the required impetus for higher investment in agriculture sector for improved productivity.

Whatever stocks which are brought to the purchase centers falling within the Government of India's specifications are purchased at the fixed support price. If the farmers get prices better than the support price from other buyers such as traders/millers *etc.*, the farmers are free to sell their produce to them. FCI and the State Government/its agencies ensure that the farmers are not compelled to sell their produce below support price.

Storage

The storage function assumes paramount importance in organization such as Food Corporation of India because of its requirement to hold huge inventory of food grains over a significant period of time. Storage plan of FCI is primarily to meet the storage requirement for holding stocks to meet the requirements of Public Distribution System and Other Welfare Schemes undertaken by the Government of India. Also, buffer stock is to be maintained for ensuring food security of the nation. Adequate scientific storage is pre-requisite to fulfill the policy objectives assigned to the Food Corporation of India for which FCI has a network of strategically located storage depots including silos all over India. Besides having own storage capacity, FCI has hired storage capacities from Central Warehousing Corporation, State Warehousing Corporations, State Agencies and Private Parties for short term as well as for guaranteed period under Private Entrepreneurs Guarantee Scheme. New Godowns are being constructed by FCI mainly through Private Participation under Private Entrepreneurs Guarantee Scheme. FCI is also augmenting and modernizing its storage capacity in the form of silos through Public Private Partnership.

Storage Capacity for Central Pool Stocks (year-wise) shown below:

State Wise Storage Capacity available with different Storage Agencies in the Country

Sl.No.	Date and Year	Grand Total (Lakh MT)
1	01.04.2011	316.1
2	01.04.2012	333.04
3	01.04.2013	377.35
4	01.04.2014	368.35
5	01.04.2015	356.63
6	01.04.2016	357.89
7	01.04.2017	352.71
8	01.04.2018	362.50

Movement

Movement plays a very important role in the working of FCI as well as in fulfilling the objectives of Food Policy and National Food Security Act. FCI undertakes movement of food grains in order to:

- Evacuate stocks from surplus regions
- Meet the requirements of deficit regions for NFSA/TPDS and Other Schemes
- Create buffer stocks in deficit regions
- Punjab, Haryana and Madhya Pradesh are the surplus States in terms of wheat procurement vis-a-vis their own consumption. Punjab, Haryana, Andhra Pradesh/Telengana, Chhattisgarh and Odisha are surplus States in terms of rice procurement vis-à-vis their own consumption. Surplus stocks of wheat and rice available in these States are moved to deficit States to meet the requirements under NFSA/TPDS and other schemes as well as to create buffer stocks. On an average 40 to 42 million tonnes of foodgrains are transported by FCI across the country in a year. FCI undertakes massive movement operation of foodgrains all over the country encompassing around 1906 FCI owned & hired depots/Slios, 557 rail-heads (owned by Indian Railways and others) and 98 FCI own sidings.

Movement Plan is Prepared on Monthly Basis Keeping in View

- Quantity available in surplus regions
- Quantity required by deficit regions
- Likely procurement
- Vacant storage capacity both in consuming as well as procuring regions
- Monthly allotment/off take of foodgrains

Mode of Transportation

- Movement of foodgrains is undertaken by Rail, Road and Waterways. Around 85 per cent of stocks are moved by rail to different parts of the country. Inter-State movement by road is mainly undertaken in those parts of the country which are not connected by rail. A small quantity is also moved by ocean vessels to Lakshadweep and A&N Islands as well as through coastal shipping and riverine movement to Kerala/Agartala (Tripura).
- FCI has 98 own Rail sidings, where foodgrain rakes are placed directly at FCI depots. Other than that, foodgrain stocks are transported 'to and fro' from the nearest rail-heads of Indian Railways.
- FCI has been able to ensure availability of sufficient foodgrain in all States by proper planning. About a decade back, nearly 90 per cent of stocks were moved Ex-North mainly from Punjab & Haryana, which has now come down to 72 per cent due to increase in procurement of rice in Andhra

Pradesh, Chhattisgarh, Odisha & West Bengal and wheat in Madhya Pradesh, Uttar Pradesh and Rajasthan.

Finance

The functions of Finance and Accounts division is primarily maintenance of Accounts and preparation of annual reports, preparation of Budget Estimates, claim subsidy from Government of India, Funds Management, Taxation, Debt collection and Financial evaluation of commercial proposals.

FCI is a statutory Organization constituted under Food Corporation's Act, 1964 and has been carrying out its operations since 1965 with an objective to trade in food grains and other Food stuffs and for matters connected therewith and incidentals thereto. At present FCI is only implementing Government of India food programme and not involved in any commercial venture. Main operation of FCI includes procurement of food grains at minimum support price declared by Government of India, store food grains so procured, transport the surplus food grains to deficit states and issue it to State Governments under Public Distribution System at a price decided by the Government of India. Since, the issue prices declared by Government of India under different schemes are much lower than the cost of food grains procured; the differential amount is reimbursed to FCI as food subsidy by the Government of India. FCI also maintains buffer stocks of food grains as mandated by the Government of India and intervene in the domestic market to control the rising prices of the food grains.

Sales

Operations

In order to achieve the food security of the country, the Sales Division looks after one of the most important operation *i.e.* distribution of food grains under TPDS/NFSA & Other Welfare Schemes. Government of India fulfils the objectives of food security through the Public Distribution System. Public Distribution System strives to meet the twin objectives of price support to the farmers for their product and supply of food grains at affordable prices. It is against the stocks procured under price support, Government releases a certain quantity of food grains in each State under the Public Distribution System. This mission of the Government of India is translated into reality by the FCI. In order to implement the food policy of Government, FCI has to fulfill certain objectives which are as follows:

- ✰ To ensure and equitable distribution of available food grains at reasonable prices to the vulnerable sections of society throughout the year.
- ✰ To maintain stability in food grains prices throughout the country during the year.
- ✰ To maintain an adequate buffer stock of food grains to deal with fluctuations in production and to meet unforeseen exigencies and natural calamities.

Functions

The functions of Sales Division are as follows:

1. Management of allocation and off take of wheat, rice, and coarse grains under different schemes of Ministry of CAF&PD at Central Issue Price which are as under:- TPDS (NFSA & Tide over/Other than NFSA) and Special additional allocations for TPDS. Other Welfare Schemes (OWS) *viz.* Mid-Day Meal, Annapurna, Welfare Institutions & Hostels, SC/ST/OBC Hostels, Wheat Based Nutrition Programme and Scheme for Adolescent Girls. Defense/Para-Military Forces (CRPF/BSF/ITBP). Open Market Sales Scheme (Domestic). Natural calamities and festivals.
2. Matters related to Stocking Norms of foodgrains for Central Pool.
3. Disposal of issuable old stocks of foodgrains through open tender.
4. Identification & declaration of Base Depot in consultation with the State Govts.
5. Policy matter related to reimbursement of Hill Transport Subsidy (HTS) to the entitled States/UTs being cost of transportation of foodgrains from base depots to Principal Distribution Centres (PDCs) on actual basis.
6. Reimbursement of Road Transportation Charges (RTC) to State/UT Govts. Where stocks are lifted from other depots instead of base depot.
7. Issues relating to National Food Security Act (NFSA).
8. Matter related to Direct Benefit Transfer (DBT) to States
9. Release of wheat at pre-determined prices in the open market from time to time to enhance the supply of wheat especially during the lean season to moderate the open market prices.
10. Disposal of pulses through normal channel and through tenders.

Stocks

Food grain stocking norms refers to the level of stock in the Central Pool that is sufficient to meet the operational requirement of food grains and exigencies at any point of time. Earlier this concept was termed as Buffer Norms and Strategic Reserve. Presently stocking norms fixed by Government of India that comprise:

Operational stocks: for meeting monthly distributional requirement under TPDS and OWS.

Food security stocks/reserves: for meeting shortfall in procurement.

Stocking norms are for a quarter and consist of operational stock for the quarter and strategic reserve to take care of short fall in production or natural calamities.

Quality Control

The Quality Control (QC) wing of FCI manned by qualified and trained personnel is entrusted with enormous task of procurement & preservation of food grains. The food grains are procured as per laid down Specifications of Government of India and inspected regularly during storage to monitor the quality. Representative samples of the stocks are drawn for physical and chemical analysis to ensure whether the quality standard meets the parameters of laid down Specifications of Government of India. Food grain samples are also referred to NABL accredited laboratories and get tested for its conformity of parameter under FSS Act also. Food Corporation of India's testing laboratories spread across the country for effective monitoring of quality of food grains providing quality assurance as per FSS Act 2006, leading to improved satisfaction level to the customers (consumers). Laboratories across the country are being upgraded with latest equipment. The IFS (Institute of Food Security) Lab, Gurgaon is in process of upgradation to a State of Art Lab.

Conclusion

It is concluded that unless some very drastic measures are taken to improve the storage capacity of food grains, the wastage of food grains cannot be curbed which otherwise could be utilized for feeding millions of poor people. Thus, warehouses and Food Corporation of India plays an ambient role in conservation of food items through buffer stocks. It also ensures National Food Security so that it can be judiciously used for future consumption throughout the country via public distribution system. Effective price support operations for safeguarding the interests of the farmers are another vital feature preventing post harvest loss.

REFERENCES

Garg, Radhey Shyam (2012). Arth Prabhand: A Journal of Economics and Management, Vol.1 Issue 3,

Patnaik, Gokul (2014). Marketing, Storage, and Extension Services State of Agriculture in India http: //www.idfresearch.org

www: //cewacor.nic.in

www: //fci.gov.in

Part C

Miscellaneous

Chapter 20

An Introduction to Biofortification and Food Fortification

Ummed Singh, C.S. Praharaj, Abhishek Bohra, Purushottam and A.K. Parihar

Introduction

Deficiency of micronutrient is in increasingly important problem worldwide especially in the face of burgeoning human population. Micronutrients are crucial for plant growth and human health. The micronutrient deficiency is popularly known as hidden hunger. Hidden hunger or micronutrient malnutrition (Zn, Fe, Cu, B, Mo and I *etc.*) could be alleviated through increasing the micronutrient density in crops that make dominant portions of the human diets worldwide. The problem of hidden hunger or micronutrient malnutrition particularly due to the deficiency of iron (Fe) and zinc (Zn) is very common among women and preschool children. Consistent efforts have been made to examine the genetic potential to increase bioavailable Fe and Zn in staple food crops such as rice, wheat, maize, pulses, common beans and cassava.

A strong correlation has been reported in the regions with zinc (Zn) and iron (Fe) deficient soils and Zn and Fe deficiency in humans (Cakmak, 2008). The severity of micronutrient deficiency is more evident in the regions (Asia and Africa) where rice and wheat are consumed as staple owing to the micronutrient-poor nature of cereal crops as compared with pulses. According to an FAO report, more than 25 per cent are hunger in sub-Saharan Africa (FAO, 2010). The anaemia based on Fe deficiency affects more than 80 per cent of the pregnant women in South Asia, and also contribute to maternal mortality and impaired mental development in children (IRRI, 2006); about 60 per cent of the under-5-year children were underweight in the

20 per cent poorest group in this region (UN, 2011). Fe deficiency affects nearly two billion people worldwide and most of these deficient people reside in developing countries (Stolzfus and Dallman, 1998). In a similar manner, anaemia is known to exert its impact on 80 per cent of India's children aged between 3 and 6 years, and more than 50 per cent of married women (15–49 years) (Krishnaswamy, 2009). More than 30 per cent of pre-school children in South and Southeast Asia and sub-Saharan Africa have ben reported to encounter Vitamin A deficiency (IRRI, 2006). Gibson (2006) reported that Zn deficiency might affect a considerable proportion (up to 70 per cent) of the population in Asia and sub-Saharan Africa; this points to the deficiency faced by nearly 2 billion people in Asia and 400 million people in sub-Saharan Africa (IRRI, 2006). More than one-third of the world's population suffers from Zn deficiency (Hotz and Brown, 2004; Stein, 2010), and Zn deficiency has been estimated to be responsible for approximately 4 per cent of the worldwide burden of morbidity and mortality in under-5-year children and a loss of nearly 16 million global disability-adjusted life years (Black *et al.* 2008; Walker *et al.* 2009).

Approaches of Biofortification

Enriching essential micronutrients and other health-promoting compounds/ elements to crops or foods in order to improve their nutritional value in edible portions of crop plants is referred to as biofortification. In other words, biofortification refers to a comparatively cost-effective, sustainable, and long-term means to enhance micronutrient density in plants. This is achieved through modern agronomic interventions or improved breeding practices (Singh *et al.* 2016). Crop breeding has been focused on achieving high yields rather than nutritional quality. The other alternatives that facilitate mitigation of the problem of micronutrient deficiency rely upon industrial fortification or pharmaceutical supplementation. This approach has the potential not only to check the growing number of malnourished people but also to improve their nutritional status. Moreover, biofortification provides a feasible means of reaching malnourished rural sections that have limited access to commercially marketed fortified foods and supplements (Singh *et al.* 2016a).

The biofortification strategy seeks to transfer the micronutrient-dense traits into those varieties that already have preferred agronomic and consumption traits, such as high yield. According to (Bouis *et al.* 2011), though biofortified staple foods may not deliver as high a level of minerals and vitamins per day as could be achieved through supplements or industrially fortified foods, these can lead towards daily adequacy of micronutrient intakes among individuals.

A brief account on different approaches that lead to biofortification is presented in this chapter. The potential approaches are as described below:

Breeding Based Interventions

Through exploring the genetic variation across crop's gene pools, plant breeding programs could cause a considerable increment in the level and bioavailability of minerals in staple crops (Welch and Graham 2005). This warrants quantification of genetic variation affecting heritable mineral traits, and assessing their interactions with the diverse environments, and finally, exploring the feasibility of breeding crops with enhanced mineral content in edible tissues without showing any ill-impact on yields or other quality traits. Breeding for increased mineral levels has several advantages over conventional interventions (*e.g.* sustainability). However, plant nutritional improvement using standard breeding procedures requires considerable time and labour. Nowadays, breeders adopt an array of modern biotechnological tools and techniques to accelerate the development of high-mineral varieties. The modern molecular tools and methods increasing employed in crop biofortification include DNA markers based quantitative trait locus (QTL) maps and marker-assisted selection (MAS) (Cakmak, 2008).

Tradition Crop Improvement Techniques

This has allowed agricultural scientists to demonstrate significant improvement in the nutritional, eating quality, and agronomic traits of major subsistence food crops. The success and pace of the conventional plant breeding is hampered by the lack of adequate genetic variation for the mineral traits and more often than not, these traits are harboured by the wild crop relative's hybridization to which is rendered difficult due the presence of strong cross ability barriers. Furthermore, the tradeoffs associated with enhanced nutritional content in plant are seen in the form of deteriorated yields. One prominent example is quality protein maize (QPM), which has taken decades of conventional plant breeding research to develop varieties acceptable to farmers. Other biofortified crops, such as the orange-fleshed sweet potatoes (OFSP) promoted through the Harvest Plus program in Africa, have been successfully selected and developed for both nutrient and (at least rainy season) yield traits (Unnevehr *et al.*, 2007).

Mutation Plant Breeding

Mutation breeding has been used extensively used in developed and developing countries to obtain varieties with improved grain quality and in some cases higher yield and other traits. Mutations are capable of generating tremendous amount of novel genetic variability, and this novel variability is created through different means such as chemical treatments or irradiation. According to the FAO/International Atomic Energy Agency (IAEA), mutation breeding attempts have led to the development of more than 2,500 varieties so far (Mutant varieties database). Most of the European and US mutants contain alterations in flowers, however majority of these mutants in Asia represent important food crops such as wheat, rice, maize, and soybeans (Singh *et al.* 2016a).

Modern Biotechnological Approaches

Unlike plant breeding, modern biotechnological tools facilitate the development of nutrient dense crops in a time saving manner. Some of the biotechnological approaches are discussed here:

Molecular Breeding

Breeding assisted by modern molecular tools is a powerful technique that attracts no significant cultural or regulatory resistance and has been embraced so far. Organic growers because have also shown increasing interest in this techniq because of its reliance of natural crop breeding processes rather than engineered gene insertions to change the DNA of plants. The capacity of this technique is enhanced several folds with the tremendous advances witnessed in the field of plant genomics, which is the study of the location and function of genes. The rapid decline in costs per data point and growing technological strength has further motivated scientists to adopt these techniques. Using this technique, plant breeders also can stack several traits into one variety such as QPM, disease resistance, and drought tolerance in maize (Pray 2006). This technique remains particular important in case of recessive traits in plants, which are otherwise difficult to address through conventional breeding or other techniques.

Genetic Engineering

Genetic engineering (GE) or recombinant DNA technology has emerged as a powerful tool to combat against mineral deficiency, and this technology enables introduction of the genes directly into breeding varieties. The genes are often not present in the crop's gene pool (Zhu *et al.* 2007).The GE can be employed to (a) improve the efficiency with which minerals are mobilized in the soil, (b) reduce the level of anti-nutritional agents (c) enhance the level of nutritional enhancer compounds such as inulin. The GE technique offers greater speed and reach. The *transgenic* trait is added without normal biological reproduction, but once introduced in plants in a single copy the gene shows simple inheritance during plant reproduction. The choice of which technology to use when biofortifying crops come down to a calculation by breeders of how to get the best results most quickly, given their budget constraint. Conventional plant breeding requires less investment in labs or highly trained human resources (molecular biologists) than either marker-assisted selection or genetic engineering, and it faces lower and less costly regulatory hurdles.

Nutritional Genomics

Recently, availability of large-scale genome sequencing data of plants, bacteria, fungi, and animals in conjunction with the development of novel bioinformatics tools enables better understanding of metabolic pathways, which has been a crucial driving force for the genesis of nutritional genomics. As a component of the biofortification research, nutritional genomics is a new approach to the study of complex biochemical pathways in plants. It seeks to elucidate the basic, underlying mechanisms involved in the synthesis and accumulation of essential vitamins and minerals in plant tissues. Given the metabolic pathways shared among organisms during the course of evolution, the knowledge enabled by these nutritional genomics tools can be readily translated to other related organisms. This comparative knowledge base will be used to increase specific micronutrient levels in crops that will be of most benefit to the needs of the developing world. Once the genes of interest are identified, they are moved into a crop species to demonstrate proof of concept to effect the desired change in nutritional content of the target tissues (Soren *et al.* 2016).

Next Generation Sequencing (Ngs)

Presently, NGS technology is one of the remarkable developments witnessed in the field of molecular biology. NGS has permitted a much-needed resolution of trait mapping. Besides classical Sanger sequencing, a range of different second generation technology (SGT) *viz.;* Roche/454 FLX Pyro sequencer, GS FLX Titanium/ GS Junior, Genome Analyzer (Solexa/Illumina), SBS (Sequencing by Synthesis), Solid Sequencer (Applied Biosystems) and third generation sequencing (TGS) technologies *viz.;* Ion Torrent's PGM/Proton (Life science), HiSeq/MiSeq from Illumina and Oxford Nanopore technologies has gained immense popularity in recent years owing to the colossal sequencing rates and read lengths, and declining sequencing costs and decreasing procedural complexity.

Tissue Cultures

Modern tissue culture techniques can allow scientists to reproduce plants from a single cell. These techniques are now used extensively to produce disease-free planting material of clonally propagated crops such as bananas. When tissue culture is combined with embryo rescue techniques, plant breeders can use the genes from wild and weedy relatives of a crop, which would normally not cross with the cultivated crop. This allows breeders to increase genetic variability of the cultivated crop and then bring in valuable traits of the wild and weedy relatives. These techniques have allowed scientists to cross Asian and African rice varieties and develop Nerica rice varieties with agronomic traits, such as higher yield and resistance to water stresses that have met with growing success in Africa. Tissue culture is an important tool for propagation of roots and tubers, such as potatoes and cassava, and both of these crops are part of current biofortification research.

Microbiological Approaches

Large quantities of N fertilizers are commonly used to obtain high yields in many crops. But these chemical fertilizers compromise on human and animal health and pollute the environment. Microbes represent a promising input, which possess an array of mechanisms to sequester macro and macronutrients from soil or water, which can be also made available to plants (Prasanna *et al.* 2012). Therefore, the application of microbial biofertilizers needs to be advocated as a possible strategy not only to increase nutrient concentrations in edible crops, but also to improve yields on infertile or nutrient poor soils. Apart from breeding programs worldwide (Kamran *et al.* 2014), including the *Harvest Plus*, biofortification of crops with micronutrients can also be achieved through microbial inoculants.

Nutrient Sequestration and Mobilization by Microbes

In agriculture, artificial inoculation with specific microorganisms has been a common practice that is being followed to obtain higher grain yield. The "core microbiome" concept is gaining interest among researchers for developing suitable inoculants as nutrient management. The rhizosphere is a metabolically active site in which microorganisms sequester, mobilize and make available, both macro and micronutrients to the plants. Pfeffer *et al.* (2013) illustrated the heritable variation in the rhizosphere microbial community composition while analyzing the diversity and heritability of field grown maize inbreds. Biofortification of crops can therefore be achieved through the application of microbial inoculants/biofertilizers which mobilize/solubilize the essential mineral micronutrients in the soil and make it easily available to the plant.

Plant Growth Promoting Agents

Cyanobacteria or Blue Green Algae represent one of the types of PGP agents which can also be key players in nutrient sequestration, improving nutrient use efficiency and crop yields (Prasanna *et al.* 2015). Besides being diazotrophs, cyanobacteria can also improve the plant growth by sequestering nutrients and improving their mobilization into plants. Their ability to synergistically interact with other microorganisms led to development of consortia and the novel concept of cyanobacteria based biofilms as biofertilizers (Nain *et al.* 2010). The use of cyanobacterial inoculants (*Anabaena-Azotobacter*biofilm) and *Anabaena* sp. -*Providencia*sp. In maize hybrids is capable to enhance the activity of defense enzymes *viz.*; peroxidase, PAL and PPO in roots, which shows positive correlation with Zn concentration in the flag leaf. Additionally, cyanobacterial inoculation enhances the Zn mobilization to flag leaf in maize hybrids, without any negative effects on plant vigor and yields (Prasanna *et al.* 2015a).

Moreover, plant growth promoting rhizobacteria (PGPR) which include beneficial bacteria that colonize plant roots and enhance plant growth by a wide variety of mechanisms. Secretion of phytosiderophores by microorganisms and plants in restricted spatial and temporal windows represent an efficient strategy for uptake of iron and other micronutrients by plants from the rhizosphere. Several mechanisms are involved in the sequestration and transformation by microorganisms in soil such as production of acids, alkalis *etc.* Microorganisms differ in competing with higher plants for micronutrients. Species of Azotobacter differed in their competitiveness with wheat plants in extracting Fe and Zn (Shivay *et al.* 2010). Consequently, biofortification of crops through application of PGPRs can be therefore considered as a possible supplementary measure, which along with breeding varieties, can lead to increased micronutrient concentrations in wheat crop, besides improving yield and soil fertility.

AM Fungi

Most plants, including all major grain crops and almost all vegetables and fruits, are associated with mycorrhizal fungi that improve the uptake of essential mineral elements from soils and, therefore, enhance plant growth and productivity. These symbiotic fungi, therefore, change, directly or indirectly, the mineral nutrition of plant products that are also essential for humans. However, role of mycorrhizas on element biofortification may be piloted through agricultural practices. Mycorrhizas can potentially offer a more effective and sustainable element biofortification to curb global human malnutrition.

Physiological Approaches

Loading of micronutrients into the phloem is essentially required for translocation to seeds. The micronutrients pass through pods prior to transport to seeds. Mineral micronutrients are considered to remain in chelated forms during phloem transport. Iron transport protein (ITP), a ligand, was identified from phloem of Ricinuscommunis. One potential phloem chelator NA has been identified that has the capacity to bind Cu, Co, Fe(II) and Fe(III), Mn, Ni, and Zn. The chelator NA is a precursor to phytosiderophores in grasses but NA is present in dicots as well and is considered to be a vital chelator of micronutrients for homeostasis during growth, and for translocation within vegetative parts of the plant and also in phloem transport of micronutrients to seeds. The pressure gradient is generated by loading nutrients such as carbohydrates and potassium into phloem at source tissues to provide osmotic potential that draws in water. The unloading of these nutrients at the seed decreases the pressure and maintains a gradient. Micronutrients will move with the major nutrients, *i.e.* K+, Cl−, and sugars in the phloem. The NA is a potential chelator of Fe in the phloem and form a stable complex at alkaline pH, that can also chelate Cu and Zn element. Other potential phloem chelators are unspecified oligopeptides compounds containing Nitrogen, suggesting that phloem transport of N and mineral micronutrients may be directly related. There is a relationship between N and micronutrient transport to seeds in seeds of several grass and dicot species. Thus, improved fertilization practices with N and Zn fertilizers could be a large step toward biofortification (Basu 2016). Seeds contain filial tissues (embryo and endosperm, aleurone) surrounded by maternal tissues (seed coat). The developing seed is connected to the maternal plant by a single vascular trace.

Several genes may be involved in translocation of micronutrients into xylem including FRD3, FPN1, HMA2, HMA4, HMA5, and MTP3. The over expression of these genes often increased movement of micronutrients into the shoot systems and increased flux into shoots providing additional micronutrients for seed biofortification. Potential methods for characterizing these new genes that affect translocation of micronutrients to seeds include quantitative trait locus (QTL) mapping which includes linkage and association mapping. Blair *et al.* (2011) characterized variations in Fe and Zn concentrations of seed coats of common bean in a recombinant inbred population and identified some of the underlying genetic loci responsible for seed coat accumulation of these metals, which could be useful for future biofortification efforts. The constitutive expression of a suite of Zn-deficiency inducible responses through the over expression of bZI19 and bZIP23 transcription factors can be used to increase Zn accumulation in edible portion of the crop (Assuncao *et al.*, 2010). The Zn accumulation in roots include transport proteins in the plasma membrane and tonolast of root cells that facilitate the uptake and sequestration of Zn in the vacuole, together with enzymes involved in the synthesis of compounds that bind Zn2+ in the rhizosphere cytoplasm.

Agronomic Approaches

Farmers have applied mineral fertilizers to soil for hundreds of years in order to improve the health of their plants, but within certain limits the same strategy can also be used to increase mineral accumulation within cereal grains for nutritional purposes (Rengel *et al.* 1999). Like supplements and fortification, agronomic intervention is probably best applied in niche situations or in combination with other strategies (Cakmak 2008).

Agronomic strategies to increase the concentrations of mineral elements in edible tissues generally rely on the application of mineral fertilizers and/or improvement of the solubilisation and mobilization of mineral elements in the soil (White and Broadley 2009). When crops are grown where mineral elements become immediately unavailable in the soil, targeted application of soluble inorganic fertilizers to roots or to leaves is practised. In situations where mineral elements are not readily translocated to edible tissues, foliar applications of soluble inorganic fertilizers are made. The most effective and efficient agronomic strategies of biofortification are discussed hereunder one by one.

Foliar Fertilization

Majority of the cultivated plants are capable of absorbing soluble compounds and gases through leaves. The phenomenon could be used for delivering nutrients to the plant through foliar fertilization. Foliar feeding of nutrients through foliar spray implies that nutrients sprayed will be absorbed and exported from the point of application (leaf) to the point of utilization (growing tissues), where leaf cuticle is the major hurdle in the events or chain of utilization (Rengel*et al.* 1999). Moreover, foliar fertilization is the most common and handy in changing the micronutrient status of edible plant parts to an appreciable level. Soluble fertilizers of N, P, K, S, Zn and Fe can be dissolved in potable water/pesticides for *in-situ* application onto green leaves or plant parts of crops especially at fruiting or seed formation. Foliarly applied nutrient (Zn, Fe, urea *etc.*) are absorbed by leaf epidermis following transportation to other parts of the plant via phloem and xylem (Singh *et al.* 2015). The elements which are even partially mobile in phloem (*e.g.* zinc) having the potential to improve concentration in the grain. For example spray of zinc during grain development stage also contributed to substantial concentration of zinc in grains. However, zinc spray at crop maturity may not have marked influence on grain yield as well as concentration in the grain (Zhang *et al.* 2010).

The effectiveness of the foliar fertilization is also governed by the factors *viz.* concentration of solution, timing of spray, stages of crop development, form of fertilizers *etc.* Foliar fertilization with Zn-EDTA, Fe-EDTA and other chelates has been used in majority of the cereals and pulses. Chelates are more effective than sulphate salt. Three sprays of Zn-EDTA 0.5 per cent solution spray at the rate of 500 litres/ha at maximum vegetative growth + flowering + grain filling stages was much better than $ZnSO_4.7H_2O$ in increasing the productivity and Zn concentration in grain and as well as straw of chickpea (Shivay *et al.* 2016). The effectiveness of the form of fertilizer is also affected by the stage of crop development and timing of spray. It is pertinent to note that timing of foliar Zn or iron spray is more crucial as spraying of zinc or iron prior to appearance of symptoms may predispose the crop to yield loss.

Seed Priming

Seed priming is the practice of treating the seeds with micronutrients either by soaking in nutrient solution of a specific concentration for a specific time or duration. Priming, on-farm priming, invigorations (the term used interchangeably for seed priming or coating) and seed enhancements are the similar practice of delivering nutrients from seed to seed for improving nutrient concentration in grains (Shivay *et al.* 2016). Here, we have discussed about the prospects and potential of seed priming with zinc, boron, molybdenum, manganese, copper, cobalt and iron *etc.* for the improvement of growth attributes, yield and micronutrient enrichment of grain.

Table 1. Zinc and Boron Enrichment of Pulses Seeds through Priming

Crop	*Fertilizer*	*Rate of Application and Tme for Priming*	*Increase in Grain Yield Over Control (Per cent)*	*Increase in Zn/B Content in Grain over Control (Per cent)*	*Reference*
Wheat	Zinc sulphate	0.3 per cent Zn, 10 hrs	14.0	12.0	Harris *et al.*, 2008
Rice	Zinc sulphate	2.2g/kg seed, 10 hrs	17.92	-	Slaton *et al.*, 2001
Chickpea	Zinc sulphate	0.05 per cent Zn, 10 hrs	19.0	29.0	Harris *et al.*, 2008
Lentil	Zinc sulphate	0.004 per cent Zn, 12hrs	-	11.7	Johnson *et al.*, 2005
Cowpea	Zinc sulphate	0.004 per cent Zn, 12hrs	-	11.7	Johnson *et al.*, 2005
Chickpea	Boric acid	0.008 M, 8 hrs	-	9.0	Johnson *et al.*, 2005
Lentil	Boric acid	0.008 M, 12 hrs	-	15.7	Johnson *et al.*, 2005

Adapted from Farooq *et al.* (2012)

Table 2. Molybdeneum and Cobalt Enrichment of Pulses Seeds through Seed Priming

Crop	*Fertilizer*	*Rate of Application and Tme for Priming*	*Increase in Grain Yield Over Control (Per cent)*	*Increase in Nodule Number Over Control (Per cent)*	*Increase in Mo/Co Content in Grain Over Control (per cent)*	*Reference*
Oat	Cobalt sulphate	0.001 per cent, 1 hrs	11.17	-	-	Saric and Saclragic 1969
Common bean	Sodium molybdate	1.0 ppm 1 hrs	12.7	12.3	-	Mohandas 1985
Chickpea	Sodium molybdate	0.0026 M 8 hrs	-	-	7.4	Johnson *et al.*, 2005
Lentil	Sodium molybdate	0.0026 M 12 hrs	-	-	-	Johnson *et al.*, 2005
Common bean	Cobalt nitrite	2.0 ppm 1 hrs	0.5	14.8	-	Mohandas 1985

Seed Coating

Seed coating, the practice of applying finely ground solid or liquid which contains dissolved or suspended solids to form a uniform continuous layer covering the seed coat (Scott, 1989). Seed coatings with desired nutrients normally entails the sequential coatings of seed with layers of adhesives followed by finely ground nutrients after proper sieving of the material, which leads to desired increase in seed size. Under this practice beneficial inputs like microorganism, plant growth regulators, chemicals and nutrients are adhered/applied around the seed with some adhesive/gummy materials. Seed coatings with trace elements *viz.*, molybdenum, iron, zinc, manganese and boron have been found more effective. In alkaline soil, application of iron, zinc and manganese have considerable significance, where availability of these elements is decreased. Further, molybdenum is commonly used in legumes with lime especially when sown in acid soils. Varieties of chelated and mineral forms of trace elements have been used in seed coatings. As evident from the findings that the effectiveness of the seed coating depends largely on chemical used, soil type, soil health or fertility status, coating time, coating agent, ratio of chemical to seed *etc.* Application of zinc through seed coating improved zinc concentration in the seed besides improving seed emergence, plant growth and leaf area (Singh, *et al.* 2016). Seed pelleting of cowpea with $ZnSO_4$ (250 mg/kg seed) produced higher grain yield, seed weight and seed yield.

Soil Application of Fertilizers

Substantial increase in micronutrient concentration in seed due to soil application of respective micronutrients (zinc, iron, boron, molybdenum *etc.*) also brings multifarious agronomic benefits for crop production. Under potentially zinc deficient soils, application of zinc to the plants is very effective in reducing the uptake and accumulation of phytate-P in plants. Such agronomic disadvantage of zinc fertilization may leads to better bioavailability of zinc in the human body. In addition to this, the seeds containing better zinc concentration also have the ability to withstand adverse climatic abnormalities (Cakmak, 2008). Among the strategies of increasing zinc or iron concentration in the crop seeds, applying zinc or iron through soil application appears to be one of the important strategy. Although, nutrient recovery efficiency is reported low due to soil application over foliar and other methods, but fertilizer strategy as soil application could be a rapid solution to tackle the micronutrient deficiency or hidden hunger. Among sources of fertilizers, zinc sulphate, iron sulphate, borax and ammonium molybdate are most widely fertilizers as basal. For inorganic source of zinc, zinc sulphate is the most widely adapted and preferred source. As zinc sulphate is highly soluble and cheaper. The other sources of zinc are ZnO, Zn-EDTA and Zn-oxysulphate. The crop response or recovery efficiency is higher with Zn-EDTA than inorganic zinc fertilizer, but due to higher price its application is limited (Mortvedt, 1991) in crops.

Nutri - Farms Scheme

The idea of nutri-farms is based on bio-fortification technology, in which a crop is made rich in a nutrient either through conventional plant breeding or using genetic modification (Singh *et al.* 2016a). Wherein, enriched biofortified crops (Rich in Vitamin A, Zn, Fe and other desired nutrient) are promoted to grow at the large scale on community basis in the village or area or region. The staple food crops are given preference for the purpose, as majority of the people eat them daily.

Nutri-farm is designed to give concurrent attention to the three major problems as follows:

1. First, we have to help farm families overcome under-nutrition as a result of calorie deprivation.
2. Second, protein hunger is becoming serious due to the inadequate consumption of pulses and milk (in the case of vegetarians) and eggs, fish and meat (in the case of non-vegetarians).

Third, there is widespread hidden hunger, caused by the deficiency of micro-nutrients like iron, iodine, zinc, Vitamin A, Vitamin B12, *etc.* in the diet. Here the role of nurti-farms comes. The technology has to be need based and a lot of under-utilised crops, in addition to our staple crops, like wheat and rice, can be brought under it. We cannot ask people to change their food habits suddenly. You can't force them to take jowar and bajra instead of wheat and rice and hence bio-fortification would play a key role. The allocation of funds for agricultural research has to be more since it needs a huge boost. There are different technologies for biofortification and ICAR will certainly look for promising technologies, like genomic research, where micro-nutrient providing genes could be inserted in plants. This can help fight iron and beta carotene deficiencies.

The development of Nutri-farms or biofortification projects which can encourage development of novel food items using modern technological tools is a step in the right direction for mitigating malnutrition. The Nutri-Farms scheme will add nutritional dimensions to the farm sector (Swaminathan 2013). The role of agribusinesses and food companies need to respond with innovative proposals to take Bio-fortification idea forward. The full potential of Indian agriculture will be realized if these measures are supplemented with further reforms in the Agricultural Produce Marketing Act by the States and the Forward Contracts Regulation Act by Government of India.

According to the International Food Policy Research Institute, success hinges on three things: effective breeding, sufficient nutrients must remain after cooking and, third, farmers must adopt them widely. The Nutri-farm Scheme where the government intends to cultivate new crop varieties rich in micro-nutrients such as iron-rich bajra, protein-rich maize and zinc-rich wheat *etc.*, will definitely have an answer to both under as well as over nutrition problems of people," explained Chief Nutritionist Ishi Khosla of The Whole Foods. Nutri-farms ensure bio-fortified food crops that are enriched in critical micronutrients. These foods essentially provide the critical nutrients such as iron and zinc besides supplementing protein and essential vitamins. The nutri-farm movement considers the followings.

Nutri-Farms scheme will be implemented in 100 high malnutrition burden districts of 9 states namely Assam, Bihar, Chhattisgarh, Jharkhand, Madhya Pradesh,

Orissa, Rajasthan, Uttar Pradesh &Uttarakhand. The cereals crops namely Rice, Maize, Pearl millets, Finger Millet, Wheat and horticulture crops *viz.* Sweet potato and Moringa (drumstick) are identified for production of nutri-rich foods under pilot Scheme (Table 3).

Table 3: Target Nutrient and Varieties of Crops being Promoted under Nutri-farm

Crop	*Target Nutrient*	*Varieties*
Wheat	-carotene, Iron & Manganese	HI-8663
	Zinc, Copper & Manganese	VL-892
Rice	Fe	MSE-9, Kalanamak, Karjat-4, Mitta Triveni, Varsha
	Fe & Zinc	Chittmutyalu
	Fe & Zinc	Udayagiri
	Zinc	Poornima, Ranbeer Basmati, Pant Sugandha-17, ADT-43,Jyoti, Ratna, Type-3, Kesari
Maize	Lysine and tryptophan	QPM-9, HQPM-1, HQPM-4, HQPM-5, HQPM-7, Vivek, Saktiman-1, Saktiman-2, Saktiman-3, Saktiman-4
Pearl Millet	Fe	ICTP-8203 FE 10-2 (DHANSHAKTI), 86 M 86, Ajit,
	Fe & Zinc	Hybrid Pusa- 415, Pusa Composite
Finger millet	Fe & Zinc	PRM-1, VL-315, VL-324

Source: Purushottam *et al.*, 2016.

Alleviating Hidden Hunger through Food Fortification

The term 'hidden hunger' has been used to describe the micronutrient malnutrition inherent in human diets that are adequate in calories but lack vitamins and/or mineral elements. Majority of world diets are deficient in Fe, Zn, Ca, Mg, Cu, Se or I, which affects human health and longevity. Mineral malnutrition can be addressed by increasing the amount of fish and animal products in diets, mineral supplementation, food fortification and/or increasing the bioavailability of mineral elements in edible crops. However, strategies to increase dietary diversification, mineral supplementation and food fortification have not always proved successful. For this reason, the biofortification of crops through the application of mineral fertilizers, combined with breeding varieties with an increased ability to acquire mineral elements, has been advocated (White and Broadley 2009). The strategies of food fortification are listed below.

Food Fortification

Food fortification is one of the most cost-effective long-term strategies for mineral nutrition (Horton 2006). Fortification of dairy products such as bread and milk with different minerals (and vitamins) has been successful in industrialized countries However, this strategy is difficult to implement in developing countries because it relies on a strong food processing and distribution infrastructure. Fortification takes place during food processing and increases the product price.

These factors make fortified products unaffordable to the most impoverished people living in remote rural areas.

Industrial Fortification

The marketed supply of a widely consumed staple food can be fortified by adding micronutrients at the processing stage, and historically this is how micronutrient deficiencies have been addressed in the developed world. Concentration in the food industry also tends to strengthen compliance and quality assurance. Consumption of wheat flour products is growing around the world, even where wheat it is not a traditional food staple, opening new fortification opportunities at the milling stage. Public support for traditional fortification has recently been enhanced by new promotion and coordination efforts: Micronutrient Initiative (based in Canada), Flour Fortification Initiative (based in Emory University), Mid Day Meal Scheme (India) and the Global Alliance for Improved Nutrition (GAIN, based in Geneva). Certain kinds of fortification may be impractical for some important food staples (*e.g.*, VA fortification of milled rice), or may introduce off-colors or flavors (*e.g.*, VA fortification of white maize).Industrial fortification will only apply to marketed supplies and therefore may not reach those among the poor who obtain food outside of commercialized channels.

Dietary Diversification

Education is an important element in ensuring that improvements in income result in better maternal and child health. However, dietary diversification is constrained by resource availability for poor households and seasonal availability of fruits and vegetables. Promotion of home gardens is often touted, but the poor have a high opportunity cost for their labor and often limited land. Increased production of fruits and vegetables for household use reduces resources available for other income-earning or food-production activities. This type of effort is also relatively expensive and difficult to sustain on any large scale.

Food Supplementation

Supplementation is the best short-term intervention to improve nutritional health, involving the distribution of pills or mineral solutions for immediate consumption. This helps to alleviate acute mineral shortages but is unsustainable for large populations and should be replaced with fortification at the earliest opportunity. In industrialized countries, with few mineral malnutrition problems, supplementation is focused on a small subset of the population with specific deficiencies resulting from medical conditions. In developing countries, where acute and chronic deficiencies are commonplace, supplementation is highly recommended to complement the diet (fortified or otherwise) of the entire population (Nantel and Tontisirin 2002).

Conclusion

In general, majority of the agricultural systems prevalent in the developing world do not provide enough micronutrients to meet the human dietary needs. The substantial impact of micronutrient malnutrition has reflected on the work capacity of people and their economic development. Finding sustainable solution to this global nutritional crisis and to address micronutrient deficiencies warrants urgent implementation of several approaches. Towards this end, food supplementation, dietary diversification and industrial fortification programs have shown positive results in treating the micronutrient malnutrition. Breeding staple foods that are dense in minerals and vitamins provides a low-cost, sustainable strategy for reducing levels of micronutrient malnutrition. We further need to explore the prospects of novel breeding techniques, omics technology, microbe mediated approaches, ferti-fortification and nutri-farm movement to improve the micronutrient quality of staple crops.

REFERENCES

Assuncao, A.G.L., Herrero, E., Lin, Y.F., Huettel, B., Talukdar, S., Smaczniak, C., Immink, R.G.H., Van Eldik M., Fiers, M., Schat, H., and Aarts, M.G.M. (2010). Arabidopsis thaliana transcription factors bZIP19 and bZIP23 regulate the adaptation to zinc deficiency. *Proc Natl Acad Sci,* USA 107: 10296–10301.

Basu, P.S. (2016). Physiological Processes Toward Movement of Micronutrients from Soil to Seeds in Biofortification Perspectives. In: Biofortification of Food Crops (Eds., Ummed Singh *et al.* Springer (India) Pvt. Ltd., New Delhi). Pp. 317-329. (DOI: 10.1007/978-81-322-2716-8_23).

Black, R.E., Allen, L.H., Bhutta, Z.A., Caulfield, L.E., de Onis M., Ezzati, M., Mathers, C. and Riviera, J. (2008). Maternal and child health consequences. *Lancet* 371: 243–260.

Blair, M.W., Astudillo, C., Rengifo, J., Beebe, S.E. and Graham, R. (2011). QTL analyses for seed iron and zinc concentrations in an intra-genepool population of Andean common beans (*Phaseolus vulgaris* L.). *Theor Appl Genet* 122: 511–521.

Bouis, H.E., Hotz, C., McClafferty, B., Meenakshi, J.V. and Pfeiffer, W.H. (2011). Biofortification: A new tool to reduce micronutrient malnutrition. *Food and Nutrition Bulletin* 32 (Supplement 1): 31S-40S.

Cakmak, I. (2008). Enrichment of cereal grains with zinc: agronomic or genetic biofortification. *Plant Soil* 302: 1–17.

FAO. (2010). The State of Food Insecurity in the World: Addressing Food Insecurity in Protracted Crises. The Food and Agriculture Organization of the United Nations, Rome, pp 58.

Farooq, M., Wahid, A. and Siddique, K.M.H. (2012). Micronutrient application through seed treatments–a review. *Journal of Soil Science and Plant Nutrition* 12(1): 125–142.

Harris, D., Rashid, A., Miraj, G., Arif, M. and Yunas, M. (2008). 'On-farm' seed priming with zinc in chickpea and wheat in Pakistan. *Plant and Soil* 306: 3–10.

Horton, S. (2006). The economics of food fortification. *Journal of Nutrition* 136: 1068–1071.

Hotz, C. and Brown, K.H. (2004). Assessment of the risk of zinc deficiency in populations and options for its control. *Food Nutrition Bulletin* 25: S91–S204.

IRRI, (2006). Bringing Hope, Improving Lives: Strategic Plan 2007–2015. International Rice Research Institute, Manila, Philippines, pp 61.

Johnson, S.E., Lauren, J.G., Welch, R.M. and Duxbury, J.M. (2005). A comparison of the effects of micronutrient seed priming and soil fertilization on the mineral nutrition of chickpea (*Cicer arietinum*), lentil (*Lens culinaris*), rice (*Oryza sativa*) and wheat (*Triticum aestivum*) in Nepal. *Experimental Agriculture* 41: 427–448.

Kamran, A., Kubota, H., Yang, R.C., Randhawa, H.S. and Spaner, D. (2014). Relative performance of Canadian spring wheat cultivars under organic and conventional field conditions. *Euphytica* 196(1): 13-24. DOI: 10.1007/s10681-013-1010-3

Krishnaswamy, K. (2009). The problem and consequences of double burden–a brief overview. In: Symposium on Nutrition Security for India–Issues and Way Forward, 3–4 August, New Delhi, Programme and Abstracts. Indian National Science Academy, New Delhi, pp 5–7.

Mohandas, S. (1985). Effect of presowing seed treatment with molybdenum and cobalt on growth, nitrogen and yield in bean (*Phaseolus vulgaris* L.). *Plant and Soil* 86: 283–285.

Mortvedt, J.J. (1991). Micronutrient fertilizer strategy. In: Mortvedt JJ, Cox FR, Shuman LM, Welch RM (eds) *Micronutrients in Agriculture*. SSSA Book Series No. 4. Madison, WI. pp 33-44.

Nain, L., Rana, A., Joshi, M., Jadhav, S.D., Kumar, D., Shivay, Y.S., Paul, S. and Prasanna, R. (2010). Evaluation of synergistic effects of bacterial and cyanobacterial strains as biofertilizers for wheat. *Plant and Soil* 331: 217-230.

Nantel, G. and Tontisirin, K. (2002). Policy and sustainability issues. *Journal of Nutrition* 132: S839–S844.

Pfeiffer, J.A., Spor, A., Koren, O., Jin, Z., Tringe, S.G., Dangi, J.L., Buckler, E.S., Ley, R.E. (2013). Diversity and heritability of the maize rhizospheremicrobiome under field conditions. *Proc Natl Acad Sci* 110: 6548-6553.

Prasanna, R., Joshi, M., Rana, A., Shivay, Y.S. and Nain, L. (2012). Influence of co-inoculation of bacteria- cyanobacteria on crop yield and C- N sequestration in soil under rice crop. *World Journal of Microbiology & Biotechnology* 28: 1223-1235.

Prasanna, R., Babu, S., Bidyarani, N., Kumar, A., Triveni, S., Monga, D., Mukherjee, A.K., Kranthi, S. Gokte-Narkhedhar, N., Adak, A., Yadav, K., Nain, L. and Saxena, A.K. (2015). Prospecting cyanobacteria fortified composts as plant growth promoting and biocontrol agents in cotton. *Experimental Agriculture* 51: 42-65.

Prasanna, R., Bidyarani, N., Babu, S., Hossain, F., Shivay, Y.S. and Nain, L. (2015a). Cyanobacterial inoculation elicits plant defense response and enhanced Zn mobilization in maize hybrids. Cogent Food & Agriculture 1(998507): 1-13. DOI: 10.1080/23311932.2014.998507

Pray, C. (2006). A model for strengthening national agricultural research systems. The Asian Maize Biotechnology Network (AMBIONET) Mexico, DF: CIMMYT

Purushottam, Singh, S.K. and Riyaj, U. (2016). Nutri-farms for Mitigating Malnutrition in India. In: Biofortification of Food Crops (Eds., Ummed Singh *et al.* Springer (India) Pvt. Ltd., New Delhi). Pp. 461-477. (DOI: 10.1007/978-81-322-2716-8_33).

Rengel, Z., Batten, G.D. and Crowley, D.E. (1999). Agronomic approaches for improving the micronutrient density in edible portions of field crops. *Field Crops Research* 60: 27–40.

Saric, T. and Saciragic, B. (1969). Effect of oat seed treatment with microelements. *Plant and Soil* 31: 185-187.

Scott, J.M. (1989). Seed coatings and treatments and their effects on plant establishment. *Advances in Agronomy* 42: 43–83

Shivay, Y.S., Prasad, R. and Rahal, A. (2010). Studies on some nutritional quality parameters of organically or conventionally grown wheat. *Cereal Research Communications* 38(3): 345-352.

Shivay, Y.S., Singh, Ummed, Prasad, R. and Kaur, R.J. (2016). Agronomic Interventions for Micronutrient Biofortification of Pulses. *Indian Journal of Agronomy* 61: S161-S172.

Singh Ummed, Kumar, N., Praharaj, C.S., Singh, S.S. and Kumar, L. (2015). Ferti-fortification: an easy approach for nutritional enrichment of chickpea. *The Ecoscan* 9(3&4): 731-736.

Singh Ummed, Praharaj, C.S., Singh, S.S., Bohra, A. and Shivay, Y.S. (2015a). Biofortification of Pulses: Strategies and Challenges. pp. 50-55. In: Proceedings of The Second International Conference on Bio-Resource and Stress Management, Hyderabad, India.

Singh, Ummed, Praharaj, C.S., Chaturvedi, S.K. and Bohra, A. (2016). Biofortification: Introduction, Approaches, Limitations and Challenges. In: Biofortification of Food Crops (Eds., Ummed Singh *et al.* Springer (India) Pvt. Ltd., New Delhi). Pp. 3-18. (DOI: 10.1007/978-81-322-2716-8_1).

Singh, Ummed, Praharaj, C.S., Singh, S.S. and Singh, N.P. (2016a). Biofortification of Food Crops. Springer (India) Pvt. Ltd. ISBN 978-81-322-2714-4 (Hard Book) ISBN 978-81-322-2716-8 (eBook) DOI: 10.1007/978-81-322-2716-8

Slaton, N.A., Wilson-Jr C.E., Ntamatungiro, S., Norman, R.J. and Boothe, D.I. (2001). Evaluation of zinc treatment for rice. *Agronomy Journal.*, 93: 152-157.

Soren, K.R., Shanmugavadivel, P.S., Gangwar, P., Singh, P., Das, A. and Singh, N.P. (2016). Genomics-Enabled Breeding for Enhancing Micronutrients in Crops. In: Biofortification of Food Crops (Eds., Ummed Singh *et al.* Springer (India) Pvt. Ltd., New Delhi). Pp. 115-128. (DOI: 10.1007/978-81-322-2716-8_10).

Swaminathan, M.S. (2013). Launching a "Nutri-Farm Movement" .www.mssrf.org, May 15 Stein AJ. 2010. Global impacts of human mineral nutrition. *Plant and Soil* 335: 133–154.

Stolzfus, R.J and Dallman, P.R. (1998). Guidelines for the use of iron supplements to prevent and treat iron deficiency anaemia. International Nutritional Anaemia Consultative Group (INACG), ILSI Press, Washington, DC.

Walker, C.L.F., Ezzati, M. and Black, R.E. (2009). Global and regional child mortality and burden of disease attributable to zinc deficiency. *European Journal of Clinical Nutrition*, 63: 591–597.

Welch, R.M. and Graham, R.D. (2005). Agriculture: the real nexus for enhancing bioavailable micronutrients in food crops. *Journal of Trace Elements in Medicine and Biology*, 18: 299–307.

White, P.J. and Broadley, M.R. (2009). Biofortification of crops with seven mineral elements often lacking in human diets – iron, zinc, copper, calcium, magnesium, selenium and iodine. *New Phytol.*, 82: 49–84.

UN. (2011). The Millennium Development Goals Report 2011. United Nations, New York, NY, pp 67.

Unnevehr, L., Pray, C. and Paarlberg, R. (2007). Addressing Micronutrient Deficiencies: Alternative Interventions and Technologies. *Ag Bio Forum* 10(3): 124-134.

Zhang,Y, Shi, R., Rezaul, K.M.D., Zhang, F. and Zou, C. (2010). Iron and zinc concentrations in grain and flour of winter wheat as affected by foliar application. *Journal of Agricultural and Food Chemistry* 58: 12268–12274.

Zhu, C., Naqvi, S., Gomez-Galera, S., Pelacho, A.M., Capell, T. and Christou, P. (2007). Transgenic strategies for the nutritional enhancement of plants. *Trends in Plant Science,* 12(12): 548–555.

Chapter 21

Diversified Agriculture and Nutritional Security

Ramjee Lal Meena, B. Jirli, Pramod Kumar Prajapati, Mohammad Hashim and Pankaj Kumar Mandal

Introduction

The technologies fail to achieve their goals if they fail to reach to the last person in the population. That's why Norman E Borlaug, the famous Nobel laureate and the founder of World Food Prize (WFP) narrated it visionary by saying ***"Take it to the farmers"***. The attempt has been to eradicate poverty and malnutrition by means of using HYVs to cover the entire human beings under the umbrella of health and prosperity. No doubt, unlike other countries of the world, India is bestowed with the diverse climatic conditions supporting a good number of diversified flora and fauna, yet the sufficient and proper nutritional availability is still under question. India, although being an agrarian nation, is witnessing a rapid change in terms of industrialization, health, education *etc.* These have lead to a situation where the people are migrating from agriculture sector to non agriculture sector in search of a better living. This has imposed an extra burden of feeding the expanding population. But fortunately, in-spite of all odds, India has shown a positive remarkable progress in this sector also. Since Independence India has given due attention and importance to agriculture sector by means of introducing a good number of programmes, projects and by taking relevant steps. The introduction of HYVs have accelerated the efforts of making the India self sufficient in food grain production. The food grains production has risen from 50 million tonnes in 1950-1951 to about 272 million tonnes in 2016-2017 (almost five folds).

The concern regarding sufficient and timely availability of food to every citizen has brought 193 countries of the world on a common platform to adopt Agenda 2030 for sustainable development on September. This Agenda, which consists of 17 Sustainable Development Goals and 169 associated main targets, demands global leaders to come together and fight for a common cause of eradicating poverty and hunger in all its dimensions and to shift to a sustainable development path integrating economic, social and environmental dimensions. The specific goals of Agenda 2030 can be defined as follows: *End hunger, achieve food security and improved nutrition, and promote sustainable agriculture* (SDG 2). This goal is not absolute but comprehensive emphasizing outcomes, covering in large part all four dimensions of food security (food availability, access, utilization and stability) and nutrition to end all forms of malnutrition by 2030.

Concept of Food and Nutritional Security

Food security is an umbrella term consisting of both food and nutritional security. With the advancement of science, we now know that nutritional security is much more essential then merely food security. Nutritional security can be achieved only be consuming diversified forms of flora and fauna consisting of essential macro and micro nutrients like proteins, Carbohydrate, fat, vitamins, minerals *etc.* M S Swaminathan has defined food security as "livelihood security for the households and all members within, which ensures both physical and economic access to balanced diet, safe drinking water, environmental sanitation, primary education and basic healthcare". Looking at the importance of availability of food to every citizen, Indian government has taken a concrete step and provided right of food to their citizen which is embedded under Article 21 of the Constitution of India guarantees a fundamental right for food which is essential because India, with a huge population of over 1.3 billion, has seen tremendous growth in the past two decades. Nearly 14.5 per cent of the population is undernourished in India consisting of 51.4 per cent of women between the age range of 15 to 49 years are anaemic. Also it has been reported that about 38.4 per cent of the children of below five years of age are stunted (too short for their age), while 21 per cent suffer from wasting, meaning their weight is too low for their height. India occupied a shocking 97th rank out of 118 countries of the world in 2016 Global Hunger Index *(FAO estimates in 'The State of Food Security and Nutrition in the World, 2017)*

Components of the Right to Food

Availability

Availability refers to the possibilities either for feeding oneself directly from productive land or other natural resources, or for well functioning distribution, processing and market systems that can move food from the site of production to where it is needed in response to demand.

Stability

Stability refers to stable food supply which should be stable over time and space.

Accessibility

Availability refers to access to adequate amount of food materials both economically and physically.

Sustainability

It refers to the process of ensuring availability of natural resources for future generation while meeting the demands of the present generation.

Adequacy

It refers to the situation where required amount of food items are available to everyone in sufficient quantity.

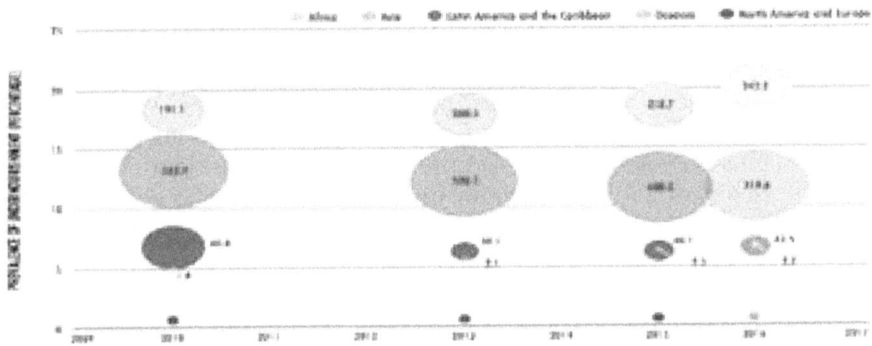

Food Security

The introduction of the term "food security" was recorded in the international development literature of the 1960s and 1970s, referring the ability of a country or region to assure adequate food supply for its current and projected population. A comprehensive definition of it was given in World Food Summit in 1996 as a situation in which *"all people, at all times, have access to sufficient, safe and nutritious food to meet their dietary needs and food preferences for an active and healthy life"* .

Nutrition Security

"Nutrition Security may be defined as the availability of physical, economic and social access to balanced diet, clean drinking water, safe environment, and health care (preventive and curative) for every individual. Thus we can see that nutritional security is much more important than just food security which is not just emphasizing the calorie need but the sources of different types of minerals, vitamins, carbohydrates *etc.*

Diversified Agricultural and Nutritional Security

As mentioned earlier, India is gifted with diversified climatic condition, having 3 out of 12 biodiversities center of the world. This has leaded a diversification not only across but within the crops, livestock, and forestry and fishery, mushroom, bee keeping, and poultry sectors also. India is on its path to assure proper nutritional security which has been aided by Green Revolution because of which food grains has increased from 144.10 million tons to 277.49 million tonnes of food grains but almost all the increase in foodgrain production has come from wheat and rice while the production of coarse grains (millets) and pulses have declined. Declines here, thus, affect food security of the poor qualitatively.

Food Grain Production and Food per Capita Food Availability/person/day

Years	*Rice*	*Wheat*	*Pulses*	*Foodgrains*	*Food Availability gm/head/day*
1951-52	58.00	24.00	22.10	144.10	394.9
1961-62	73.4	28.9	25.2	171.1	468.7
1971-72	70.3	37.8	18.7	171.1	468.8
1981-82	72.2	47.3	13.7	166	454.8
1991-92	80.9	60	15.2	186.2	510.1
2001-02	69.5	49.6	10.9	151.9	416.2
2011-12	102.75	88.31	42.08	250	453.6
2017-18	111.25	97.1	24.51	279.5	518.1

Source: Agricultural Statistics at a Glance, 2017-18.

Conclusions and Recommendations

The changes in living styles of the people, their preference, over population, migration, high cost of food materials have risen the threat of non availability, non affordability as well as non consumption of available nutritional sources (particularly in urban areas)has risen the threat of malnourished society. Thus, there is an urgent need of making the people aware of the nutritional importance and also assuring proper, adequate and affordable availability of nutritious sources at near and far, to rich and poor, without any gender or regional boundaries.

REFERENCES

http: //akvopedia.org/wiki/Food_ per cent26_Nutrition_Security_Portal

https: //secure.mygov.in/group/food-security/

http: //niti.gov.in/writereaddata/files/document_publication/Indias per cent20performance per cent20in per cent20Global per cent20Hunger per cent20Index.pdf

http: //www.fao.org/in-action/food-security-at-the-top-of-indias-agenda/en/

https: //pdfs.semanticscholar.org/de50/33f8ec908962bea0d2ab6ecb6e940108d494.pdf

(PDF) Diversification of Indian Agriculture Available from:

https: //www.researchgate.net/publication/46535062_Diversification_of_Indian_Agriculture_Composition_Determinants_and_Trade_Implications [accessed Aug 19 2018].

UNICEF/WHO/The World Bank, Levels and Trends in Child Malnutrition, (2018), p.3

FAO, The State of Food Security and Nutrition in the World, (2017) p. 7

http: //www.fao.org/3/a-i3448e.pdf

https: //www.indiafoodbanking.org/hunger

http: //agri.ckcest.cn/ass/a7ffdeca-ff39-431e-9068-0dc918ee9463.pdf

https: //knoema.com/atlas/sources/UNICEF

www.iosrjournals.org

http: //naasindia.org/Policy per cent20Papers/policy per cent2007.pdf

Sheereen, Z. and Banu S. (2016) Agriculture Diversification and Food Security Concerns in India, Journal of Agriculture and Veterinary Science (IOSR-JAVS) e-ISSN: 2319-2380, p-ISSN: 2319-2372. Volume 9, Issue 11, PP 56-63.

https: //www.researchgate.net/publication/280112251_Agricultural_Diversification_Food_Self_Sufficiency_and_Food_Security_in_Ghana

Chapter 22

Human Development: Towards Bridging Inequalities

G.K. Sinha

Introduction

The human development approach, developed by the economist Mahbub-Ul-Haq, is anchored in the Nobel laureate Amartya-Sen's work on human capabilities, often framed in terms of whether people are able to "be" and "do" desirable things in life. Examples include Beings: well fed, sheltered, healthy; Doings: work, education, voting, participating in community life.

Human development is about enlarging human choices– focusing on the richness of human lives rather than simply the richness of economies. Critical to this process is work, which engages people all over the world in different ways and takes up a major part of their lives. Of the world's 7.3 billion people, 3.2 billion are in jobs, and others engage in care work, creative work, voluntary work, or other kinds of work or are preparing themselves as future workers. Some of this work contributes to human development, and some does not. Some work even damages human development.

Human development is a process of enlarging people's choices–as they acquire more capabilities and enjoy more opportunities to use those capabilities. But human development is also the objective, so it is both a process and an outcome. Human development implies that people must influence the process that shapes their lives. In all this, economic growth is an important means to human development, but not the goal.

Human development is development of the people through building human capabilities, for the people by improving their lives and by the people through active participation in the processes that shape their lives. It is broader than other approaches, such as the human resource approach; the basic needs approach and the human welfare approach.

Human Development Report

Twenty-five years ago the first Human Development Report presented the concept of human development, a simple notion with far-reaching implications. For too long, the world had been preoccupied with material opulence, pushing people to the periphery. The human development framework, taking a people-centered approach, changed the lens for viewing development needs, bringing the lives of people to the forefront. It emphasized that the true aim of development is not only to boost incomes, but also to maximize human choices– by enhancing human rights, freedoms, capabilities and opportunities and by enabling people to lead long, healthy and creative lives. The human development concept is complemented with a measure– the Human Development Index (HDI) – that assesses human well-being from a broad perspective, going beyond income

Measuring Human Development

The Human Development Index (HDI) is a composite index focusing on three basic dimensions of human development: to lead a long and healthy life, measured by life expectancy at birth; the ability to acquire knowledge, measured by mean years of schooling and expected years of schooling; and the ability to achieve a decent standard of living, measured by gross national income per capita. The HDI has an upper limit of 1.0. To measure human development more comprehensively, the Human Development Report also presents four other composite indices. The Inequality adjusted HDI discounts the HDI according to the extent of inequality. The Gender Development Index compares female and male HDI values. The Gender Inequality Index highlights women's empowerment. And the Multidimensional Poverty Index measures non-income dimensions of poverty.

Human Development in India

When India became independent in 1947, Jawaharlal Nehru stressed the importance of the task that lay ahead of ending poverty, ignorance, disease, and inequality of opportunity. As the 1st Five Year Plan (FYP) was launched, it however did not spell out any specific c planning strategy linking sectoral investment proposals to the objective of the plan. But in the 2nd FYP the principles of 'socialistic pattern of society' underlay the planning strategy and emphasized social gain. It put stress on raising standards of living by raising national income through a rapid industrialization process with focus on heavy industry. This was expected to generate employment opportunities and reduce inequalities in society through trickling down to the poorer sections in society. The plan also placed emphasis on comprehensive village planning, taking the more vulnerable parts of the population, such as landless farmers and artisans into account. It was, in addition, pointed out that national planning should always be carried out in a manner, which takes the programmes of the national, State and district plans into consideration. During the 60's, 70's and 80's, most of the focus was, however, put on accelerating economic growth, savings and investments. This was nothing unique to India, but was the dominating approach to development in most developing countries, as the belief in the trickle-down effect to solve the issue of poverty was strong. The inadequacy of tackling poverty through this strategy was recognized by the Government of India: "The equity objective was sought to be pursued through redistribution of assets. But, land reforms could not be implemented effectively. The problem of poverty could not be tackled through growth, which itself was slow over a long period of time."8

During the 1990s, India introduced economic reforms, aiming at liberalizing the economy through various initiatives. As stated in the 8th FYP (1992-1997): "The Eighth Plan is being launched at a time, which marks a turning point in both international and domestic economic environment.

Human Development Measurements

Five indices are used by the Human Development Reports to measure progress on human development. The first Human Development Report in 1990 introduced a new way of measuring development by combining indicators of life expectancy, educational attainment and income into a composite human development index, the HDI. The components are measured by four variables: GDP per capita, (PPP USD), literary rates (per cent), combined gross enrollment ratio, (per cent) and life expectancy at birth (years). The composite index results in a figure between 0 and 1, of which 1 indicates high level of human development and 0 being no level of human development. Countries are consequently given a specific rank dependent on their success in achieving HD, presented yearly in the Global HDRs. In 1995, the Gender-related Development Index (GDI) and Gender Empowerment Measure

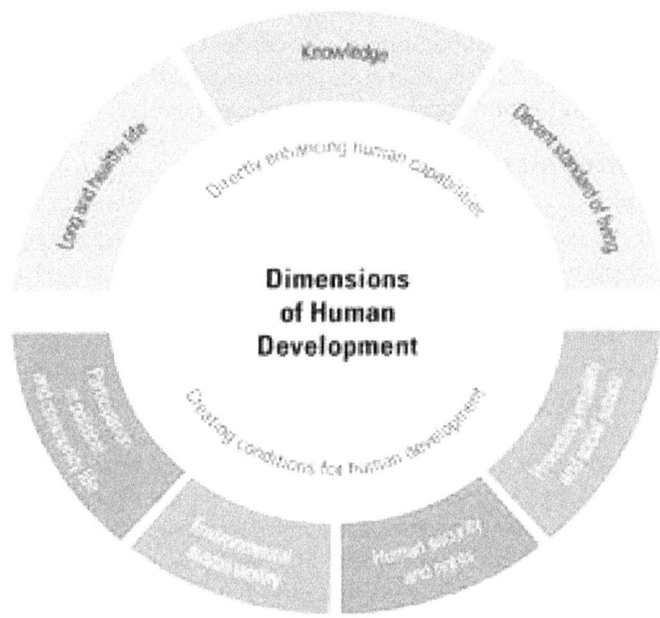

Source: www.hdr.undp.org

(GEM) were added to the reports, as a response to the criticism that HDI did not capture gender inequalities. The GDI measures the same variables as the HDI, but calculates the components separately for women. The methodology used imposes a penalty for inequality, such that the GDI falls when the achievement levels of both women and men in a country diverge or when the disparity between their achievements increases. The greater the gender disparity, the lower a country's GDI. The GDI is simply the HDI discounted, or adjusted downwards, for gender inequality. The GEM measures women's empowerment in public life through assessing the share of seats in parliament held by women, of female legislators, senior officials and managers, of female professional and technical workers, and gender disparities in earned income, reflecting economic independence. While GDI focuses on expansion of capabilities, GEM is concerned with the use of those capabilities to take advantage of the opportunities in life.

Table 1: India's Human Development Index Trends

Year	*Life Expectancy at Birth*	*Expected Years of Schooling*	*Mean Years of Schooling*	*GNI Per Capita (2011 PPP$)*	*HDI Value*
1980	53.9	6.4	1.9	1,255	0.362
1985	55.8	7.3	2.4	1,446	0.397
1990	57.9	7.7	3.0	1,754	0.428
1995	60.4	8.3	3.5	2,046	0.462
2000	62.6	8.5	4.4	2,522	0.496
2005	64.5	9.9	4.8	3,239	0.539
2010	66.5	11.1	5.4	4,499	0.586
2011	66.9	11.7	5.4	4,745	0.597
2012	67.3	11.7	5.4	4,909	0.600
2013	67.6	11.7	5.4	5,180	0.604
2014	68.0	11.7	5.4	5,497	0.609

Source: Human Development Report-2015.

It is evident from the Table 1 that India's HDI value for 2014 is 0.609– which put the country in the medium human development category– positioning it at 130 out of 188 countries and territories. Between 1980 and 2014, India's HDI value increased from 0.362 to 0.609, an increase of 68.1 per cent or an average annual increase of about 1.54 per cent.

Table had shown India's progress in each of the HDI indicators. Between 1980 and 2014, India's life expectancy at birth increased by 14.1 years, mean years of schooling increased by 3.5 years and expected years of schooling increased by 5.3 years. India's GNI per capita increased by about 338.0 per cent between 1980 and 2014.

Expenditure on Social Infrastructure

Social infrastructure with its positive externalities has a significant role in the conomic development of a country. It is empirically proven and widely recognized that education and health impact the growth of an economy. Investing in human capital by way of education, skill development, training and provision of health care facilities enhances the productivity of the workforce and welfare of the population. In India, the proportion of economically active population (15-59 years) has increased from 57.7 per cent to 63.3 per cent during 1991 to 2013, as per Sample Registration System (SRS) data for 2013.

It is evident from the Table 2 as a proportion of the Gross Domestic Product (GDP), expenditure on education has hovered around 3 per cent during 2008-09 to 2014-15. Similarly, there has not been any significant change in the expenditure on health as a proportion of GDP and it has remained stagnant at less than 2 per cent during the same period. During 2013-14, out of the total expenditure on social services, 11.6 per cent was spent on education, while 4.6 per cent was spent on health.

At the state level, in 2013-14, the total state capital expenditure on education was R110, 894 million. Of this, Tamil Nadu had the highest share in the expenditure of about 12 per cent, followed by Uttar Pradesh with a share of 8.67 per cent and Gujarat with 6.67 per cent However, in terms of per student expenditure, Sikkim and Goa spent more than R2000, while Tamil Nadu spent about R726 states such as Rajasthan and Madhya Pradesh spent as little as R37 and R40 per student respectively.

Though the expenditure on social sectors in India has not reflected an increasing trend, an increase in expenditure per-se may not always guarantee appropriate outcomes and achievements. The efficiency of expenditure incurred so far can be assessed by the performance of social sectors through various social indicators. An overall assessment of social sector expenditures in terms of achievements shows that wide gaps still exist in educational and health outcomes and there

Gender Inequality

The 2010 HDR introduced the Gender Inequality Index (GII), which reflects gender-based inequalities in three dimensions – reproductive health, empowerment, and economic activity. Reproductive health is measured by maternal mortality and adolescent birth rates; empowerment is measured by the share of parliamentary seats held by women and attainment in secondary and higher education by each gender; and economic activity is measured by the labour market participation rate for women and men. The GII can be interpreted as the loss in human development due to inequality between female and male achievements in the three GII dimensions. India has a GII value of 0.563, ranking it 130 out of 155 countries in the 2014 index. In India, 12.2 per cent of parliamentary seats are held by women, and 27.0 per cent of adult women have reached at least a secondary level of education compared to 56.6 per cent of their male counterparts. For every 100,000 live births, 190 women die from pregnancy related causes; and the adolescent birth rate is 32.8 births per 1,000 women of ages 15-19. Female participation in the labour market is 27.0 per cent compared to 79.9 for men.

In comparison, Bangladesh and Pakistan are ranked at 111 and 121 respectively on this index.

Table 2: Trends in Social Services Expenditure by Centre and States Government

Item/Year	*2008-09*	*2009-10*	*2010-11*	*2011-12*	*2012-13*	*2013-14*	*2014-15 RE*	*2015-16 BE*
As Percentage to GDP								
Total Expenditure	28.4	28.6	27.6	27.4	27.0	26.2	28.1	27.0
Expenditure on Social Services	6.8	6.9	6.8	6.6	6.6	6.5	7.0	6.7
of which:								
i) Education	2.9	3.0	3.1	3.1	3.1	3.0	3.1	3.0
ii) Health	1.3	1.4	1.3	1.2	1.3	1.2	1.3	1.3
iii) Others	2.6	2.5	2.4	2.2	2.2	2.3	2.6	2.4
As Percentage to Total Expenditure								
Expenditure on Social Services	23.8	24.1	24.7	24.0	24.4	24.8	24.9	24.9
of which:								
i) Education	10.1	10.6	11.4	11.4	11.6	11.6	24.9	11.2
ii) Health	4.6	4.8	4.7	4.6	4.7	4.6	4.8	4.9
iii) Others	9.0	8.7	8.6	8.0	8.2	8.6	9.1	8.9
As Percentage to Social Services								
i) Education	42.6	44.1	46.1	47.7	47.5	46.7	44.0	44.9
ii) Health	19.5	19.7	19.0	19.0	19.1	18.6	19.3	19.5
iii) Others	37.9	36.1	34.9	33.3	33.4	34.7	36.7	35.6

Source: Economic Survey-2015-16.

Notes: 1. Social services includes, education, sports, art and culture; medical and public health, family welfare; water supply and sanitation; housing; urban development; welfare of Scheduled Castes (SC), Scheduled Tribes (ST) and Other Backward Castes (OBC); labour and labour welfare; and social security and welfare, nutrition, relief on account of natural calamities. 2. Expenditure on 'Education' pertains to expenditure on 'Education, Sports, Arts and Culture'. 3. Expenditure on 'Health' includes expenditure on 'Medical and Public Health', 'Family Welfare' and 'Water Supply and Sanitation'. 4. Data for states from 2013-14 onwards is provisional and pertain to budgets of 25 state governments. 5. GDP data from 2011-12 is based on new base year 2011-12.

Table 3: India's GII for 2014 Relative to Selected Countries and Groups

Country	*GII Value*	*GII Rank*	*Maternal Mortality Ratio*	*Adolescent Birth Rate*	*Female seats in Parliament (per cent)*	*Population with at Least some Secondary Education (per cent)*		*Labour force Participation Rate (per cent)*	
						Female	*Male*	*Female*	*Male*
India	0.563	130	190	32.8	12.2	27.0	56.6	27.0	79.9
Bangladesh	0.503	111	170	80.6	20.0	34.1	41.3	57.4	84.1
Pakistan	0.536	121	170	27.3	19.7	19.3	46.1	24.6	82.9
South Asia	0.536	–	183	38.7	17.5	29.1	54.6	29.8	80.3
Medium HDI	0.506	–	168	43.4	18.8	34.8	55.3	37.5	79.8

Source: Human Development Report-2015.

Note: Maternal mortality ratio is expressed in number of deaths per 100,000 live births and adolescent birth rate is expressed in number of births per1, 000 women ages 15-19.

It is evident from the table that the India's GII value higher than South Asia Region and neighbouring countries like Pakistan and Bangaldesh. Maternal mortality ratio was also high *i.e.* 190 in comparison of region and neighbours. Whereas percentage of female seats in parliament was low *i.e.* 12.2 in India in comparison of its neighbours *i.e.* 20.0 per cent 19.7 per cent in Bangladesh and Pakistan respectively. Population with at least some secondary education in females was 27.0 per cent of females it's higher than Pakistan and lower than Bangladesh while percentage of male *i.e.* 56.6 which higher than Bangladesh and Pakistan. Likewise Labour force participation rate India was 27.0 per cent in females which is higher than Pakistan while lower than Bangladesh while percentage of male *i.e.* 79.9 per cent which is lower than its neighbors *i.e.* 84.1 and 82.9 per cent respectively.

McKinsey Global Institute has released a new report, The Power of Parity: Advancing Women's Equality in India, with data showing poor levels of gender parity in Indian society. India's global Gender Parity Score or GPS is 0.48, where a score of 1 would be ideal. India's score represents an "extremely high" level of gender inequality, which compares poorly with 0.71 for Western Europe and 0.74 for North America and Oceania.

The flip side to the abysmal ranking, the report suggests, is that a focus on improving gender parity can help India reap rich rewards of economic growth. "India could boost its GDP by $ 0.7 trillion in 2025, the largest relative boost of all 10 regions analyzed by MGI. This translates into 1.4 per cent per year of incremental GDP growth for India," it says.

According to the report, about 70 per cent of the GDP increase can happen by raising India's female labour force participation rate by 10 percentage points, from 31 per cent now to 41 per cent in 2025. That would mean jobs for 68 million women over the next 10 years. Gender disparity has been a major issue in India's pursuit for achieving the goal of universal elementary education. In order to overcome the problems faced by girls, several measures have been initiated across the country.

Female education has long been acknowledged to have strong correlations with other dimensions of human and social development. As Mehrotra (2006) notes, low levels of education significantly affect the health and nutritional status of women. For instance, in the case of India, he notes that chances of suffering from the diseases caused by malnutrition decrease steadily with increased levels of education. Height and Body Mass Index (BMI) vary with level of education and illiterate women are reportedly at more risk of having lower height and BMI (leading to higher deficiency of iron and other nutrients). Similarly, he noted that while 56 per cent of illiterate women suffer from anaemia, the percentage declines to 40 per cent in the case of the women who have completed at least high school (Mehrotra, 2006: 914).

While gender inequalities intensify with poverty, caste inequalities and geographical location (particularly in underdeveloped rural areas), particular gender-differentiated ideologies cut across all social groups, explaining why in all social groups, girls lag behind boys in access to and participation in education. These include specific views on the appropriate roles to be played by women in family and society, and the underlying controls placed on female mobility and chastity. These gender-specific ideologies are responsible for the continued wide gaps in female secondary schooling enrolment. Further, responsibilities for securing domestic water and fuel place tremendous time burdens on women, often shared with younger girls in the family who could otherwise be in school or at rest or play. Investments in water supply, sustainable energy and renewable sources of fuel all can have significant impact on female education.

Table 4: Sex-wise Literacy Rates of Youth and Adult and Gender Parity Index

Country	*Reference*	*Youth (15-24) Literacy Rates*			*Adult (15+) Literacy Rates*		
		Female	*Male*	*GPI*	*Female*	*Male*	*GPI*
Afghanistan	2011	32.11	61.88	0.52	17.61	45.42	0.39
Argentina	2012	99.42	99.06	1.00	97.95	97.88	1.00
Bangladesh	2012	81.91	78.01	1.05	55.05	62.46	0.88
Bhutan	2005	67.96	80.04	0.85	38.68	65.05	0.59
Brazil	2012	99.02	98.22	1.01	91.63	91.02	1.01
China	2010	99.59	99.69	1.00	92.71	97.48	0.95
India	2006	74.36	88.41	0.84	50.82	75.19	0.68
Maldives	2006	99.36	99.24	1.00	98.43	98.37	1.00
Mexico	2012	99.03	98.72	1.00	93.18	95.39	0.98
Nepal	2011	77.47	89.24	0.87	46.71	71.11	0.66
Pakistan	2011	63.14	78.04	0.81	41.98	66.99	0.63
Russian Federation	2010	99.76	99.66	1.00	99.65	99.73	1.00
South Africa	2012	99.27	98.50	1.01	92.59	94.96	0.97
Sri Lanka	2010	98.59	97.69	1.01	89.96	92.58	0.97

GPI: Gender Parity Index is the Female to Male ratio.

Source: Women & Men in India- 2015.

Table 4 reveals that the literacy ratio in females between aged15 to 24 was highest in Argentina, Brazil, China, Maldives, Mexico, Russian Federation, South Africa and Sri Lanka *i.e.* above 99percent whereas in India its 74.36 per cent with 0.84GPI which is lowest in BRIC nation. Likewise in male literacy ratio were 88.41 in India while Argentina, Brazil, China, Maldives, Mexico, Russian Federation, South Africa, and Sri Lanka were 99 per cent and GPI was 1.00 or more than 1.00.

Employment and Workforce

A focus on employment is particularly important in the context of rapid changes in the Indian economy in which rewards to formal sector work have rapidly outstripped rewards to other activities. For a barely literate manual worker, a monthly salaried job as a waiter in a roadside restaurant is far more remunerative, on an average, than seasonal agricultural work. However, if the same worker is able to find a job as a waiter in a government run canteen or cafe, his salary will most likely outstrip his earnings in a privately owned cafe. Two forces are at work here.

First, movements from agricultural work to non-farm regular employment increase income by reducing underemployment. Second, employment in government or the public sector further boosts salaries. Another important theme of this segment is gender discrimination in employment. Women are less likely to participate in the work force than men. When women work, they are largely concentrated in agriculture and the care of the livestock. Even when they engage in wage work, they work fewer days per year and at a considerably lower pay than men. Even education fails to bridge the gender gap in labour force participation. Educated women seem to be less likely to be employed than their less educated sisters. The progressive decline in labour force participation with higher levels of education stops only at college graduation. However, college graduates form a very small segment of the female population. Finally, regional inequalities in employment are pervasive. Both employment opportunities and wage rates vary dramatically by state. In some cases, state variations in employment mirror state development levels. There are informative exceptions in the hill states for rural non-farm work that demonstrate the potential for combining agricultural and non-agricultural employment. And the vast statewide variations in gender inequalities in employment are not at all related to state levels of development.

Table 5: Total Workers in India - 2011 Census

Population/Workers	*Persons*		*Male*	*Female*
Population	Total	1,21,05,69,573	62,31,21,843	58,74,47,730
Workers		48,17,43,311	33,18,65,930	14,98,77,381
Percentage of Workers		39.79	53.26	25.51
Population	Rural	83,34,63,448	42,76,32,643	40,58,30,805
Workers		34,85,97,535	22,67,63,068	12,18,34,467
Percentage of Workers		41.83	53.03	30.02
Population	Urban	37,71,06,125	19,54,89,200	18,16,16,925
Workers		13,31,45,776	10,51,02,862	2,80,42,914
Percentage of Workers		35.31	53.76	15.44

Source: GOI, Statistical Profile on Women Labour-2012-13

Note: Workers Include Both Main Workers and Marginal Workers

Table 5 reveals that the share of workers was 39.79 per cent of in our total population in which 53.26 was male while 25.51 was female workers during 2011 Census. Likewise 41.83 per cent workers in rural area in which share of female workers were 30.02 per cent whereas 53.03 per cent were male workers. Similarly in urban area 35.31 per cent of total workers were 15.44 was female workers and 53.76 per cent share was male workers. The difference between male and female workers in urban area was huge in comparison of rural area.

Table 6 reveals that the total number of persons in the labour force under usual status method was highest in the year 2011-12 *i.e.* 483.7 million whereas lowest was 407.0 million in the year 1999-2000. Similarly under current daily status (CDS) method it was highest in the year 2011-12 *i.e.* 440.4 million while lowest persons in the labour force *i.e.* 363.3 million during 1999-2000. Likewise highest persons and person days employed under usual status method was 472.9 million during 2011-12 whereas lowest persons and person days employed during 1999-2000 *i.e.* 398.0 million. Under current daily status method highest number of persons and person days employed was 415.7 million during 2011-12 while lowest number *i.e.* 336.9 million during 1999-2000.

Table 6: Employment and Unemployment Scenario in India

Method	*1999- 2000*	*2004- 2005*	*2009- 2010*	*2011-12*
	Persons in the Labour Force (in millions)			
Usual Status	407.0	469.0	468.8	483.7
Current Daily Status	363.3	417.2	428.9	440.4
	Persons and Person Days Employed (in millions)			
Usual Status	398.0	457.9	459.0	472.9
Current Daily Status	336.9	382.8	400.8	415.7
	Unemployment Rate (in per cent)			
Usual Status	2.2	2.3	2.0	2.2
Current Daily Status	7.3	8.2	6.6	5.6

Source: Economic Survey2014-15 Vol.2, p.135.

Health Scenario

An essential component of human development, health of a population is vital for a nation's- economic growth and stability. Good health calls for disease control and nutrition, and an efficient medical care system. Incidentally, while health is a state subject, population stabilization is on the Concurrent List. 'Health for All' (HFA) by the year 2000 through emphasis on primary health care was the main slogan in the Alma Ata Declaration of 1978 of which India too was a signatory. Accordingly, the National Health Policy was adopted by Parliament in 1983. Within the HFA strategy, health for the underprivileged is to be promoted consciously and consistently through community-based systems.

The National Health Policy, 2002 (NHP-2002) outlines improvement in the health status of the population as one of the major thrust areas in social development programme. It focuses on the need for enhanced funding and an organizational restructuring of the national public health initiatives in order to facilitate more equitable access to the health facilities.

Another theme to emerge from the IHDS data is the dominant position of the private sector in medical care. In the early years following independence, discourse on health policy was dominated by three major themes: providing curative and preventive services delivered by highly trained doctors, integrating Indian systems of medicine (for example, Ayurvedic, homeopathic, unani) with allopathic medicine, and serving hard to reach populations through grassroots organization and use of community health care workers.

This discourse implicitly and often explicitly envisioned a health care system dominated by the public sector. Public policies have tried to live up to these expectations. A vast network of Primary Health Centres (PHCs) and sub-centres, as well as larger government hospitals has been put in place, along with medical colleges to train providers. Programmes for malaria, tuberculosis control, and immunization are but a few of the vertically integrated programmes initiated by the government. A substantial investment has been made in developing community-based programmes, such as Integrated Child Development Services, and networks of village-level health workers. In spite of these efforts, growth in government services has failed to keep pace with the private sector, particularly in the past two decades.

Table 7: Demographic and Health Status Indicators

Sl.No.	*Parameters*	*1951*	*1981*	*1991*	*2001*	*2013*
1	Crude Birth Rate (per 1000 Population)	40.8	33.9	29.5	25.4	21.4 (2013)
2	Crude Death Rate (per 1000 population)	25.1	12.5	9.8	8.4	7.0 (2013)
3	Total Fertility Rate	6.0	4.5	3.6	3.1	2.3 (2013)
4	Maternal Mortality Ratio (per 100,000 live births)	NA	NA	398	301	167 (2011-13)
5	Infant Mortality Rate (per 1000 live births)	146	110	80	66	40 (2013)
6	Expectation of Life at Birth (in years) Person	NA	55.4	59.4	63.4	67.5
	Male	37.1	55.4	50.0	62.3	65.8
	Female	36.1	55.7	59.7	64.6	69.3 (2009-13)

Crude Birth Rate: Number of Births per 1000 population in a given year.

Crude Death Rate: Number of Deaths per 1000 population in a given year.

Source: Govt. of India, Health & Family Welfare Statistics India-2015.

The Table 7 had shown the demographic and health status indicators have shown significant improvements over time. In the year 1951 crude birth rate were 40.8 it was decreasing continuously *i.e.* 33.9, 29.5, 25.4 and 21.4 during 1981, 1991, 2011 and 2013 respectively. Likewise crude death rates were 25.1 in the year 1951 it was 12.5, 9.8, 8.4, and 7.0 in the year 1981, 1991, 2001, and 2013 respectively. Similarly maternal mortality ratio was also decreasing since *i.e.* 398 in 1991 whereas in the year 2011-13 it was 167. Infant mortality rate was also declining during *i.e.*146 in 1951 and40 in 2013. Life expectancy rate was also improve significantly in both male and female it was 55.4 in the year 1981 it increased continuously *i.e.* 67.5 year.

Defining Universal Health Coverage

The Universal Health Coverage (UHC) index has been developed by the World Bank to measure the progress made in health sectors in selected countries of the World. India ranks 143 among 190 countries in terms of per capita expenditure on health ($146PPPin2011). It has 157th position according to per capita government spending on health which is just about $44PPP. India's performance on the indicator on treatment of diarrhoea needs improvement in terms of enhancing the coverage. The impoverishment indicator reflects the financial risk protection coverage, with a higher percentage reflecting better coverage.

The Planning Commission(2011) adopted the following definition of Universal Health Coverage (UHC): Ensuring equitable access for all Indian citizens, resident in any part of the country, regardless of income level, social status, gender, caste or religion, to affordable, accountable, appropriate health services of assured quality (promotive, preventive, curative and rehabilitative) as well as public health services addressing the wider determinants of health delivered to individuals and populations, with the government being the guarantor and enabler, although not necessarily the only provider, of health and related services.

This definition incorporates the different dimensions of universal health assurance: health care, which includes ensuring access to a wide range promotive, preventive, curative, and rehabilitative health services at different levels of care; health coverage, that is inclusive of all sections of the population, and health protection, that promotes and protects health through its social determinants. These services should be delivered at an affordable cost, so that people do not suffer financial hardship in the pursuit of good health. The foundation for UHC is a universal entitlement to comprehensive health security and an all-encompassing obligation on the part of the State to provide adequate food and nutrition, appropriate medical care, access to safe drinking water, proper sanitation, education, health-related information, and other contributors to good health.

Ten principles have guided the formulation of our recommendations for introducing a system of UHC in India: (i) universality; (ii) equity; (iii) non-exclusion and non-discrimination; (iv) comprehensive care that is rational and of good quality; (v) financial protection; (vi) protection of patients' rights that guarantee appropriateness of care, patient choice, portability and continuity of care; (vii) consolidated and strengthened public health provisioning; (viii) accountability and transparency; (ix) community participation; and (x) putting health in people's hands.

According to the International Labour Organization, nearly 50 countries have attained universal or near universal coverage. Conspicuous gaps still exist, however, particularly in Asia, Africa and the Middle East.

Escalating health care costs, inadequate public spending, and weak health care delivery systems in low and middle income countries have been barriers to UHC in the past. Today there is greater international recognition of the need for health systems to adopt sustainable financing mechanisms that permit population-wide coverage and the efficient delivery of a wide range of health services. The 2005 World Health Assembly (WHA) urged member states to pursue UHC, ensuring equitable distribution of quality health care infrastructure and human resources, to protect individuals seeking care against catastrophic healthcare expenditure and possible impoverishment. It also highlighted the importance of taking advantage, where appropriate, of opportunities that exist for collaboration between public and private providers and health-financing organizations, under strong overall government stewardship.

Table 8: Universal Health Coverage (UHC) Index Values (in per cent) for Select Countries

Country	Year	Immuni-zation	Diarrhoea Treatment	Inpatient Admission	Impoverish-ment	UHC
Brazil	1998	67.6	44.1	92.3	94.1	82.0
	2006	67.6	43.0	85.9	93.6	81.6
India	1998	26.0	10.3	71.6	85.2	51.6
	2006	34.5	22.1	71.6	83.7	56.9
Indonesia	2000	49.4	49.7	27.9	93.4	47.3
	2009	59.5	48.2	27.9	93.1	47.0
Philippines	1998	68.5	23.0	64.7	95.7	66.2
	2008	72.6	54.3	64.7	95.0	75.2
South Africa	1997	61.7	53.0	96.6	95.9	78.8
	2003	61.7	53.0	96.6	96.7	79.3
Vietnam	1997	46.7	45.2	86.7	84.6	57.8
	2008	30.6	65.9	86.7	66.1	61.1

Source: Economic Survey 2015-16; World Bank Group Policy Research Working Paper 7470 on Measuring Progress towards Universal Health Coverage--With an Application to 24 Developing Countries, November, 2015.

Multidimensional Poverty Index (MPI)

The 2010 HDR introduced the MPI, which identifies multiple deprivations in the same households in education, health and living standards. The education and health dimensions are each based on two indicators, while the standard of living dimension is based on six indicators. All of the indicators needed to construct the MPI for a household are taken from the same household survey. The indicators are weighted to create a deprivation score, and the deprivation scores are computed for each household in the survey. A deprivation score of 33.3 per cent (one-third of the weighted indicators), is used to distinguish between the poor and non-poor. If the household deprivation score is 33.3 per cent or greater, the household (and everyone in it) is classified as multidimensionally poor. Households with a deprivation score greater than or equal to 20 per cent but less than 33.3 per cent are near multidimensional poverty. Finally, households with a deprivation score greater than or equal to 50 per cent live in severe multidimensional poverty.

The recent survey data that were publically available for India's MPI estimation refer to 2005/2006. In India 55.3 per cent of the population (631,999 thousand

Table 9: The Most Recent MPI for India Relative to Selected Countries

	Survey Year	*MPI Value*	*Head-count (per cent)*	*Intensity of Deprivations (per cent)*	*Population Share (per cent)*			*Contribution to Overall Poverty of Deprivations in (per cent)*		
					Near Poverty	*In Severe Poverty*	*Below Income Poverty Line*	*Health*	*Education*	*Living Standards*
India	2005/2006	0.282	55.3	51.1	18.2	27.8	23.6	32.5	22.7	44.8
Bangladesh	2011	0.237	49.5	47.8	18.8	21.0	43.3	26.6	28.4	44.9
Pakistan	2012/2013	0.237	45.6	52.0	14.9	26.5	12.7	32.3	36.2	31.6

Source: Human Development Report-2015.

people) are multidimensionally poor while an additional 18.2 per cent live near multidimensional poverty (208,588 thousand people). The breadth of deprivation (intensity) in India, which is the average of deprivation scores experienced by people in multidimensional poverty, is 51.1 per cent. The MPI, which is the share of the population that is multi-dimensionally poor, adjusted by the intensity of the deprivations, is 0.282. Bangladesh and Pakistan have MPIs of 0.237 and 0.237 respectively.

Table 9 compares income poverty, measured by the percentage of the population living below PPP US$1.25 per day, and multidimensional poverty. It shows that income poverty only tells part of the story. The multidimensional poverty headcount is 31.7 percentage points higher than income poverty. This implies that individuals living above the income poverty line may still suffer deprivations in education, health and other living conditions. Table 9 also shows the percentage of India's population that lives near multidimensional poverty and that lives in severe multidimensional poverty. The contributions of deprivations in each dimension to overall poverty complete a comprehensive picture of people living in multidimensional poverty in India. Figures for Bangladesh and Pakistan are also shown in the table for comparison.

Table 10: State-Wise Poverty Situation during 2011-12 (in per cent)

Population below Poverty Line	*Rural Poverty*	*Urban Poverty*	*Total Poverty*
Less than 10	Goa, Punjab, Himachal Pradesh, Kerala, Sikkim	Goa, Sikkim, Himachal Pradesh, J&K, Mizoram, Kerala, Andhra Pradesh, Tamil Nadu, Meghalaya, Maharashtra, Punjab, Tripura	Goa, Kerala, Himachal Pradesh, Sikkim, Punjab, Andhra Pradesh
10 to 20	Andhra Pradesh, Haryana, Meghalaya, Rajasthan, J&K, Nagaland, Tripura, Tamil Nadu, Uttarakhand	Gujarat, Haryana, Uttarakhand, Rajasthan, West Bengal, Karnataka, Nagaland, Odisha	J&K, Haryana, Uttarakhand, Tamil Nadu, Meghalaya, Tripura, Rajasthan, Gujarat, Maharashtra, Nagaland, West Bengal
20 to 30	Gujarat, West Bengal, Maharashtra, Karnataka	Arunachal Pradesh, Assam, Madhya Pradesh, Chhattisgarh, Jharkhand, Uttar Pradesh	Mizoram, Karnataka, Uttar Pradesh
30 to 40	Arunachal Pradesh, Manipur, Madhya Pradesh, Assam, Uttar Pradesh, Bihar, Odisha, Mizoram,	Bihar, Manipur	Madhya Pradesh, Assam, Odisha, Bihar, Arunachal Pradesh, Manipur, Jharkhand, Chhattisgarh
Above 40	Jharkhand, Chhattisgarh		

Source: Based on NITI Aayog estimates, 2011-12

Poverty estimates based on the Tendulkar Committee methodology using household consumption expenditure survey data collected by the NSSO in its 68th round (2011-12) shows that the incidence of poverty declined from 37.2 per cent in

2004-05 to 21.9 per cent in 2011-12 for the country as a whole, with a sharper decline in the number of rural poor. While the rural poverty ratio declined from 41.8 per cent in 2004-05 to 25.7 per cent in 2011-12, the urban poverty ratio declined from 25.7 per cent in 2004-05 to 13.7 per cent in 2011-12. The rural poverty ratio still remains much higher than the urban. The high rural poverty can be attributed to lower farm incomes due to subsistence agriculture, lack of sustainable livelihoods in rural areas, impact of rise in prices of food products on rural incomes, lack of skills, underemployment, and unemployment. Table gives a picture of state-level rural and urban poverty.

Table 10 reveals that the states having less than 10 per cent poverty in both areas *i.e.* urban and rural were Goa, Kerala, Himachal Pradesh, Sikkim, Punjab, and Andhra Pradesh. Similarly states having 10 to 20 per cent population below poverty line *i.e.* Jammu & Kashmir, Haryana, Tamil Nadu, Meghalaya, Tripura, Rajasthan, Gujarat, Maharashtra, Nagaland and West Bengal. Likewise Mizoram, Karnataka, and Uttar Pradesh states 20 to 30 per cent population living below poverty line. Whereas 30 to 40 per cent population of Madhya Pradesh, Assam, Odisha, Bihar, Arunachal Pradesh, Manipur, Jharkhand and Chhattisgarh states living below poverty line.

Inequality Dimensions

Inequality has been an important issue in development debates. Several philosophers and economists have discussed about inequality. Tendulkar (2010) draws a distinction between inequity and inequality. He examines the path breaking work of Simon Kuznets who indicates that inequalities rise with economic growth upto a point and then decline. This is the so called Kuznets inverted 'U' shape curve. Initially economic growth increase overall inequality as the rural-urban transformation takes place and labour moves from low productivity agriculture to high productivity urban industrial and service sector activities. Tendulkar says that even if measured inequality increases, there may not be increasing feeling of inequity as people observe high mobility and can aspire to move upwards like others.

Recently, Credit Suisses (CS) and Oxfam have released reports on global wealth and inequality. According to CS report, the top percentile of wealth holders now own over half of the world's wealth and the richest decile 87.7per cent. The richest 1 per cent owns half of all the wealth in the world. Oxfam report released ahead of the annual World economic Forum in Davos in 2015,shows that the combined wealth of the richest 1 per cent will overtake that of the other 99 per cent in 2016 unless the current trend of rising inequality is checked. The share of global wealth of richest 1 per cent rose from 44 per cent in2009 to 48 per cent in 2014 and at this rate it will be more than 50 per cent in 2016.Credit Suisse report on India reveals that the richest 1 per cent owned 53 per cent of the country's wealth while the share of the top 10 per cent was 76.3 per cent. In other words, 90 per cent of Indians own a less than 25 per cent of the country's wealth. Generally inequality is examined with consumption distribution as income distribution data is not available. Inequality in consumption may be an under estimate as NSS data may not be capturing the consumption of the rich adequately. Inequality in income would be much higher than that of consumption. It may be noted that if we consider access to education and other public services like health, electricity, drinking water, the inequalities could be much higher.

Conclusion

Human development focuses on improving the lives people lead rather than assuming that economic growth will lead, automatically, to greater wellbeing for all. Income growth is seen as a means to development, rather than an end in itself.

The principal objective of human development is the attainment of higher standard of living for the people. This requires a more equitable distribution of development benefits and opportunities, better living environment and empowerment of the poor and marginalized. There is special need to empower women who can act as catalysts for change. In making the development process inclusive, the challenge is to formulate policies and programmes to bridge regional, social, and economic disparities in as effective and sustainable a manner as possible.

From a human development perspective, work, rather than jobs or employment is the relevant concept. A job is a narrow concept with a set of pre-determined time-bound assigned tasks or activities, in an input-output framework with labour as input and a commodity or service as output. Yet, jobs do not encompass creative work (*e.g.* the work of a writer or a painter), which go beyond defined tasks; they do not account for unpaid care work; they do not focus on voluntary work. Work thus is a broader concept, which encompasses jobs, but goes beyond by including the dimensions mentioned above, all of which are left out of the job framework, but are critical for human development.

Human development is about giving people more freedom to live lives they value. In effect this means developing people's abilities and giving them a chance to use them. For example, educating a girl would build her skills, but it is of little use if she is denied access to jobs, or does not have the right skills for the local labour market. Three foundations for human development are to live a long, healthy and creative life, to be knowledgeable, and to have access to resources needed for a decent standard of living. Many other things are important too, especially in helping to create the right conditions for human development. Once the basics of human development are achieved, they open up opportunities for progress in other aspects of life.

REFERENCES

Desai, Sonalde B. (2010). Human Development in India: Challenges for a Society in Transition, Oxford University Press, New Delhi

Govt. of India, Economic Survey various issue

Govt. of India, Health & Family Welfare Statistics India-2015

Govt. of India, Women and Man in India-2015

High Level Expert Group Report on Universal Health Coverage for India (2011). Planning Commission, New Delhi

Indian Express various issue

Mahadev, S. (2016). Economic Reforms Poverty and Inequality, Indira Gandhi Institute of Development Research, Mumbai

Mehrotra, Santosh and Gandhi, Ankita (2012). India's Human development in 2000s: Towards Social Inclusion, Institute of Applied Manpower Research, New Delhi

Mukherjee, Scchidanand (2014). Three Decades of Human Development across Indian State: Inclusive Growth or Perpetual Disparity, National Institute of Public Finance and Policy, New Delhi

Stewart, Frances (2013). Capabilities and Human Development: Beyond the Individual–the Critical Role of Social Institutions and Social Competencies, UNDP, New York

The Times of India various issue

UNDP (2010). Human Development in India: Analysis to Action, New Delhi

UNDP, Human Development Report various issue

www.hdr.undp.org

www.mospi.gov.in

www.planningcommission.nic.in

Chapter 23

Mushroom: Nutritional Value and Health Benefits

M.K. Yadav, Ram Chandra, Preeti Dubey, S.K. Yadav and Pratibha Srivastava

Introduction

Mushrooms have been considered as ingredient of gourmet cuisine across the globe; especially for their unique flavor and have been valued by humankind as a culinary wonder. More than 2,000 species of mushrooms exist in nature, but around 25 are widely accepted as food and few are commercially cultivated. Mushrooms are considered as a delicacy with high nutritional and functional value, and they are also accepted as nutraceutical foods; they are of considerable interest because of their organoleptic merit, medicinal properties, and economic significance. The global food and nutritional security of growing population is a great challenge, which looks for new crop as source of food and nutrition. In this context, mushrooms find a favor which can be grown even by landless people, that too on waste material and could be a source for proteineous food. Use of mushrooms as food and neutraceutical have been known since time immemorial, as is evident from the description in old epics Vedas and Bible. Earlier civilizations had also valued mushrooms for delicacy and therapeutic value. In the present time, it is well recognized that mushroom is not only rich in protein, but also contains vitamins and minerals, whereas, it lacks cholesterol and has low calories. Furthermore, it also has high medicinal attributes like immune- modulating, antiviral, antitumor, antioxidants and hepatoprotective properties. With the growing awareness for nutritive and quality food by growing health conscious population, the demand for food including mushrooms is quickly rising and will continue to rise with increase in global population which will be 8.3 million by 2025 and flexible income (Singh, 2011).

Literally the term 'Nutraceutical' is a concurrence of nutrition and pharmaceuticals. In fact Nutraceutical are the natural substances or food supplements that assert to have medicinal effects on human health. The International medical boards and community medicines has described the term Nutraceutical as any substance which may be considered as food or a part of food and provides some medical or health benefits including prevention, treatment and cure of some diseases (Barros *et al.*, 2008). It is needless to say that a bulk of our present population relies on drug resources of plant origin and the herbal medicines are increasingly being used not only by developing countries but also by the developed countries as supplement or alternative to their health care system. Preparation of drugs and medicines from natural resources is the most eco-friendly method of new drug discovery. Considering the growing demand of herbal medicines, the fast growing mushrooms have received remarkable interest in recent decades with realization to the fact that they are good sources of delicious food with high nutritional attributes and some having medicinal values as well. Mushrooms are nutritionally functional food and a source of medicines and most remarkably can produce mycopharmaceuticals and myconutraceutical (Chang and Miles, 1989). Many edible mushrooms like *Pleurotus sp., Agaricus bisporus, Agaricus campestris, Volvariella volvacea, Lentinus edodes, Grifola sp., Schizophyllum communis, Auricularia auriculata etc.* enjoy a high demand of premium price because of their high nutritional value as well as their pharmaceutical potential. Keeping this fact in mind, the present topic, "Mushroom: Nutritional Value and Health Benefits" has been chosen giving emphasis on the pharmaceutical properties of some edible mushrooms which can safely be used as Nutraceutical.

Nutritional Importance of Mushrooms

Nutritional Value

The nutritional value of edible mushrooms is due to their high protein, fiber, vitamin and mineral contents, and low-fat levels (Mattila *et al.*, 2001 and Barros *et al.*, 2008). They are very useful for vegetarian diets because they provide all the essential amino acids for adult requirements; also, mushrooms have higher protein content than most vegetables. Besides, edible mushrooms contain many different bioactive compounds with various human health benefits. It is important to remark that the growth characteristics, stage and postharvest condition may influence the chemical composition and the nutritional value of edible mushrooms. Also, great variations occur both among and within species (Kalac, 2013 and Reis *et al.*, 2012). Mushrooms contain a high moisture percentage that ranges between 80 and 95 g/100 g, approximately. As above mentioned, edible mushrooms are a good source of protein, 200–250 g/kg of dry matter; leucine, valine, glutamine, glutamic and aspartic acids are the most abundant. Mushrooms are low-calorie foods since they provide low amounts of fat, 20–30 g/kg of dry matter, being linoleic (C18:2), oleic (C18:1) and palmitic (C16:0) the main fatty acids. Edible mushrooms contain high amounts of ash, 80–120 g/kg of dry matter (mainly potassium, phosphorus, magnesium, calcium, copper, iron, and zinc). Carbohydrates are found in high proportions in edible mushrooms, including chitin, glycogen, trehalose, and mannitol; besides, they contain fiber, -glucans, hemicelluloses, and pectic substances. Additionally, glucose, mannitol, and trehalose are abundant sugars in cultivated edible mushrooms, but fructose and sucrose are found in low amounts. Mushrooms are also a good source of vitamins with high levels of riboflavin (vitamin B2), niacin, folates, and traces of vitamin C, B1, B12, D and E. Mushrooms are the only nonanimal food source that contains vitamin D and hence they are the only natural vitamin D ingredients for vegetarians. Wild mushrooms are generally excellent sources of vitamin D2 unlike cultivated ones; usually cultivated mushrooms are grown in darkness and UV-B light is needed to produce vitamin D2 (Guillamon *et al.*, 2010).

Numerous bioactive polysaccharides or polysaccharide-protein complexes from medicinal mushrooms appear to enhance innate and cell-mediated immune responses and exhibit antitumor activities in animals and humans. A wide range of these mushroom polymers have been reported previously to have immunotherapeutic properties by facilitating growth inhibition and destruction of tumor cells. Several of the mushroom polysaccharide compounds have proceeded through clinical trials and are used extensively and successfully in Asia to treat various cancers and other diseases. A total of 126 medicinal functions are thought to be produced by selected mushrooms. Mushrooms are usually eaten for their culinary properties, providing a flavoring and garnish for other foods. They are cultivated with special technique and usually consumed by the rich people because the price of mushroom is usually much higher than that of the most common vegetables. This may give one the impression that mushrooms constitute a luxury food and that their promotion would only benefit relatively rich people. Actually, mushrooms are rich in protein and contain several vitamins and mineral salts and should thus be considered as high protein vegetables to enrich all human diets.

Table 1: Composition of Common Edible Mushrooms (G/100 Fresh Weight)

Mushroom	*Moisture*	*Protein*	*Fat*	*Carbo-hydrate*	*Fibre*	*Ash*	*Calories*
A. bisporus	90.1	2.9	0.3	5.0	0.9	0.8	36
V. volvacea	90.1	2.1	1.0	4.7	1.1	1.0	36
P. sajor-caju	90.2	2.5	0.2	5.2	1.3	0.6	35
Cabbage	91.9	1.8	0.1	4.6	1.0	0.6	27
Cauliflower	90.0	2.6	0.4	4.0	1.2	1.0	30
Potato	74.7	1.6	0.1	22.6	0.4	0.6	97

Source: Sohi, 1988.

Mushrooms have been found effective against cancer, cholesterol reduction, stress, insomnia, asthma, allergies and diabetes (Bahl, 1983). Due to high amount of proteins, they can be used to bridge the protein malnutrition gap. Mushrooms as functional foods are used as nutrient supplements to enhance immunity in the form of tablets. Due to low starch content and low cholesterol, they suit diabetic and heart patients. One third of the iron in the mushrooms is available in mineral form. Their polysaccharide content is used as anticancer drug. Even, they have been used to combat HIV effectively (Namba, 1993 and King, 1993).

Vitamins

Mushrooms are good source of vitamins (Table 2) such as vitamin 'B1' (Thiamine), vitamin 'B2' (Riboflavin), niacin, biotin and vitamin 'C' (Ascorbic acid). Mushrooms are one of the best sources of vitamins especially Vitamin B (Chang and Buswell, 1996; Mattila *et al.*, 2000). Vitamin content of edible mushrooms has been reported by Esselen and Fellers (1946), and Litchfield (1964).

Table 2: Vitamin Content of Edible Mushrooms

Mushrooms	*Content mg/100g D. wt.*			
	Thiamine	*Riboflavin*	*Niacin*	*Ascorbic Acid**
Agaricus bisporus	1.1	5.0	55.7	81.9
Lentinus edodes	7.8	4.9	54.9	0.0
Volvariella volvacea	0.32-0.35	1.63-2.97	64.8	20.2
Pleurotus spp.	1.16-4.8	4.7*	46.1	0.0

Adapted from Eli V. Crisan and Anne Sands (Chang and Hayes, 1978)

Source: Chadha and Sharma, 1995.

Proteins

The value of protein is determined by the kinds of amino acids that form protein. Mushrooms contain all the essential amino acids as well as the most commonly occurring non-essential amino acids and amides. Mushrooms are rich in lysine and tryptophan, the two essential amino acids that are deficient in cereals. The most nutritious mushrooms are almost equal in nutritional value to meats and milk. Protein is the main body building constituent of our food. Protein content of mushrooms depends on the composition of the substratum, size of pileus, harvest time and species of mushrooms (Bano and Rajarathnam, 1982).

Table 3: Essential amino acid (per cent crude protein) in edible mushrooms

Amino Acid	*Agaricus bisporus*	*Pleurotus sajor-caju*	*Volvariella volvacea*
Leucine	7.5	7.0	4.5
Isoleucine	4.5	4.4	3.4
Valine	2.5	5.3	5.4
Tryptophan	2.0	1.2	1.5
Lysine	9.1	5.7	7.1
Threonine	5.5	5.0	3.5
Phenyl alanine	4.	5.0	2.6
Methionine	0.9	1.8	1.1
Histidine	2.7	2.2	3.8

Sources: Bano and Rajarathnam, 1982; Li and Chang, 1982.

Verma *et al.* (1987) reported that mushrooms are very useful for vegetarian because they contain some essential amino acids which are found in animal proteins. The digestibility of *Pleurotus* mushrooms proteins is as that of plants (90 per cent) whereas that of meat is 99 per cent (Bano and Rajarathnam, 1988). Rai and Saxena (1989) observed decrease in the protein content of mushroom on storage. The protein conversion efficiency of edible mushrooms per unit of land and per unit time is far more superior compared to animal sources of protein (Bano and Rajarathnam, 1988).

Minerals

Mushrooms are also good source of minerals such as, potassium, phosphorus, sodium, calcium, and contain low but available form of Iron. Sodium and potassium ratio is very high which idea is for patient of hypertension.

Table 4: Minerals Content in Oyster Mushrooms

Species	*K*	*P*	*Mg*	*Na*	*Ca*	*Fe*	*Cd*	*Zn*	*Cu*	*Pb*	*Hg*
	(mg/100g Dry wt.)				*(ppm)*						
P. sajor-caju	3260	760	221	60	20	12.4	0.3	29	12.2	3.2	0
P. eous	4570	1410	242	78	23	9.0	0.4	82.7	17.8	1.5	0
P. flabellatus	3760	1550	292	75	24	12.4	0.5	56.6	21.9	1.5	0
P. florida	4660	1850	192	62	24	18.4	0.5	11.5	15.8	1.5	0

Source: Rai, 1995).

The fruiting bodies of mushrooms are characterized by a high level of well assimilated mineral elements. Major mineral constituents in mushrooms are K, P, Na, Ca, Mg and elements like Cu, Zn, Fe, Mo, Cd form minor constituents (Bano and Rajarathanum, 1982; Bano *et al.*, 1981). K, P, Na and Mg constitute about 56 to 70 per cent of the total ash content of the mushrooms (Li and Chang, 1982) while potassium alone forms 45 per cent of the total ash. Abou-Heilah *et al.*, (1987) found that content of potassium and sodium in *A. bisporous* was 300 and 28.2 ppm respectively. *A. bisporus* ash analysis showed high amount of K, P, Cu and Fe (Anderson and Fellers, 1942). Varo *et al.* (1980) reported that *A. bisporus* contains Ca (0.04 g), Mg (0.16), P (0.75 g), Fe (7.8 g), Cu (9.4 mg), Mn (0.833 mg) and Zn (8.6 mg) per kilogram fresh weight.

Medicinal Mushroom

Medicinal mushrooms or extracts from mushrooms that are used as possible treatments for diseases. Some mushroom compounds, including polysaccharide, glycoprotein and proteoglycons modulate immune system responses and inhibit tumor growth. Some medicinal mushroom isolates that have been identified also show cardiovascular, antiviral, antibacterial, antiparasitic, anti-inflammatory, and antidiabetic properties. Currently, several extracts have widespread use in Japan, Korea and China, as adjuncts to radiation treatments and chemotherapy.

Table 5: Various Pharmaceutical Components Isolated from Several Mushrooms

Pharmacodynamics	*Component*	*Species*
Antibacterial effect	Hirsuitic acid	Many species
Antibiotic	E-B, Methoxyacrylate	*O. radicata*
Antiviral effect	Polysaccharide, Protein	*L. edodes* and *Polyporaceal*
Cardiac tonic	Volvatoxin, Flammutoxin	*Volvariella, Flammulina*
Decrease cholesterol	Eritadenine	*Collybia velutipes*
Decrease blood pressure	Triterpene	*G. lucidum*
Decrease level of blood	Peptide, glycogen, Ganoderan Glucon.	*G. lucidum*
Antithrombus	5' AMP, 5'GMP	*Psalliota hartensis*
Inhibition of PHA	r-GHP	*P. hartensis, L. edodes*
Anti tumour	B-glucan RNA complex	*Hysizygus marmoreus etc.*
Increase secretion of bile	Armillarisia A	*Armillaria tabescens*
Analgesic & sedative effect	Marasmic acid	*Marasmices androsaceus*

Source: Chadha and Sharma 1995.

Historically, mushrooms have long had medicinal uses, especially in traditional Chinese medicine. Mushrooms have been a subject of modern medical research since the 1960s, where most modern medical studies concern the use of mushroom extracts, rather than whole mushrooms. Only a few specific mushroom extracts have been extensively tested for efficacy. Polysaccharide-K and lentinan are among the mushroom extracts with the firmest evidence. The available results for most other extracts are based on *in vitro* data, effects on isolated cells in a lab dish, animal models like mice, or underpowered clinical human trials. Studies show that glucan-containing mushroom extracts primarily change the function of the innate and adoptive immune system, functioning as bioresponse modulator, rather than by directly killing bacteria, viruses, or cancer cells as cytocidal agents. In some countries, extracts like, Polysaccharide-K schizophyllan, polysaccharide peptide and lentinan are government-registered adjuvant cancer therapies.

Pharmaceuticals worth $700 million are produced annually in Japan from *Lentinus, Coriolus, Schizophyllum* and *Ganoderma*. According to Mizuno *et al.* (1990) Mushroom extracts have a high amount of retene that has an antagonistic effect on some form of tumor. Some mushroom extracts induce formation of interferon, a defense mechanism against viral infection and have hypo-cholesteroemic activity (lowering cholesterol levels). Further, compounds extracted from mushroom have antifungal and antibacterial properties. The low fat content and cholesterol free mushroom diets comfortable for patients of hyperlipemia, blood pressure and hypertension and for diabetes due to lack of sugar in mushroom. Mushroom diets are also effective against diarrhea patient due to presence of fiber and its more digestibility and tumors due to some medicinal properties.

Nutraceuticals

In addition to the nutritional components found in edible mushrooms, some have been found to comprise important amounts of bioactive compounds. The content and type of biologically active substances may vary considerably in edible mushrooms; their concentrations of these substances are affected by differences in strain, substrate, cultivation, developmental stage, age, storage conditions, processing, and cooking practices (Barros *et al.*, 2008).

The bioactive substances found in mushrooms can be divided into secondary metabolites (acids, terpenoids, polyphenols, sesquiterpenes, alkaloids, lactones, sterols, metal chelating agents, nucleotide analogs, and vitamins), glycoproteins and polysaccharides, mainly -glucans. New proteins with biological activities have also been found, which can be used in biotechnological processes and for the development of new drugs, including lignocellulose-degrading enzymes, lectins, proteases and protease inhibitors, ribosome-inactivating proteins, and hydrophobins (Erjavec, *et al.*, 2012).

In China, many species of edible wild-grown mushrooms, that is *Tricholoma matsutake, Lactarius hatsudake, Boletus aereus* are appreciated as food and also in traditional Chinese medicine. The rich amount of proteins, carbohydrates, essential minerals, and low energy levels contributes to considering many wild-grown mushrooms as good food for the consumer, which can virtually be compared with meat, eggs, and milk (Wang, *et al.*, 2014). There are more than 100 species of mushroom fungi that are commonly used in traditional medicine.

Ganoderma lucidum mushroom is considered as a symbol of happy augury, good fortune, good health, longevity and even immortality in china. It can reduce the blood pressure, cholesterol and blood sugar level and inhibit platelets aggregation. It has cancer curing ability attributed to various polysaccharides. *Lentinus edodes* (shiitake) can be used both for edible and medicinal purposes such as treatment of colds, measles in children, bronchitis inflammation, stomachache, headache, dizziness, dropsy, small-pox, mushroom poisoning, blood pressure and blood cholesterol level and also valuable for stimulating the immune system to increase the body's ability towards of cancerous tumors.

Cordyceps sinensis contains substances called cordycepin, cordycepic acid, with therapeutic applications like the effects of increased oxygen utilization, ATP production, and stabilization of blood sugar metabolism. Besides, it has antibacterial function, reduces asthma, and lowers blood pressure. On the other hand, it has been reported as organ protector, as well as with a protective effect for heart, liver, and kidney diseases. It is a unique fungus that grows on the insect larvae. It is most used for treating of debility after illness, weakness, spitting of blood caused by tuberculosis, chronic coughing and asthma caused by senility, night sweating, spontaneous sweating and anemia and malignant tumor. *Tremella fulciformis* is used as a cough syrup for chronic bronchitis and other cough related conditions such as asthma and dry cough. It is able to help strengthen the vital internal organs such as sense organ, kidney, lungs, stomach and ulcer treatment. It is also used for the weakness after childbirth, constipation, abnormal menstruation, blood in sputum, dysentery, gastritis and removes facial freckles. *Auricularia auricular* is considered for bleeding, especially excessive uterine bleeding, abdominal and tooth pain, anti-mutagenic, antiulcer with little effect on the gastric acid secretion, anticoagulant.

Coriolus versicolar is traditionally used as tea or dried powder to treat the infection or inflammation of the upper respiratory, urinary and digestive tract linear diseases, weakness and tumors. It is successful effective anticancer drug. A protein bound polysaccharides is capable of modifying the host biological response by stimulating the immune system. *Hericum eninaceous,* and *Flammulina voluptipes* are used for improving of spleen and stomach ailments, calming nerves and mind. They have been used for the controlling blood pressure, sugar, cholesterol, chronic gastritis, cancerous carcinoma and esophageal carcinoma. *Grifola frondosa* is promoted as anticancer agent, particularly on human gastric carcinoma, such effect results from the induction of cell apoptosis and could significantly accelerate the anticancer activity. *Antrodia cinnanomea* is a medicinal mushroom and it has been used to treat food and drug intoxication, diarrhea, abdominal pain, hypertension, skin itching, and cancer. *Panellus serotinus* is helps to prevent the development of nonalcoholic fatty liver disease.s

Conclusion

Several mushroom species have been pointed out as sources of bioactive compounds, in addition to their important nutritional value. The inclusion of whole mushrooms into the diet may have efficacy as potential dietary supplements. The production of mushrooms and the extraction of bioactive metabolites is a key feature for the development of efficient biotechnological methods to obtain these metabolites. It has been shown by a wide range of studies that mushrooms contain components with outstanding properties to prevent or treat different type of diseases. Powder formulations of some species have revealed the presence of essential nutrients. They present a low fat content and can be used in low-calorie diets, just like the mushrooms fruiting bodies. Some formulations could be used as antioxidants to prevent oxidative stress and thus ageing.

Future studies into the mechanisms of action of mushroom extracts will help us to further delineate the interesting roles and properties of various mushroom phytochemicals in the prevention and treatment of some degenerative diseases. In view of the current situation, the research of bioactive components in edible wild and cultivated mushrooms is yet deficient. There are numerous potential characteristics and old and novel properties, provided by mushrooms with nutraceutical and health benefits, which deserve further investigations.

REFERENCES

Abou-Heilah, A.N., Kasionalsim, M.Y. and Khaliel, A.S. (1987). Chemical composition of the fruiting bodies of *Agaricus bisporus*. *Int. J. Expt. Bot.*, 47: 64-68.

Anderson, E.E. and Fellers, C.R. (1942). The food value of mushrooms (*A. compestris*). *Proc. Am. Soc. Hort. Sci.*, 41: 301.

Bahl, N. (1983). Medicinal value of edible fungi. In: *Proceeding of the International Conference on Science and Cultivation Technology of Edible Fungi. Indian Mushroom Science II*, 203-209.

Bano, Z. and Rajarathanam, S. (1982). *Pleurotus* mushrooms as a nutritious food. In: *Tropical mushrooms –Biological Nature and cultivation methods*, (Chang, S.T. and Quimio, T.H., Eds.). *The Chinese University press*, Hongkong, 363-382.

Bano, Z. and Rajarathanum, S. (1988). *Pleurotus* mushroom part II. Chemical composition nutritional value, post harvest physiology, preservation and role as human food. *Crit. Rev. Food Sci. Nutr.*, 27: 87-158.

Bano, Z., Bhagya, S. and Srinivasan, K.S. (1981). Essential amino acid composition and proximate analysis of Mushroom, *Pleurotus florida*. *Mushrooms News Letter Trop.*,1: 6-10.

Barros, L., Correia, D. M., Ferreira, I. C. F. R., Baptista, P. and Santos-Buelga, C. (2008). Optimization of the determination of tocopherols in *Agaricus sp*. edible mushrooms by a normal phase liquid chromatographic method, *Food Chemistry*, 110(4): 1046–1050.

Chadha, K.L. and Sharma, S.R. (1995). Mushroom Research in India. History, Infrastructure and Achievements. In: *Advances in Horticulture* (Chadha, K.L. and Sharma, S.R., Eds.), Malhotra Publish House, New Delhi, 13: 1-33.

Chang, S.T. and Buswell, J.A. (1996). Mushroom nutriceuticals. *World J. Microb. Biotech.*, 12: 473-476.

Chang, S.T. and Hayes, W.A. (1978). The biology and cultivation of edible mushrooms. *New York: Academic Press.*

Chang, S.T. and Miles, P.G. (1989). Edible mushroom and their cultivation. *Florida, CRC Press.*

Erjavec, J., Kos, J., Ravnikar, M., Dreo, T. and Sabotic, J. (2012). Proteins of higher fungi-from forest to application, *Trends in Biotechnology*, 30(5): 259–273.

Esselen, W.B. and Fellers, C.R. (1946). Mushrooms for food and flavor. *Bull. Mass. Agric. Exp. Sta.*, 434.

Guillamon, E., Garcia-Lafuente, A. and Lozano, M. (2010). Edible mushrooms: role in the prevention of cardiovascular diseases, Fitoterapia, 81(7): 715–723.

Kalac, P. (2013). A review of chemical composition and nutritional value of wild-growing and cultivated mushrooms, *J. Sc. Food and Agri.*, 93(2): 209–218.

King, T.A. (1993). Mushrooms, the ultimate health food but little research in U. S to prove it. *Mushroom News*, 41: 29-46.

Li, G.S.F. and Chang, S.T. (1982). Nutritive value of *Volvariella volvacea, In: Tropical mushrooms –Biological nature and cultivation methods* (Chang, S.T. and Quimio, T.H., Eds.) *Chinese university press Hong Kong*, 199-219.

Litchfield, J.H. (1964). Nutrient content of morel mushroom mycelium: B vitamin composition. *J. Food Sci.*, 29: 690-691.

Mattila, P., Konko, K. and Eurola, M. (2001). Contents of vitamins, mineral elements, and some phenolic compounds in cultivated mushrooms, *Journal of Agricultural and Food Chemistry*. 49 (5): 2343–2348.

Mattila, P.K., Konko, M., Eurola, J., Pihlava, J., Astola, L., Vahteristo, V., Hietaniemi, J., Kumpulainen, N., Valtonen, V. and Piironen, V. (2000). Contents of vitamins, mineral elements and some phenolic compounds in the cultivated mushrooms. *J. Agric. Food Chem.*, 49: 2343-2348.

Mizuno, T., Inagaki, R., Kanao, T., Hagiwara, T., Nakamura, T., Hohshimura, R., Suniyat and Asukura (1990). Antitumor activity and some properties of water insoluble Heteroglycans from "Himematsutake", the fruing body of *Agaricus blazei* Murill. *Agricultural and Biological Chemistry,* 54(11): 2889- 2896.

Namba, H. (1993). Shiitake mushroom the king mushroom. *Mushroom News,* 41: 22-25.

Rai, R.D. (1995). Nutritional and medicinal values of mushrooms. *In: Advances in Horticulture.* (Chadha K.L. and Sharma S.R., eds.), Malhotra Publishing House, New Delhi, 537-551.

Rai, R.D. and Saxena, S. (1989). Biochemical changes during the post harvest storage of button mushroom (*Agaricus bisporus*). *Curr. Sci.,* 58: 508-510.

Reis, F. S., Barros, L., Martins, A. and Ferreira, I. C. F. R. (2012). Chemical composition and nutritional value of the most widely appreciated cultivated mushrooms: an inter-species comparative study, *Food and Chemical Toxicology,* 50(2): 191–197.

Singh, H.P. (2011). *Mushrooms, Cultivation, Marketing and Consumption.* RMCU, DMR, Solan.

Sohi, H.S. (1988). Mushroom culture in India, Recent research findings. *Indian Phytopath.,* 41: 313-326.

Varo, P., Lahelman, O. Nuurtamo, M. Saari, E. and Koivistoinen, P. (1980). Mineral element composition of Finish Food. VII Postal, Vegetables, fruits, berries, nuts and mushrooms. *Acta Agric. Scandinavica Supplement,* 22: 107-113.

Verma, R.N., Singh, G.B. and Bilgrami, K.S. (1987). Fleshy fungal flora of N. E. H. India- I. Manipur and Meghalaya. *Indian Mush. Sci.,* 2: 414- 421.

Wang, X. M., Zhang, J. and Wu, L.-H. (2014). A mini-review of chemical composition and nutritional value of edible wild-grown mushroom from China, *Food Chemistry,* 151: 279–285.

Chapter 24

Nutraceuticals: Health-Giving Intervention in Food Product for Medicinal Application

Dilip Kumar and D.C. Rai

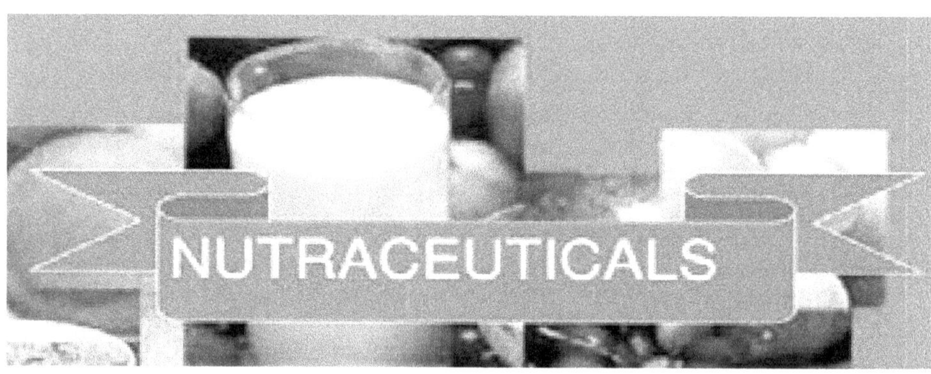

Introduction

The term nutraceutical was coined from nutrition and pharmaceutical in 1989 by Stephen Defelice, founder and chairman of foundation for innovation in medicine, an American organization which encourages medical health. According to him "a nutraceutical is any substance that is a food or a part of food and provides medical or health benefits, including the prevention and treatment of disease". Such products may range from isolated nutrients, dietary supplements and specific diets to genetically engineered designer foods and herbal products. It is natural that people's focus is shifting to positive approach for prevention of diseases to stay healthy. Milk is also unique food providing a variety of essential nutrients necessary to properly fuel to the body. Milk contains several bioactive components inherently, whereas several others are produced when milk is subjected to processing (fermentation *etc.*) and specific treatment conditions. Several milk components, namely lactose, fat, protein (casein and whey protein) and minerals serve as the parent molecules for nutraceuticals.

Food as medicine: Considered a father of Western medicine, Hippocrates advocated the healing effects of food. The Indians, Egyptians, Chinese, and Sumerians are just a few civilizations that have provided evidence suggesting that foods can be effectively used as medicine to treat and prevent disease. Ayurveda, the 5,000 year old ancient Indian health science, have mentioned benefits of food for therapeutic purpose. Documents hint that the medicinal benefits of food have been explored for thousands of years. Hippocrates, considered by some to be the father of Western medicine, said that people should "Let food be thy medicine."

WHO estimates that 60 per cent of the cardiac patients in the world will be Indians by 2030. Asia is expected to have 190 million diabetes cases, more than half of them are in India and China. Scientists say that the percentage of overweight/ obese people in India is on track to rise from 9 per cent in 1995 to 24 per cent in 2025.

Area Covered by Nutraceutical Products

All therapeutic areas such as anti-arthritic, pain killers, cold and cough, sleeping disorders, digestion and prevention of certain cancers, osteoporosis, blood pressure, cholesterol, depression and diabetes have been covered by nutraceuticals (Figure 1).

Hippocrates highlighted around 2000 year ago "Let food be your medicine and medicine be your food". Nutraceuticals are foods or food ingredients thatprovide medical or health benefits. This emerging class of products blurs the line between food and drugs . They do not easily fall into the legal categories of food or drug and often inhabit a grey area between the two.

Within European Union (EU) law the legal categorization of a nutraceutical is, in general, made on the basis of its accepted effects on the body. Thus, if the substance contributes only to the maintenance of healthy tissues and organs it may be considered to be a food ingredient. If, however, it can be shown to have a modifying effect on one or more of the body's physiological processes, it is likely to be considered to be a medicinal substance. Within European Medicines law a nutraceutical can be defined as a medicine for two reasons:

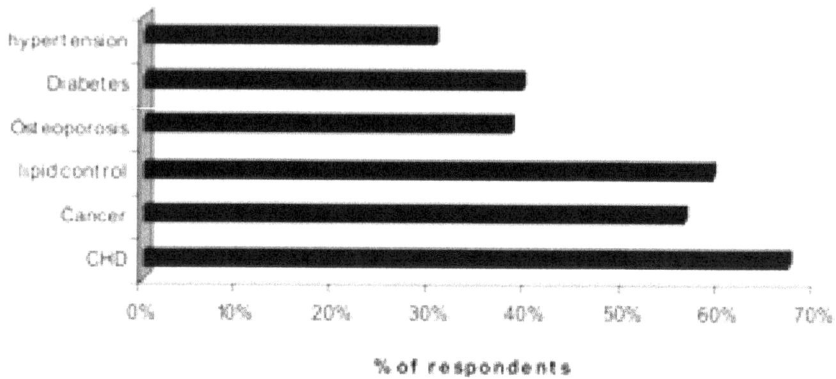

Figure 1: Percentage Area Covered by Nutraceutical Products

Bridging the Gap Between Food and Medicine.

1. It can used for the prevention, treatment or cure of a condition or disease or
2. It can be administered with a view to restoring, correcting or modifying physiological functions in human beings.

Nutraceutical Categories

Dietary Supplements Including Botanicals

- ☆ Vitamins, minerals, co-enzyme Q, carnitine
- ☆ Gingsing, Gingko Biloba, Saint John's Wort, Saw Palmetto

Dietary supplements, such as the vitamin B, are typically sold in pill form. A dietary supplement is a product that contains nutrients derived from food products that are concentrated in liquid or capsule form. The Dietary Supplement Health and Education Act (DSHEA) of 1994 defined generally what constitutes a dietary supplement. "A dietary supplement is a product taken by mouth that contains a "dietary ingredient" intended to supplement the diet. The "dietary ingredients" in these products may include: vitamins, minerals, herbs or other botanicals, amino acids, and substances such as enzymes, organ tissues, glandulars, and metabolites. Dietary supplements can also be extracts or concentrates, and may be found in many forms such as tablets, capsules, softgels, gelcaps, liquids, or powders."

Functional Foods

A food product that is part of usual diet but has beneficial effects that go beyond the traditional nutritional effects. Functional foods are designed to allow consumers to eat enriched foods close to their natural state, rather than by taking dietary supplements manufactured in liquid or capsule form. Functional foods have been either enriched or fortified, a process called nutrification. This practice restores the nutrient content in a food back to similar levels from before the food was processed. Sometimes, additional complementary nutrients are added, such as vitamin D to milk.

Examples

- Yogurts - Probiotics for intestinal health.
- Foods/cereals/snacks enriched with soluble fibres, vitamins and minerals.
- Omega-3 milk in prevention of heart disease
- Canola oil with lowered triglycerides for cholesterol reduction
- Oats, bran, psyllium and lignin's for heart disease and colon cancer
- Prebiotics - oligofructose for control of intestinal flora
- Stanols (Benecol) in reduction of cholesterol adsorption

Medicinal Foods

- Health bars with added medications
- Transgenic cows and lacto-ferrin for immune enhancement
- Transgenic plants for oral vaccination against infectious diseases

Table 1: Fortified Foods with their Ayurvedic Nutraceuticals

Fortified Foods	*Ayurvedic Nutraceuticals*
Calcium enriched edli	Antioxidants and bone density enhancer
Probiotic fortified yogurt	Curcumin
Buttermilk	Green tea extract
Omega -3-fortified health drinks and baby foods	Fish oil, Brahmi, Senna, Lutein Sugar free ayurvedic supplement
Dal and Atta noodles	Low calorie sweeteners

Table 2: Nutraceutical Ingredients with their Therapeutic Applications

Nutraceutical Ingredients	*Therapeutic Applications*
Probiotics, Prebiotics	Bone and Joint Health
Vitamins, Antioxidants	Cancer Risk Reduction
Soya based ingredients	Cardiovascular Health
Minerals	Maternal and Infant Health
Nutritional lipids and oil	Immune system
Fibers and carbohydrates	Energy and Eye Health
Dairy base ingredients	Skin Health, Respiratory Health, Weight Management, Cognitive and Mental function Cholesterol Reduction

Table 3: Nutraceuticals with their Therapeutic Benefits

Name of Nutraceutical	*Therapeutic Benefits*
Natural Lycopene	Reducing risk of prostate and cervical cancers. Supporting cardiovascular health.
Natural Purified Lutein Esters	Dietary supplement, Functional foods, Antioxidants.
Garlic	Cholesterol lowering, Cardiac diseases, Diabetic support
Green Tea	Cancer prevention, Weight management, Lowering cholesterol
Gymnema, Momordica	Diabetic control
Glucosamine	Arthritis treatment
Ginkgo Biloba	Allergy relief
Digestive Enzymes	Digestive support
Ginseng	Immunomodulator
Phycocyanin Powder	Antioxidan

Table 4: Examples of Functional Food Components

Components		*Source*	*Potential benefits*
1) Polyphenols	Anthocyanidine	Fruits	Neutralizes free radicals, reduce risk of cancer
	Catechin	Tea, Babul pods, Mustard cake, Rapeseed, Shorea robusta seeds.	
	Flavanones	Citrus	
	Flavones	Fruit, Vegetables, Soybean	
2) Saponins		Soybean, GNC, Lucerne	Lower cholesterol level Anti-cancer
3) Probiotics/ Prebiotics/ Symbiotic	Lactobacillus	Yoghurt	Improve GI health
	Fructo-oligosaccharides	White grain, Onions	
4) Phytoestrogens	Diadzein	Soybean, Flax, Maize	Reduce menopause symptoms, improve bone health
	Genistein	Berseem clover, Lucerne	
	Lignans	Flax, Rye, Vegetables	Reduce cancer and heart disease
5) Dietary Fibres	Insoluble fibres	Wheat bran	Reduce breast, colon cancer
	Glucan	Oats	
	Whole grains	Cereal grains	Reduce Cordial Vascular Disease
6) Carotenoids	-Carotene, Zeaxanthin, Lutein	Clover, Lucerne, Oats, Corn, Carrots, Vegetables, Fruits	Neutralize free radicals, Healthy vision

Classification of Nutraceuticals

Regarding the promise of nutraceuticals, they should be considered in two ways:

1. Potential nutraceuticals
2. Established nutraceuticals

A potential nutraceutical is one that holds a promise of a particular health or medical benefit; such a potential nutraceutical only becomes an established one after there are sufficient clinical data to demonstrate such a benefit. It is disappointing to note that the overwhelming majority of nutraceutical products are in the 'potential' category, waiting to become established. The food products used as nutraceutical are categorized as.

- ✰ Probiotic
- ✰ Prebiotic
- ✰ Dietary fiber
- ✰ Omega 3 fatty acid
- ✰ Antioxidant

Benefits of Nutraceuticals

From the consumers' point of view, functional foods and nutraceuticals may offer many benefits:

- ✰ May increase the health value of our diet.
- ✰ May help us live longer.
- ✰ May help us to avoid particular medical conditions.
- ✰ May have a psychological benefit from doing something for oneself.
- ✰ May be perceived to be more "natural" than traditional medicine and less likely to produce unpleasant side-effects.
- ✰ May present food for populations with special needs (*e.g.* nutrient-dense foods for the elderly).

New Concepts with Nutraceuticals

Farmaceuticals

According to a report written for the United States Congress entitled "Agriculture: A Glossary of Terms, Programs, and Laws", "(Farmaceuticals) is a melding of the words farm and pharmaceuticals. It refers to medically valuable compounds produced from modified agricultural crops or animals (usually through biotechnology). Proponents believe that using crops and possibly even animals as pharmaceutical factories could be much more cost effective than conventional methods. Nutraceutical a portmanteau of the words 'nutrition' and 'pharmaceutical', is a food or food product that reportedly provides health and medical benefits, including the prevention and treatment of diseases.

Aquaceuticals

Recently, manufacturers of bottled water-considered a conventional food by FDA-have begun following suit. The introduction of "aquaceuticals" offers the bottled water industry an avenue for growth and a means of attracting new segments of the health conscious consuming public. Functional foods, including aquaceuticals, are a cross between dietary supplements and conventional foods/beverages; they are foods that contain ingredients intended to promote health. The result is a line of products that may have an effect on the structure or function of the body. Recent additions to the functional food marketplace have included juices, snacks, bars and soup products.

Cosmeceuticals

Cosmeceuticals is the combination of cosmetics and pharmaceuticals. Cosmeceuticals are cosmetic products with bioactive ingredients purported to have medical or drug-like benefits. The "cosmeceutical" label applies only to products applied topically, such as creams, lotions and ointments. Products which are similar in perceived benefits but ingested orally are known as nutricosmetics. Examples: anti-aging creams and moisturizers.

Regulatory Bodies

- Foods for Specified Health Use (FOSHU)-Japan
- Joint Health Claims Initiative (JHCI) - UK
- Codex Alimentarius - FAO and WHO
- Functional Food Science in Europe (FUFOSE) - European Union
- Food Safety and Standard Act 2006-India

Conclusion

Nutraceuticals are available in the form of isolated nutrients, dietary supplements and specific diets to genetically engineered foods, herbal products and processed foods such as cereals, soups and beverages. Nutraceuticals provide all the essential substances that should be present in a healthy diet for the human and it provides energy and nutrient supplements to body, which are required for maintaining optimal health. Nutraceuticals are widely used in the food and pharmaceutical industries. Some nutraceuticals are useful in maintaining healthy prostate function, remedy for restlessness and insomnia. Nutraceuticals, such as glucosamine and chondroitin sulfate, offer possible chondro-protective effects against joint injury and it used in many more ways.

REFERENCES

Adelaja Adesoji O and Schilling Brian J. (1999). Nutraceutical: blurring the line between food and drugs in the twenty-first century. The Magazine of Food, Farm and Resource Issues; 14: 35-40.

Brower, B. (1998). Nutraceuticals: poised for a healthy slice of the market. *Nat Biotechnology*, 16: 728-33.

Consumer Association of Canada. Available from: http: //www.consumermanitoba.ca/resources.html. Accessed on date June 18, 2018.

De Felice L and Stephen. (1995). The nutraceutical revolution, its impact on food industry. *Trends in Food Sci. and Tech*; 6: 59-61.

Dietary Supplement Health Education Act (DSHEA) of (1994). Public Law 103–417, available from FDA website: http: //www.fda.gov.

Felice De L Stephen. (1995). The nutraceutical revolution, its impact on food industry. *Trends in Food Sci. and Tech*, 6: 59-61.

Freedonia group (2009). World Nutraceutical Ingredients, industry study with forecast for 2013 and 2018: 1- 487. (www.freedoniagroup.com)

Jack, D.B. (1995). Keep taking the tomatoes - the exciting world of nutraceuticals. *Mol Med Today*, 1(3): 118-21.

Kokate, C.K, Purohit, A.P and Gokhale, S.B. (2002). Nutraceutical and Cosmaceutical. Pharmacognosy, 21st edition, Pune, India: Nirali Prakashan, p 542-549.

Om P Gulati, and Peter Berry Ottaway. (2006). Legislation relating to nutraceuticals in the European Union with a particular focus on botanical-sourced products. *Toxicol.* 221: 75–87.

Patil, C.S. (2011). Current trends and future prospective of nutraceuticals in health promotion, BIOINFO *Pharmaceutical Biotechnology* 1 (1); 01-07.

Richardson, D.P. (1996). Functional foods–shades of grey: an industry perspective. Nutr. Rev.; 54: 174–180.

Rishi, R.K. (2006). Nutraceutical: borderline between food and drug. Pharma Review Available from: http: //www.kppub.com/articles/herbal-safety-pharmareview-004/nutraceutica-borderline-between-food-and-drugs.html.

Sami Labs-pioneer in nutraceuticals. The Hindu Newspaper; Aug 05, 2002, Avaliable from: http: //www.hinduonnet.com/thehindu/biz/2002/08/05/stories/2002080500040200.htm.

Stephen, D.A (2012) report of National Nutraceutical Centre, Nutraceuticals India 2012 webinar;: 1-22.

Chapter 25

Colostrum as a Functional Food

Arun Kumar and Raman Seth

Introduction

The world wide trend towards the development of health promoting foods offers new vistas for applications which contain specific antibody ingredients derived from immunized cows. Colostrum based preparations have remarkable potential to contribute to human health care as a part of health promoting diet and as an alternative or a supplement to the medical treatment of specified human diseases. Colostrum is a form of milk produced by the mammary glands of mammals in late pregnancy. Colostrum contains antibodies to protect the newborn against diseases, as well as being lower in fat and higher in protein than ordinary milk. It contains many essential nutrients and bioactive components of human cellular and humoral immune defence such as growth factors, immunoglobulins (Igs), lactoperoxidase (Lp), lysozyme (Lys), lactoferrin (Lf), cytokines, vitamins, peptides and oligosaccharides, which are of increasing relevance to human health by curing cancer, cardiovascular diseases, diabetes, autoimmune diseases, allergies, bacterial, viral and parasitic infections, cold as well as flu. Colostrum has been processed into products designed for pharmaceutical and nutraceutical purposes. Internationally, colostrum derived products have become valuable niche products and are currently being sold into highly competitive markets with current focus on protein components because of their physiological effects and hence possess commercial value. Colostrum based products that are available in the market include health supplements, infant supplements, beverages, lozenges and sports drink and tablets.

Bioactive Components of Colostrum

Growth Factors

Growth factors are so called because historically they have been identified by their ability to stimulate the growth of various cell lines in vitro but, in reality, the functions of these peptide based molecules are considerably more diverse. Concentration ranges of growth factors reported for bovine colostrum and milk is as follows:

Growth Factors	*Concentration (ng/ml)*	
	Colostrum	*Milk*
Insulin like growth–I	100 - 2000	5-100
Insulin like growth factor-II	150 – 600	5-100
Transforming growth factor - 1	10 – 50	< 5
Transforming growth factor- 2	150 -1150	10-70
Epidermal growth factor	4 – 325	1- 150
Betacellulin	< 5	< 5
Platelet-derived growth factor	NA	NA
Fibroblast growth factor	NA b	< 1

Epidermal Growth Factors (EGF)

EGF is a 53–amino acid peptide produced by the salivary glands and the Brunners glands of the duodenum in adults. EGF acts as a "luminal surveillance peptide" in the adult gut, readily available to stimulate the repair process at the sites of injury. The EGF in colostrum and milk may therefore play a role in preventing bacterial translocation and stimulating gut growth in suckling neonates by playing an important role in cell differentiation rather than cell proliferation and in stimulating mucus secretion. It increases DNA synthesis, mitotic activity and enhances brush-border membrane enzyme activity when injected into suckling mice.

Transforming Growth Factor (TGF)

TGF- α is a 50–amino acid molecule that is produced within the mucosa throughout the gastrointestinal tract. In contrast with EGF, TGF- is produced within the mucosa throughout the gastrointestinal tract. Systemic administration of TGF- stimulates gastrointestinal growth and repair, inhibits acid secretion, stimulates mucosal restitution after injury, and Increases gastric mucin concentrations. Up-regulation of TGF- expression has been shown to occur in the gastrointestinal mucosa at sites of injury as well as in the liver after partial hepatectomy, supporting a role for TGF- in mucosal growth and repair. Other findings support the role of TGF- in maintaining epithelial continuity but suggest that the relative importance of peptides such as this might vary from one region of the gut to another.

Transforming Growth Factor β (TGF β)

This family of molecules is structurally distinct from TGF- α and, in most systems, actually inhibits proliferation. There are 5 different isoforms of TGF- β and their major site of expression in the normal gastrointestinal tract is in the superficial zones. TGF-β has many diverse functions. It is a potent chemo- attractant for neutrophils and stimulates epithelial cell migration at wound sites. It is therefore, likely to be a key player in stimulating restitution, the early phase of the repair process during which surviving cells from the edge of a wound migrate over the denuded area to reestablish epithelial continuity. TGF- β and TGF- β -like molecules are present in high concentrations in both bovine milk (1–2 mg/L) and colostrum (20– 40 mg/L).

Platelet-Derived Growth Factor (PDGF)

PDGF is an acid-stable molecule that was originally identified from platelets but is also synthesized and secreted by macrophages. It consists of 2 disulfide linked polypeptides: chain A (14 kDa) and chain B (17 kDa). The dimer, therefore, exists in 3 isoforms (AA, AB, and BB) that binds to tyrosine kinase–type receptors. PDGF is a potent mitogen for fibroblasts and arterial smooth muscle cells and administration of exogenous PDGF has been shown to facilitate ulcer healing when administered orally to animals. Although PDGF is present in bovine milk and colostrum, most of the PDGF-like mitogenic activity in bovine milk is actually derived from bovine colostral growth factors which shares homology with growth factors.

Insulin-like Growth Factors (IGF)

The most abundant and best characterized growth factors in bovine colostrum are insulin-like growth factors (IGF-1 and IGF-2). Both IGF-I and -II are single chain polypeptides with 70 and 67 amino acid residues and molecular weights of about 7.6 and 7.5 kDa, respectively. Each IGF molecule contains three disulphide bridges. They are similar in structure to proinsulin and it is possible that they also exert insulin-like effects at high concentrations. Bovine colostrum contains much higher concentrations of IGF-I (500 mg/L with lower concentrations in mature bovine milk (10 mg/L). These growth factors are relatively stable to both heat and acidic conditions. They therefore survive the harsh conditions of both commercial milk processing and gastric acid to maintain their biological activity. IGF-I and IGF-II promote cell proliferation and differentiation. They also exert insulin-like effects at high concentrations.

Immune Factors

Immunoglobulins

Immunoglobulins (Igs) are a family of globular proteins with antimicrobial and other protective bioactivities. They exist at different concentrations in milk and colostrum. Immunoglobulin-rich fractions are usually prepared by removing fat and casein followed by concentration, sterilization and sometimes lyophilization or spray-drying. Bovine colostrum contains large amounts of secretory IgA, which protects against viruses (*e.g.* poliovirus, influenza A virus and herpes simplex virus) and bacteria such as *E. coli*, Salmonella and Streptococcus. IgG antibodies express multifunctional activities, including complement activation, bacterial opsonisation and agglutination, and act by binding to specific sites on the surfaces of most infectious agents or products, either inactivating them or reducing infection. In bovine colostrum and milk, immunoglobulin G (IgG; subclasses IgG1 and IgG2) is the major immune component, although low levels of IgA and IgM are also present IgG1 constitutes approximately 80 per cent of the total Ig content of bovine milk. The concentration of immunoglobulins in colostrum and normal milk is as follows:

Immunoglobulin	*Colostrum (g/L)*	*Normal Milk (g/L)*
IgG1	52. 0 – 87.0	0.31 – 0.40
IgG2	1.6 – 2.1	0.03 – 0.08
IgM	3.7 – 6.1	0.03 – 0.06
IgA	3.2 – 6.2	0.04 – 0.06

Lactoferrin

Lactoferrin is an 80 kDa iron-binding glycoprotein present in colostrum and shows antiviral, antibacterial, anti-inflamatory properties. The concentration of lactoferrin in bovine colostrum and mature milk is about 1.5-5 mg/mL and 0.1 mg/mL respectively. Lactoferrin has been implicated in the treatment of diseases like cancer, HIV, herpes, chronic fatigue *etc.* Lactoferrin's affinity for iron is very high (about 260 times that of blood serum transferrin). Lactoferrin has been shown to inhibit the growth of several microbes, including *E.coli, Salmonella typhimurium, Shigella dysenteria, Listeria monocytogenes, Streptococcus mutans, Bacillus stearothermophilus* and *Bacillus subtilis.* It has been proposed that the antimicrobial effect of lactoferrin is based on its capacity to bind iron, which is essential for the growth of bacteria. Lactoferrin exerts its antimicrobial activity by modifying bacterial cell membranes.

Lysozyme

Lysozyme is a well-known antibacterial and lytic enzyme present in many mammalian body fluids, including colostrum. The concentration of lysozyme in colostrum and in normal milk is about 0.14-0.7 and 0.07-0.6 mg/L, respectively. The natural substrate of the enzyme is the peptidoglycan layer of the bacterial cell wall and its degradation results in lysis of the bacteria. Some recent results suggest that the antibacterial activity of lysozyme is not only due to its enzymatic activity, but also due to its cationic and hydrophobic properties. The presence of lactoferrin enhances the antibacterial activity of lysozyme against E.coli, which also supports the hypothesis that lactoferrin damages the outer membrane of Gram-negative bacteria.

Lactoperoxidase

Lactoperoxidase is a major antibacterial enzyme in colostrum. Bovine colostrum and milk contain about 11-45 mg/L and 13-30 mg/L lactoperoxidase, respectively. It is a basic glycoprotein containing a heme-group with Fe^{3+} and catalyzes the oxidation of thiocyanate (SCN-) in the presence of hydrogen peroxide (H_2O_2), producing a toxic intermediate oxidation product. This product inhibits bacterial metabolism via the oxidation of essential sulphydryl groups in proteins. The lactoperoxidase system is also toxic to other Gram-positive and Gram negative bacteria such as *Pseudomonas* sp., *Salmonella typhimurium, Listeria monocytogenes,*

Streptococcus mutans, Staphylococcus aureus and psychrotrophic bacteria in milk. Apart from this, the lactoperoxidase system inactivates polio virus and human immunodeficiency virus type I in vitro. Bovine lactoperoxidase also contains a site with high affinity for calcium.

Proline-Rich Polypeptides (PRP)

PRP is a hormone that regulates the thymus gland, stimulating an underactive immune system or down-regulating an overactive immune system as seen in autoimmune disease(Multiple sclerosis, rheumatoid arthritis, lupus, scleroderma, chronic fatigue syndrome, allergies, *etc.*). PRP stimulates immature thymocytes to turn into functionally active T-cells. Furthermore, PRP acts as an immune-regulator by changing surface markers and functions of cells .It is a mixture of peptides (polypeptide-clostrinin) derived from colostrum which could help to slow down the progression of Alzheimer's Disease by reducing the build-up of beta-amyloid, a toxic protein that accumulates in the brains of Alzheimer's sufferers.

Processing of Colostrum

Some of the possible pathogens like *Campylobacter spp., Escherichia coli, Listeria monocytogenes, Mycoplasma spp., Mycobacterium avium ssp. paratuberculosis, Salmonella spp. etc.* can be transmitted through colostrum and milk. Best method for reducing or eliminating bacterial pathogens in colostrum is to heat fresh bovine colostrum .Pasteurization is the process of heating liquids in order to destroy harmful organisms such as bacteria, protozoa, molds, and yeasts .Unlike sterilization, pasteurization is not intended to kill all microorganisms. Instead, pasteurization aims to achieve a reduction in the number of viable organisms, reducing their number and hence are unlikely to cause disease. There are two common methods of pasteurizing: *i.e.* batch pasteurization(63°C/30 min) and HTST continuous flow pasteurization(72°C/15s).

The stability of bovine IgG towards thermal treatment has been widely studied using different techniques From these studies it was suggested that IgG denaturation involves an initial reversible unfolding of native structure with loss of globular configuration, which can proceed further to irreversible denaturation and aggregation via hydrophobic and disulfide interactions. Studies on the effects of heat treatment on structure and solubility of the Immunoglobulin fraction of whey reported that IgG are among the most heat-stable whey proteins, which is attributed to their high content of disulfide bonds and components such as fat, lactose, salts and other proteins which helped in the stabilization of antibodies during thermal treatment.

Depending on consumer acceptance and marketing strategies, colostrum may be processed into colostrum powder(marketed in the form of tablets, capsules or as dietary supplements), liquid colostrum whey based nutritional beverage, immunoglobulin concentrates, lactoferrin rich concentrates *etc.* The technologies applied in processing of colostrum are basically patented but they have to be adapted keeping in view the temperature sensitivity of the components in colostrum. The initial steps are same for each end products. Frozen colostrum is thawed, defatted and pasteurized at 63-70°C(batch process). Membrane filtration like microfiltration can be applied when a liquid sterile end product is desired. Also membrane filtration is acceptable when specific component rich fractions are to be prepared.

To lower the microbial load in colostrum, hygienic conditions should be maintained and storage should be done at freezing temperature. Freezing prevents nutrient breakdown during storage. Colostrum stored under frozen conditions, had no appreciable changes in pH, acidity, fat, total solids, total nitrogen and non protein nitrogen. There was no loss of vitamin A and less than a 6 per cent loss of carotene in colostrum stored at -20.5°C for six months.

Clinical Applications of Colostrums

Bovine colostrum is attributed with curative powers has been proved by recent clinical studies which helped in rediscovering the potential benefits of the bioactive components of colostrums. However, colostrums exert a wide range of beneficial health effects which are discussed below.

Cancer

Bovine colostrums contain a hormone like substance called leptin which regulates the incorporation and usage of fat in the body. Growth hormone alongwith IGF-1 was observed to have a role in the repair and functioning of heart muscles .Also, the cytokines like Interleukins 1, 6 and 10, Interferon G and Lymphokines found in colostrum are involved in the treatment of cancer. Colostrum lactalbumin has been found to cause the selective death (apoptosis) of cancer cells, leaving the surrounding non-cancerous tissues unaffected. Lactoferrin has similarly been reported to possess anti-cancer activity. The mixture of immune and growth factors in colostrum can inhibit the spread of cancer cells.

Diabetes

Juvenile diabetes (Type I, Insulin Dependent) is thought to be brought about through an autoimmune mechanism, possibly initiated by an allergic reaction to the protein GAD found in cow's milk. Colostrum contains several factors, which can offset this and other allergies. Human trials reported that IgF-1 stimulates glucose utilization, effectively treating acute hyperglycemia and lessening a Type II diabetic dependence on insulin.

Infective Diarrhoea

Hyperimmune milk or colostrum preparations have been shown to be beneficial in the prevention and treatment of infection and to increase weight gain in both clinical and veterinary practice. The use of whole hyperimmune colostrum rather than specific antibodies purified from milk or other sources has the added value of potentially stimulating the repair process (due to the presence of growth factors) as well as facilitating the eradication of the infection by mechanisms involving nonspecific antibacterial factors in colostrum and milk.

Heart Disease

Colostrum PRP plays a role in reversing heart disease very much like it does with allergies and autoimmune diseases. Additionally, IgF-1 and GH in colostrum can lower LDL-cholesterol while increasing HDL-cholesterol concentrations. Colostrum growth factors promote the repair and regeneration of heart muscle and the regeneration of new blood vessels for collateral coronary circulation.

Anti-inflammatory

Inflammation is associated with strenuous exercise and anti-inflammatories are the most commonly prescribed class of drug to athletes. Inflammation is typically centered in the joints and in the digestive tract. Inflammation is a protective response to an injury, invading foreign substance, or an internally produced substance (*e.g.* in auto-immune disorders like rheumatoid arthritis). Colostrum reduces the need for damaging medication, and because it is a natural food, unlike NSAIDS, it has absolutely no negative side-effects and has a multitude of benefits.

Increased Brain Function

Phospholipids, components of alpha lipid, help in increasing brain function and have been associated with improved memory. They have also shown to elevate moods and reduce the symptoms of depression.

Allergies and Autoimmune Diseases

PRP from colostrum can work as a regulatory substance of the thymus gland. It has been demonstrated to improve or eliminate symptomatology of both allergies and autoimmune diseases (Multiple Sclerosis, rheumatoid arthritis, lupus, and myasthenia gravis). PRP inhibits the overproduction of lymphocytes and T-cells and reduces the major symptoms of allergies and autoimmune disease: pain, swelling and inflammation.

Weight Loss

IgF-1 is required by the body to metabolize fat for energy through Krebs cycle. With aging, less IgF-1 is produced in the body. Inadequate levels are associated with an increased incidence of Type II diabetes and difficulty in losing weight despite a proper nutritional intake and adequate exercise. Colostrum provides a good source of IgF-1 as a complementary therapy for successful weight loss.

Respiratory Disease

Respiratory system disorders are best fought by body's own defense mechanism. But as a person ages, the immune system loses its ability to regulate itself and this mainly occurs because the thymus gland begins to shrink. T lymphocytes are generated from bone marrow and mature in thymus. Some of these cells called killer T-lymphocytes generate cell mediated responses. Helper or suppressor T lymphocytes regulate the quality and strength of immune response. Several investigators have reported that the thymus can be restored to normal function by growth factors in colostrum.

Wound Healing

Several colostrum components stimulate wound-healing. Nucleotides, EGF, TGF and IGF-1 stimulate skin growth, cellular growth and repair by direct action on DNA and RNA. These growth factors facilitate the healing of tissues damaged by ulcers, trauma, burns, surgery or inflammatory disease. The tissues benefitted by colostrums wound healing properties are skin, muscle, cartilage, bone and nerve cells. Powdered colostrum can be applied topically to gingivitis, sensitive teeth, ulcers, cuts, abrasions and burns after they have been cleaned and disinfected.

Autoimmune diseases

Colostrum contains components, such as cytokines, which are known to be responsible for regulating this inflammatory response, and has potential as an anti-inflammatory aid. Proline-Rich-Polypeptide (PRP) in colostrum can work as a regulatory substance of the thymus gland. It has been demonstrated to improve or eliminate symptoms of both allergies and auto-immune diseases (Multiple Sclerosis, rheumatoid arthritis, lupus, and myasthenia gravis). PRP inhibits the overproduction of lymphocytes and T-cells and reduces the major symptoms of allergies and autoimmune disease.

Athletic Stress

Exhaustive workouts and athletic competition can temporarily depress the immune system, decreasing the number of T-lymphocytes and NK cells. Athletes are therefore, more prone to develop infections, including Chronic Fatigue Syndrome. Many of colostrum's immune factors can help significantly reduce the number and severity of infections caused by both physical and emotional stress.

Leaky Gut Syndrome

One of the major benefits of colostrum supplementation is enhanced gut efficiency due to the many immune enhancers that control clinical and subclinical GI infections. Colostral growth factors also play a role by keeping the intestinal mucosa sealed and impermeable to toxins. Healing leaky gut syndrome reduces toxic load and helps in the reversal of many allergic and autoimmune conditions.

Conclusion

It is anticipated that colostrums based preparations may have remarkable potential to contribute to human wellness not only in the form of health promoting diet but also as a supplement to the medical treatment of specified human diseases as a part of health promoting food. Bovine colostrum virtually contains all compounds of human cellular and humoral immune defense. They are ideal sources of these defense molecules for industrial production because of their ready availability and safety as compared with blood derived analogues. The ongoing success of colostrums based products speaks for itself. However, the challenge for manufacturers still remains that how to process colostrums in a cost effective manner.

REFERENCES

Appelmelk, B.J., An Y.Q, Geerts, M, Thijs B.G., de Boer, H.A., MacLaren, D.M., de Graaff, J. & Nuijens, J.H. (1994). Lactoferrin is a lipid A-binding protein. *Infection and Immunity*, 62:2628-2632.

Blum, J. W. and Hammon, H. (2000). Colostrum effects on the gastrointestinal tract, and on nutritional, endocrine and metabolic parameters in neonatal calves. *Livest. Prod. Sci.*, 66: 151.

Chen, C.C., Tu, Y.Y., and Chang, H.M. (2000). Thermal stability of bovine milk immunoglobulin G (IgG) and the effect of added thermal protectants on the stability. *J. Food Sci.*, 65: 188.

Donovan, S.M. and Odle, J. (1994). Growth factors in milk as mediators of infant development. *Ann. Rev of Nutr.*, 14: 147.

Elfstrand, L., Lindmark-Mansson, H., Paulsson, M., Nyberg, L. and Akesson, B. (2002). Immunoglobulins, growth factors and growth hormone in bovine colostrum and the effects of processing. *Int. Dairy J.*, 12: 879.

Gapper, L., Copestake D., Otter, D., and Indyk, H. (2007). Analysis of bovine immunoglobulin G in milk, colostrum and dietary supplements: a review Anal *Bioanal Chem.*, 389:93.

Godden, S.M., Smith, S., Feirtag, J.M., Green, L.R., Wells, S.J. and Fetrow, J.P. (2003). Effect of on-farm commercial batch pasteurization of colostrum on colostrum and serum immunoglobulin concentrations in dairy calves. *J. Dairy Sci.*, 86: 1503.

Korhonen, H., Marnila, P. and Gill, H.S. (2000). Bovine milk antibodies for health. *Br J Nutr.*, 84: S135.

Kurokowa, M., Lynch, K. and Podolsky, D.K. (1987). Effects of growth factors on an intestinal epithelial cell line: transforming growth factor beta inhibits proliferation and stimulates differentiation. *Biochemical and Biophysical Research Communications* 142:775-182.

Mehra, R., Marnila, P. and Korhonen, H. (2006). Milk immunoglobulins for health promotion. *Intl. Dairy J.*, 16.1262-1271.

Muntjewerf, A.K. and Korhonen H. (2002). Bovine immunoglobulins: Technological and applications. *Bull. Int. Dairy Fed.*, 375:59-64.

Pakkanen, R. and Aalto J. (1997). Growth factors and antimicrobial factors of bovine colostrum. *Intl Dairy J.*, 7:285–297.

Roitt, I.M. (1999). Essential Immunology. Blackwell science Ltd. 9th Edition.

Wan, L., Almers, W., and Chen, W. (2005). Emerging processing technologies for functional foods. *Australian journal of Dairy Technology.*, 60:167-169.

Wilson, J. (1997). Immune system breakthrough: Colostrum. *J Longevity Res.*, 3:7–10.

Index

D

E

F

G

H

I

J

K

L

M

N

O

P

Q

R

S

T

U

V

W

www.ingramcontent.com/pod-product-compliance
Ingram Content Group UK Ltd.
Pitfield, Milton Keynes, MK11 3LW, UK
UKHW021447280726
14060UKWH00001BA/286

9 789390 371532